a metrix Can have only one inverse

*Elementary Linear Algebra*

# Elementary Linear Algebra

**HOWARD ANTON**

**Drexel University**

JOHN WILEY & SONS, INC.    New York · London · Sydney · Toronto

Copyright © 1973, by John Wiley & Sons, Inc.

All rights reserved. Published simultaneously in Canada.

*Library of Congress Cataloging in Publication Data:*

Anton, Howard.
Elementary linear algebra.

1. Algebras, Linear.    I. Title.
QA251.A57        512′.5        72-5511
ISBN 0-471-03247-6

Printed in the United States of America

10 9 8 7 6 5 4 3 2 1

*To*
*Pat*
*and my children*
*Brian and Davey*

# *Preface*

This textbook gives an elementary treatment of linear algebra that is suitable for students in their freshman or sophomore year. Calculus is *not* a prerequisite. I have, however, included a number of exercises for students with a calculus background; these are clearly marked: "For students who have studied calculus."

My aim in writing this book is to present the fundamentals of linear algebra in the clearest possible way. Pedagogy is the main consideration. Formalism is secondary. Where possible, basic ideas are studied by means of computational examples (over 200 of them) and geometrical interpretation.

My treatment of proofs varies. Those proofs that are elementary and have significant pedagogical content are presented precisely, in a style tailored for beginners. A few proofs that are more difficult, but pedagogically valuable, are placed at the end of the section and marked "Optional." Still other proofs are omitted completely, with emphasis placed on applying the theorem. Whenever a proof is omitted, I try to motivate the result, often with a discussion about its interpretation in 2-space or 3-space.

It is my experience that $\Sigma$-notation is more of a hindrance than a help for beginners in linear algebra. Therefore, I have generally avoided its use.

It is a pedagogical axiom that a teacher should proceed from the familiar to the unfamiliar and from the concrete to the abstract. The ordering of the chapters reflects my adherence to this tenet.

Chapter 1 deals with systems of linear equations, how to solve them, and some of their properties. It also contains the basic material on matrices and their arithmetic properties.

Chapter 2 deals with determinants. I have used the classical permutation approach. In my opinion, it is less abstract than the approach through $n$-linear alternating forms and gives the student a better intuitive grasp of the subject than does an inductive development.

Chapter 3 introduces vectors in 2-space and 3-space as arrows and develops the analytic geometry of lines and planes in 3-space. Depending on the background of the students, all but the first section of this chapter can be omitted without a loss of continuity (see the guide for the instructor that follows this preface).

Chapter 4 and Chapter 5 develop the basic results about real finite dimensional vector spaces and linear transformations. I begin with a study of $R^n$ and proceed slowly to the general concept of a vector.

Chapter 6 deals with the eigenvalue problem and diagonalization. I have also included two sections concerned with identifying conic sections and quadric surfaces. These sections serve as an introduction to quadratic forms and also give the beginning student a solid application of eigenvalues and eigenvectors. Some instructors may choose to omit these two sections and cover topics in Chapter 7 instead.

Chapter 7 gives an introduction to the numerical methods of linear algebra; it does not require access to computing facilities since the exercises are designed for hand computation. This chapter gives the student a basic understanding of how certain linear algebra problems are solved practically. Too many students complete their linear algebra studies with the naive belief that eigenvalues are found in practice by solving the characteristic equation. Some instructors may wish to use this section in conjunction with a programming course.

I have included a large number of exercises. Each exercise set begins with routine "drill" problems and progresses toward more theoretical problems. Answers to all the computational problems are given at the end of the text.

Since there is more material in this book than can be covered in a one-semester or one-quarter course, the instructor will have to make a selection of topics. To help in this selection, I have provided a guide to the instructor.

I express my appreciation for the helpful guidance provided by the reviewers: Professor Joseph Buckley of Western Michigan University, Professor Harold S. Engelsohn of Kingsborough Community College, Professor Lawrence D. Kugler of the University of Michigan, and Professor Robert W. Negus of Rio Hondo Junior College. I am particularly grateful to Professor William F. Trench of Drexel University, who read the entire manuscript. His suggestions immeasurably improved both the style and content of the text. For her accurate and skillful typing of the manuscript, I thank Miss Susan R. Gershuni; thanks are also due Professor Dale Lick who encouraged my work in its early stages. Finally, I express my appreciation to my editor, Frederick C. Corey, and the production staff of John Wiley and Sons for making this project most enjoyable.

**Howard Anton**

# A Guide For The Instructor

Chapter 3 can be omitted without loss of continuity if the students have previously studied about lines, planes, and geometric vectors in 2-space and 3-space. Depending on the available time and the background of the students, the instructor may wish to add all or part of this chapter to the following suggested core material:

|  |  |
|---|---|
| Chapter 1 | 8 lectures |
| Chapter 2 | 5 lectures |
| Chapter 4 | 10 lectures |
| Chapter 5 | 9 lectures |
| Chapter 6 | 5 lectures (omit 6.4 and 6.5) |

This schedule is rather liberal; it allows a fair amount of classroom time for discussion of homework problems, but assumes that little classroom time is devoted to the material marked "optional". The instructor can build on this core as time permits by including lectures on optional material, Chapter 3, Sections 6.4 and 6.5, and Chapter 7.

# Contents

*xi*

*Elementary Linear Algebra*

# Chapter 1 | Systems of Linear Equations and Matrices

## 1.1 INTRODUCTION TO SYSTEMS OF LINEAR EQUATIONS

In this section we introduce basic terminology and discuss a method for solving systems of linear equations.

A line in the $xy$-plane can be represented algebraically by an equation of the form

$$a_1 x + a_2 y = b$$

An equation of this kind is called a linear equation in the variables $x$ and $y$. More generally, we define a **linear equation** in the $n$ variables $x_1, x_2, \ldots, x_n$ to be one that can be expressed in the form

$$a_1 x_1 + a_2 x_2 + \cdots + a_n x_n = b$$

where $a_1, a_2, \ldots, a_n$ and $b$ are real constants.

### Example 1

The following are linear equations:

$$x + 3y = 7 \qquad\qquad x_1 - 2x_2 - 3x_3 + x_4 = 7$$

$$y = \frac{1}{2}x + 3z + 1 \qquad\qquad x_1 + x_2 + \cdots + x_n = 1$$

Observe that a linear equation does not involve any products or roots of variables. All variables occur only to the first power and do not appear as arguments for trigonometric, logarithmic, or exponential functions. Each of the following *fails* to be a linear equation:

$$x + 3y^2 = 7 \qquad\qquad 3x + 2y - z + xz = 4$$

$$y - \sin x = 0 \qquad\qquad \sqrt{x_1} + 2x_2 + x_3 = 1$$

A *solution* of a linear equation $a_1x_1 + a_2x_2 + \cdots + a_nx_n = b$ is a sequence of $n$ numbers $s_1, s_2, \ldots, s_n$ such that the equation is satisfied when we substitute $x_1 = s_1$, $x_2 = s_2, \ldots, x_n = s_n$. The set of all solutions of the equation is called its *solution set*.

### Example 2

Find the solution set of each of the following:

(i) $4x - 2y = 1$        (ii) $x_1 - 4x_2 + 7x_3 = 5$

To find solutions of (i), we can assign an arbitrary value to $x$ and solve for $y$, or choose an arbitrary value for $y$ and solve for $x$. If we follow the first approach and assign $x$ an arbitrary value $t$, we obtain

$$x = t, \qquad y = 2t - \frac{1}{2}$$

These formulas describe the solution set in terms of the arbitrary parameter $t$. Particular numerical solutions can be obtained by substituting specific values for $t$. For example, $t = 3$ yields the solution $x = 3, y = 11/2$ and $t = -1/2$ yields the solution $x = -1/2, y = -3/2$.

If we follow the second approach and assign $y$ the arbitrary value $t$, we obtain

$$x = \frac{1}{2}t + \frac{1}{4}, \qquad y = t$$

Although these formulas are different from those obtained above, they yield the same solution set as $t$ varies over all possible real numbers. For example, the previous formulas gave the solution $x = 3, y = 11/2$ when $t = 3$, while these formulas yield this solution when $t = 11/2$.

To find the solution set of (ii) we can assign arbitrary values to any two variables and solve for the third variable. In particular, if we assign arbitrary values $s$ and $t$ to $x_2$ and $x_3$ respectively and solve for $x_1$, we obtain

$$x_1 = 5 + 4s - 7t, \qquad x_2 = s, \qquad x_3 = t$$

A finite set of linear equations in the variables $x_1, x_2, \ldots, x_n$ is called a *system of linear equations*. A sequence of numbers $s_1, s_2, \ldots, s_n$ is called a *solution* of the system if $x_1 = s_1, x_2 = s_2, \ldots, x_n = s_n$ is a solution of every equation in the system. For example, the system of two equations in three unknowns

$$4x_1 - x_2 + 3x_3 = -1$$
$$3x_1 + x_2 + 9x_3 = -4$$

has the solution $x_1 = 1, x_2 = 2, x_3 = -1$ since these values satisfy both equations. However, $x_1 = 1, x_2 = 8, x_3 = 1$ is not a solution since these values satisfy only the first of the two equations in the system.

Not all systems of linear equations have solutions. For example, if we multiply the second equation of the system

$$x + y = 4$$
$$2x + 2y = 6$$

by $1/2$, it becomes evident that there is no solution, since the two equations in the resulting system

$$x + y = 4$$
$$x + y = 3$$

contradict each other.

A system of equations that has no solution is said to be ***inconsistent.*** If there is at least one solution, it is called ***consistent.*** To illustrate the possibilities that can occur in solving systems of linear equations, consider a general system of two linear equations in the unknowns $x$ and $y$:

$$a_1x + b_1y = c_1 \quad (a_1, b_1 \text{ both not zero})$$
$$a_2x + b_2y = c_2 \quad (a_2, b_2 \text{ both not zero})$$

The graphs of these equations are lines; call them $l_1$ and $l_2$. Since a point $(x, y)$ lies on a line if and only if the numbers $x$ and $y$ satisfy the equation of the line, the solutions of the system of equations will correspond to points of intersection of $l_1$ and $l_2$. There are three possibilities (Figure 1.1).

(a) The lines $l_1$ and $l_2$ may be parallel, in which case there is no intersection, and consequently no solution to the system.

(b) The lines $l_1$ and $l_2$ may intersect at only one point, in which case the system has exactly one solution.

(c) The lines $l_1$ and $l_2$ may coincide, in which case there are infinitely many points of intersection, and consequently infinitely many solutions to the system.

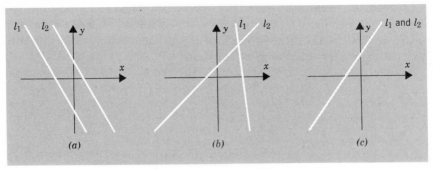

**Figure 1.1** (*a*) No solution. (*b*) One solution. (*c*) Infinitely many solutions.

Although we have only considered two equations with two unknowns here, we will show later that this same result holds for arbitrary systems; that is, *every system of linear equations has either no solutions, exactly one solution, or infinitely many solutions.*

An arbitrary system of $m$ linear equations in $n$ unknowns will be written

$$
\begin{aligned}
a_{11}x_1 + a_{12}x_2 + \cdots + a_{1n}x_n &= b_1 \\
a_{21}x_1 + a_{22}x_2 + \cdots + a_{2n}x_n &= b_2 \\
\vdots \qquad \vdots \qquad \qquad \vdots \qquad &\ \ \vdots \\
a_{m1}x_1 + a_{m2}x_2 + \cdots + a_{mn}x_n &= b_m
\end{aligned}
$$

where $x_1, x_2, \ldots, x_n$ are the unknowns and the subscripted $a$'s and $b$'s denote constants.

For example, a general system of three linear equations in four unknowns will be written

$$
\begin{aligned}
a_{11}x_1 + a_{12}x_2 + a_{13}x_3 + a_{14}x_4 &= b_1 \\
a_{21}x_1 + a_{22}x_2 + a_{23}x_3 + a_{24}x_4 &= b_2 \\
a_{31}x_1 + a_{32}x_2 + a_{33}x_3 + a_{34}x_4 &= b_3
\end{aligned}
$$

The double subscripting on the coefficients of the unknowns is a useful device that we shall employ to establish the location of the coefficient in the system. The first subscript on the coefficient $a_{ij}$ indicates the equation in which the coefficient occurs, and the second subscript indicates which unknown it multiplies. Thus, $a_{12}$ is in the first equation and multiplies unknown $x_2$.

If we mentally keep track of the location of the $+$'s, the $x$'s, and the $=$'s, a system of $m$ linear equations in $n$ unknowns can be abbreviated by writing only the rectangular array of numbers:

$$
\begin{bmatrix}
a_{11} & a_{12} & \cdots & a_{1n} & b_1 \\
a_{21} & a_{22} & \cdots & a_{2n} & b_2 \\
\vdots & \vdots & & \vdots & \vdots \\
a_{m1} & a_{m2} & \cdots & a_{mn} & b_m
\end{bmatrix}
$$

This is called the **augmented matrix** for the system. (The term *matrix* is used in mathematics to denote a rectangular array of numbers. Matrices arise in many contexts; we shall study them in more detail in later sections.) To illustrate, the augmented matrix for the system of equations

$$
\begin{aligned}
x_1 + x_2 + 2x_3 &= 9 \\
2x_1 + 4x_2 - 3x_3 &= 1 \\
3x_1 + 6x_2 - 5x_3 &= 0
\end{aligned}
$$

is

$$
\begin{bmatrix}
1 & 1 & 2 & 9 \\
2 & 4 & -3 & 1 \\
3 & 6 & -5 & 0
\end{bmatrix}
$$

The basic method for solving a system of linear equations is to replace the given system by a new system that has the same solution set, but is easier to solve. This new system is generally obtained in a series of steps by applying the following three types of operations.

**1.** Multiply an equation through by a nonzero constant.
**2.** Interchange two equations.
**3.** Add a multiple of one equation to another.

Since the rows (horizontal lines) in an augmented matrix correspond to the equations in the associated system, these three operations correspond to the following operations on the rows of the augmented matrix.

**1.** Multiply a row through by a nonzero constant.
**2.** Interchange two rows.
**3.** Add a multiple of one row to another row.

We shall now give an example that illustrates how these operations can be used to solve systems of linear equations. Since a systematic procedure for finding solutions will be derived in the next section, it is not necessary to worry about how the steps in this example were selected. The main effort at this time should be devoted to understanding the computations and the discussion.

**Example 3**

In the left column below we solve a system of linear equations by operating on the equations in the system, and in the right column we solve the same system by operating on the rows of the augmented matrix.

$$
\begin{aligned}
x + y + 2z &= 9 \\
2x + 4y - 3z &= 1 \\
3x + 6y - 5z &= 0
\end{aligned}
\qquad
\begin{bmatrix}
1 & 1 & 2 & 9 \\
2 & 4 & -3 & 1 \\
3 & 6 & -5 & 0
\end{bmatrix}
$$

Add $-2$ times the first equation to the second to obtain

Add $-2$ times the first row to the second to obtain

$$
\begin{aligned}
x + y + 2z &= 9 \\
2y - 7z &= -17 \\
3x + 6y - 5z &= 0
\end{aligned}
\qquad
\begin{bmatrix}
1 & 1 & 2 & 9 \\
0 & 2 & -7 & -17 \\
3 & 6 & -5 & 0
\end{bmatrix}
$$

Add $-3$ times the first equation to the third to obtain

Add $-3$ times the first row to the third to obtain

$$
\begin{aligned}
x + y + 2z &= 9 \\
2y - 7z &= -17 \\
3y - 11z &= -27
\end{aligned}
\qquad
\begin{bmatrix}
1 & 1 & 2 & 9 \\
0 & 2 & -7 & -17 \\
0 & 3 & -11 & -27
\end{bmatrix}
$$

Multiply the second equation by 1/2 to obtain

$$x + y + 2z = 9$$
$$y - \frac{7}{2}z = -\frac{17}{2}$$
$$3y - 11z = -27$$

Multiply the second row by 1/2 to obtain

$$\begin{bmatrix} 1 & 1 & 2 & 9 \\ 0 & 1 & -\frac{7}{2} & -\frac{17}{2} \\ 0 & 3 & -11 & -27 \end{bmatrix}$$

Add $-3$ times the second equation to the third to obtain

$$x + y + 2z = 9$$
$$y - \frac{7}{2}z = -\frac{17}{2}$$
$$-\frac{1}{2}z = -\frac{3}{2}$$

Add $-3$ times the second row to the third to obtain

$$\begin{bmatrix} 1 & 1 & 2 & 9 \\ 0 & 1 & -\frac{7}{2} & -\frac{17}{2} \\ 0 & 0 & -\frac{1}{2} & -\frac{3}{2} \end{bmatrix}$$

Multiply the third equation by $-2$ to obtain

$$x + y + 2z = 9$$
$$y - \frac{7}{2}z = -\frac{17}{2}$$
$$z = 3$$

Multiply the third row by $-2$ to obtain

$$\begin{bmatrix} 1 & 1 & 2 & 9 \\ 0 & 1 & -\frac{7}{2} & -\frac{17}{2} \\ 0 & 0 & 1 & 3 \end{bmatrix}$$

The system has now been sufficiently simplified so that the solution can be found. From the last equation we have $z = 3$. Substituting this value into the second equation and solving for $y$ gives $y = 2$. Substituting $y = 2$ and $z = 3$ into the first equation and solving for $x$ yields $x = 1$. The solution is therefore

$$x = 1, \quad y = 2, \quad z = 3$$

## EXERCISE SET 1.1

1. Which of the following are linear equations in $x_1$, $x_2$, and $x_3$?
   (a) $x_1 + 2x_1x_2 + x_3 = 2$
   (b) $x_1 + x_2 + x_3 = \sin k$
   ($k$ is a constant)
   (c) $x_1 - 3x_2 + 2x_3^{1/2} = 4$
   (d) $x_1 = \sqrt{2}x_3 - x_2 + 7$
   (e) $x_1 + x_2^{-1} - 3x_3 = 5$
   (f) $x_1 = x_3$
2. Find the solution set for:
   (a) $6x - 7y = 3$
   (b) $2x_1 + 4x_2 - 7x_3 = 8$
   (c) $-3x_1 + 4x_2 - 7x_3 + 8x_4 = 5$
   (d) $2v - w + 3x + y - 4z = 0$

3. Find the augmented matrix for each of the following systems of linear equations.

(a)
$$\begin{aligned} x_1 - 2x_2 &= 0 \\ 3x_1 + 4x_2 &= -1 \\ 2x_1 - x_2 &= 3 \end{aligned}$$

(b)
$$\begin{aligned} x_1 \quad\quad + x_3 &= 1 \\ -x_1 + 2x_2 - x_3 &= 3 \end{aligned}$$

(c)
$$\begin{aligned} x_1 \quad + x_3 \quad\quad &= 1 \\ 2x_2 - x_3 + x_5 &= 2 \\ 2x_3 + x_4 \quad &= 3 \end{aligned}$$

(d)
$$\begin{aligned} x_1 &= 1 \\ x_2 &= 2 \end{aligned}$$

4. Find a system of linear equations corresponding to each of the following augmented matrices.

(a) $\begin{bmatrix} 1 & 0 & -1 & 2 \\ 2 & 1 & 1 & 3 \\ 0 & -1 & 2 & 4 \end{bmatrix}$

(b) $\begin{bmatrix} 1 & 0 & 0 \\ 0 & 1 & 0 \\ 1 & -1 & 1 \end{bmatrix}$

(c) $\begin{bmatrix} 1 & 2 & 3 & 4 & 5 \\ 5 & 4 & 3 & 2 & 1 \end{bmatrix}$

(d) $\begin{bmatrix} 1 & 0 & 0 & 0 & 1 \\ 0 & 1 & 0 & 0 & 2 \\ 0 & 0 & 1 & 0 & 3 \\ 0 & 0 & 0 & 1 & 4 \end{bmatrix}$

5. For which value(s) of the constant $k$ does the following system of linear equations have no solutions? Exactly one solution? Infinitely many solutions?

$$\begin{aligned} x - y &= 3 \\ 2x - 2y &= k \end{aligned}$$

6. Consider the system of equations

$$\begin{aligned} ax + by &= k \\ cx + dy &= l \\ ex + fy &= m \end{aligned}$$

Discuss the relative positions of the lines $ax + by = k$, $cx + dy = l$, and $ex + fy = m$ when:
(a) the system has no solutions
(b) the system has exactly one solution
(c) the system has infinitely many solutions

7. Show that if the system of equations in Exercise 6 is consistent, then at least one equation can be discarded from the system without altering the solution set.

8. Let $k = l = m = 0$ in Exercise 6; show that the system must be consistent. What can be said about the point of intersection of the three lines if the system has exactly one solution?

9. Consider the system of equations

$$\begin{aligned} x + y + 2z &= a \\ x \quad\quad + z &= b \\ 2x + y + 3z &= c \end{aligned}$$

Show that in order for this system to be consistent, $a$, $b$, and $c$ must satisfy $c = a + b$.

**10.** Show that if two linear equations in the variables $x_1, x_2, \ldots, x_n$ each have the same solution set, then each equation is a multiple of the other.

## 1.2 GAUSSIAN ELIMINATION

In this section we give a systematic procedure for solving systems of linear equations; it is based on the idea of reducing the augmented matrix to a form that is simple enough so that the system of equations can be solved by inspection.

In the last step of Example 3 we obtained the augmented matrix

$$\begin{bmatrix} 1 & 1 & 2 & 9 \\ 0 & 1 & -\dfrac{7}{2} & -\dfrac{17}{2} \\ 0 & 0 & 1 & 3 \end{bmatrix}$$

after which we easily obtained the solution $x = 1, y = 2, z = 3$ to the given system of equations. It is possible, however, to make the solution more evident from the augmented matrix by applying a few additional row operations. To illustrate, add $-1$ times the second row of the above matrix to the first row to obtain

$$\begin{bmatrix} 1 & 0 & \dfrac{11}{2} & \dfrac{35}{2} \\ 0 & 1 & -\dfrac{7}{2} & -\dfrac{17}{2} \\ 0 & 0 & 1 & 3 \end{bmatrix}$$

Now add $-(11/2)$ times the third row to the first row and $7/2$ times the third row to the second row. This gives

$$\begin{bmatrix} 1 & 0 & 0 & 1 \\ 0 & 1 & 0 & 2 \\ 0 & 0 & 1 & 3 \end{bmatrix} \tag{1.1}$$

The corresponding system of equations is

$$\begin{array}{rcl} x_1 & = & 1 \\ x_2 & = & 2 \\ x_3 & = & 3 \end{array}$$

so that the solution

$$x_1 = 1, \qquad x_2 = 2, \qquad x_3 = 3$$

is obvious from the augmented matrix.

Matrix (1.1) is an example of a matrix that is in **reduced row-echelon form.** To be of this form, a matrix must have the following properties.

1. *If a row does not consist entirely of zeros, then the first nonzero number in the row is a* 1.
2. *If there are any rows that consist entirely of zeros, then they are grouped together at the bottom of the matrix.*
3. *If row j and row j* + 1 *are any two successive rows that do not consist entirely of zeros, then the first nonzero number in row j* + 1 *occurs further to the right than the first nonzero number in row j.*
4. *Each column that contains the first nonzero number of some row has zeros everywhere else.*

A matrix having properties **1, 2,** and **3** is said to be in **row-echelon form.**

### Example 4

The following matrices are in reduced row-echelon form.

$$\begin{bmatrix} 1 & 0 & 0 & 4 \\ 0 & 1 & 0 & 7 \\ 0 & 0 & 1 & -1 \end{bmatrix}, \quad \begin{bmatrix} 1 & 0 & 0 \\ 0 & 1 & 0 \\ 0 & 0 & 1 \end{bmatrix}, \quad \begin{bmatrix} 0 & 1 & -2 & 0 & 1 \\ 0 & 0 & 0 & 1 & 3 \\ 0 & 0 & 0 & 0 & 0 \\ 0 & 0 & 0 & 0 & 0 \end{bmatrix}, \quad \begin{bmatrix} 0 & 0 \\ 0 & 0 \end{bmatrix}$$

The following matrices are in row-echelon form.

$$\begin{bmatrix} 1 & 1 & 3 & 7 \\ 0 & 1 & 6 & 2 \\ 0 & 0 & 1 & 5 \end{bmatrix}, \quad \begin{bmatrix} 1 & 1 & 0 \\ 0 & 1 & 0 \\ 0 & 0 & 0 \end{bmatrix}, \quad \begin{bmatrix} 0 & 1 & 2 & 6 & 0 \\ 0 & 0 & 1 & -1 & 0 \\ 0 & 0 & 0 & 0 & 1 \end{bmatrix}$$

The reader should check to see that each of the above matrices satisfies all the necessary requirements.

If, by a sequence of row operations, the augmented matrix for a system of linear equations can be put in reduced row-echelon form, then the solution set for the system can be obtained by inspection or, at worst, after a few simple steps. The next example will illustrate this point.

### Example 5

Suppose that the augmented matrix for a system of linear equations has been reduced by row operations to the given reduced row-echelon form. Solve the system.

$$(a) \begin{bmatrix} 1 & 0 & 0 & 5 \\ 0 & 1 & 0 & -2 \\ 0 & 0 & 1 & 4 \end{bmatrix} \qquad (b) \begin{bmatrix} 1 & 0 & 0 & 4 & -1 \\ 0 & 1 & 0 & 2 & 6 \\ 0 & 0 & 1 & 3 & 2 \end{bmatrix}$$

(c) $\begin{bmatrix} 1 & 6 & 0 & 0 & 4 & -2 \\ 0 & 0 & 1 & 0 & 3 & 1 \\ 0 & 0 & 0 & 1 & 5 & 2 \\ 0 & 0 & 0 & 0 & 0 & 0 \end{bmatrix}$    (d) $\begin{bmatrix} 1 & 0 & 0 & 0 \\ 0 & 1 & 0 & 0 \\ 0 & 0 & 0 & 1 \end{bmatrix}$

*Solution to* (a). The corresponding system of equations is

$$
\begin{aligned}
x_1 &\phantom{=} & &= & 5 \\
 & x_2 & &= & -2 \\
 & & x_3 &= & 4
\end{aligned}
$$

By inspection, $x_1 = 5$, $x_2 = -2$, $x_3 = 4$.

*Solution to* (b). The corresponding system of equations is

$$
\begin{aligned}
x_1 & & &+ 4x_4 &= & -1 \\
 & x_2 & &+ 2x_4 &= & 6 \\
 & & x_3 &+ 3x_4 &= & 2
\end{aligned}
$$

Since $x_1$, $x_2$, and $x_3$ each "begin" an equation, we call them **leading variables.** Solving for the leading variables in terms of $x_4$ gives

$$
\begin{aligned}
x_1 &= -1 - 4x_4 \\
x_2 &= \phantom{-}6 - 2x_4 \\
x_3 &= \phantom{-}2 - 3x_4
\end{aligned}
$$

Since $x_4$ can be assigned an arbitrary value, say $t$, we have infinitely many solutions. The solution set is given by the formulas

$$x_1 = -1 - 4t, \qquad x_2 = 6 - 2t, \qquad x_3 = 2 - 3t, \qquad x_4 = t$$

*Solution to* (c). The corresponding system of equations is

$$
\begin{aligned}
x_1 + 6x_2 & & &+ 4x_5 &= & -2 \\
 & x_3 & &+ 3x_5 &= & 1 \\
 & & x_4 &+ 5x_5 &= & 2
\end{aligned}
$$

Here the leading variables are $x_1$, $x_3$, and $x_4$. Solving for the leading variables in terms of the remaining variables gives

$$
\begin{aligned}
x_1 &= -2 - 4x_5 - 6x_2 \\
x_3 &= \phantom{-}1 - 3x_5 \\
x_4 &= \phantom{-}2 - 5x_5
\end{aligned}
$$

Since $x_5$ can be assigned an arbitrary value, $t$, and $x_2$ can be assigned an arbitrary value, $s$, there are infinitely many solutions. The solution set is given by the formulas

$$x_1 = -2 - 4t - 6s, \qquad x_2 = s, \qquad x_3 = 1 - 3t, \qquad x_4 = 2 - 5t, \qquad x_5 = t$$

*Solution to* (*d*). The last equation in the corresponding system of equations is

$$0x_1 + 0x_2 + 0x_3 = 1$$

Since this equation can never be satisfied, there is no solution to the system.

We have just seen how easy it is to solve a system of linear equations, once its augmented matrix is in reduced row-echelon form. Now we shall give a step-by-step procedure, called **Gauss-Jordan elimination,**[*] which can be used to reduce any matrix to reduced row-echelon form. As we state each step in the procedure, we shall illustrate the idea by reducing the following matrix to reduced row-echelon form.

$$\begin{bmatrix} 0 & 0 & -2 & 0 & 7 & 12 \\ 2 & 4 & -10 & 6 & 12 & 28 \\ 2 & 4 & -5 & 6 & -5 & -1 \end{bmatrix}$$

**Step 1.** Locate the leftmost column (vertical line) that does not consist entirely of zeros.

$$\begin{bmatrix} 0 & 0 & -2 & 0 & 7 & 12 \\ 2 & 4 & -10 & 6 & 12 & 28 \\ 2 & 4 & -5 & 6 & -5 & -1 \end{bmatrix}$$

⎣—**Leftmost nonzero column**

**Step 2.** Interchange the top row with another row, if necessary, so that the entry at the top of the column found in Step 1 is different from zero.

$$\begin{bmatrix} 2 & 4 & -10 & 6 & 12 & 28 \\ 0 & 0 & -2 & 0 & 7 & 12 \\ 2 & 4 & -5 & 6 & -5 & -1 \end{bmatrix}$$  ( The first and second rows in the previous matrix were interchanged. )

**Step 3.** If the entry that is now at the top of the column found in Step 1 is *a*, multiply the first row by $1/a$ so that the top entry in the column becomes a 1.

$$\begin{bmatrix} 1 & 2 & -5 & 3 & 6 & 14 \\ 0 & 0 & -2 & 0 & 7 & 12 \\ 2 & 4 & -5 & 6 & -5 & -1 \end{bmatrix}$$  ( The first row of the previous matrix was multiplied by 1/2. )

[*]*Carl Friedrich Gauss* (1777–1855). Sometimes called the "prince of mathematicians," Gauss made profound contributions to number theory, theory of functions, probability, and statistics. He discovered a way to calculate the orbits of asteroids, made basic discoveries in electromagnetic theory, and invented a telegraph.

*Camille Jordan* (1838–1922). Jordan was a professor at the École Polytechnique in Paris. He did pioneering work in several branches of mathematics, including matrix theory. He is particularly famous for the Jordan Curve Theorem, which states: A simple closed curve (such as a circle or a square) divides the plane into two nonintersecting connected regions.

**Step 4.** Add suitable multiples of the top row to the rows below so that in the column located in Step 1, all entries below the top become zeros.

$$\begin{bmatrix} 1 & 2 & -5 & 3 & 6 & 14 \\ 0 & 0 & -2 & 0 & 7 & 12 \\ 0 & 0 & 5 & 0 & -17 & -29 \end{bmatrix} \quad \begin{pmatrix} -2 \text{ times the first row of} \\ \text{the previous matrix was} \\ \text{added to the third row.} \end{pmatrix}$$

**Step 5.** Now cover the top row in the matrix and begin again with Step 1 applied to the submatrix that remains. Continue in this way until the *entire* matrix is in row-echelon form.

$$\begin{bmatrix} 1 & 2 & -5 & 3 & 6 & 14 \\ 0 & 0 & -2 & 0 & 7 & 12 \\ 0 & 0 & 5 & 0 & -17 & -29 \end{bmatrix}$$

└──Leftmost nonzero column in
the submatrix

$$\begin{bmatrix} 1 & 2 & -5 & 3 & 6 & 14 \\ 0 & 0 & 1 & 0 & -\frac{7}{2} & -6 \\ 0 & 0 & 5 & 0 & -17 & -29 \end{bmatrix} \quad \begin{pmatrix} \text{The first row in} \\ \text{the submatrix was} \\ \text{multiplied by } -1/2. \end{pmatrix}$$

$$\begin{bmatrix} 1 & 2 & -5 & 3 & 6 & 14 \\ 0 & 0 & 1 & 0 & -\frac{7}{2} & -6 \\ 0 & 0 & 0 & 0 & \frac{1}{2} & 1 \end{bmatrix} \quad \begin{pmatrix} -5 \text{ times the first row of the} \\ \text{submatrix was added to the} \\ \text{second row of the submatrix.} \end{pmatrix}$$

$$\begin{bmatrix} 1 & 2 & -5 & 3 & 6 & 14 \\ 0 & 0 & 1 & 0 & -\frac{7}{2} & -6 \\ 0 & 0 & 0 & 0 & \frac{1}{2} & 1 \end{bmatrix} \quad \begin{pmatrix} \text{The top row in the submatrix} \\ \text{was covered and we} \\ \text{returned again to Step 1.} \end{pmatrix}$$

└──Leftmost nonzero column
in the new submatrix

$$\begin{bmatrix} 1 & 2 & -5 & 3 & 6 & 14 \\ 0 & 0 & 1 & 0 & -\dfrac{7}{2} & -6 \\ 0 & 0 & 0 & 0 & 1 & 2 \end{bmatrix}$$

$\left(\begin{array}{l}\text{The first (and only) row}\\ \text{in the new submatrix}\\ \text{was multiplied by 2.}\end{array}\right)$

The *entire* matrix is now in row-echelon form.

**Step 6.** Beginning with the last row and working upward, add suitable multiples of each row to the rows above so that the fourth requirement in the definition of reduced row-echelon form is satisfied.

$$\begin{bmatrix} 1 & 2 & -5 & 3 & 6 & 14 \\ 0 & 0 & 1 & 0 & 0 & 1 \\ 0 & 0 & 0 & 0 & 1 & 2 \end{bmatrix}$$ $\left(\begin{array}{l}\text{7/2 times the third row of}\\ \text{the previous matrix was}\\ \text{added to the second row.}\end{array}\right)$

$$\begin{bmatrix} 1 & 2 & -5 & 3 & 0 & 2 \\ 0 & 0 & 1 & 0 & 0 & 1 \\ 0 & 0 & 0 & 0 & 1 & 2 \end{bmatrix}$$ $\left(\begin{array}{l}-6 \text{ times the third}\\ \text{row was added to}\\ \text{the first row.}\end{array}\right)$

$$\begin{bmatrix} 1 & 2 & 0 & 3 & 0 & 7 \\ 0 & 0 & 1 & 0 & 0 & 1 \\ 0 & 0 & 0 & 0 & 1 & 2 \end{bmatrix}$$ $\left(\begin{array}{l}5 \text{ times the second}\\ \text{row was added to}\\ \text{the first row.}\end{array}\right)$

The last matrix is in reduced row-echelon form.

**Example 6**

Solve by Gauss-Jordan elimination.

$$\begin{aligned} x_1 + 3x_2 - 2x_3 \qquad\quad + 2x_5 \qquad\qquad &= \phantom{-}0 \\ 2x_1 + 6x_2 - 5x_3 - 2x_4 + 4x_5 - 3x_6 &= -1 \\ 5x_3 + 10x_4 \qquad + 15x_6 &= \phantom{-}5 \\ 2x_1 + 6x_2 \qquad\quad + 8x_4 + 4x_5 + 18x_6 &= \phantom{-}6 \end{aligned}$$

The augmented matrix for the system is

$$\begin{bmatrix} 1 & 3 & -2 & 0 & 2 & 0 & 0 \\ 2 & 6 & -5 & -2 & 4 & -3 & -1 \\ 0 & 0 & 5 & 10 & 0 & 15 & 5 \\ 2 & 6 & 0 & 8 & 4 & 18 & 6 \end{bmatrix}$$

Adding $-2$ times the first row to the second and fourth rows gives

$$\begin{bmatrix} 1 & 3 & -2 & 0 & 2 & 0 & 0 \\ 0 & 0 & -1 & -2 & 0 & -3 & -1 \\ 0 & 0 & 5 & 10 & 0 & 15 & 5 \\ 0 & 0 & 4 & 8 & 0 & 18 & 6 \end{bmatrix}$$

Multiplying the second row by $-1$ and then adding $-5$ times the second row to the third row and $-4$ times the second row to the fourth row gives

$$\begin{bmatrix} 1 & 3 & -2 & 0 & 2 & 0 & 0 \\ 0 & 0 & 1 & 2 & 0 & 3 & 1 \\ 0 & 0 & 0 & 0 & 0 & 0 & 0 \\ 0 & 0 & 0 & 0 & 0 & 6 & 2 \end{bmatrix}$$

Interchanging the third and fourth rows and then multiplying the third row of the resulting matrix by $1/6$ gives the row-echelon form

$$\begin{bmatrix} 1 & 3 & -2 & 0 & 2 & 0 & 0 \\ 0 & 0 & 1 & 2 & 0 & 3 & 1 \\ 0 & 0 & 0 & 0 & 0 & 1 & \dfrac{1}{3} \\ 0 & 0 & 0 & 0 & 0 & 0 & 0 \end{bmatrix}$$

Adding $-3$ times the third row to the second row and then adding 2 times the second row of the resulting matrix to the first row yields the reduced row-echelon form

$$\begin{bmatrix} 1 & 3 & 0 & 4 & 2 & 0 & 0 \\ 0 & 0 & 1 & 2 & 0 & 0 & 0 \\ 0 & 0 & 0 & 0 & 0 & 1 & \dfrac{1}{3} \\ 0 & 0 & 0 & 0 & 0 & 0 & 0 \end{bmatrix}$$

The corresponding system of equations is

$$\begin{aligned} x_1 + 3x_2 \quad\quad + 4x_4 + 2x_5 \quad\quad\quad &= 0 \\ x_3 + 2x_4 \quad\quad\quad\quad &= 0 \\ x_6 &= \frac{1}{3} \end{aligned}$$

(We have discarded the last equation, $0x_1 + 0x_2 + 0x_3 + 0x_4 + 0x_5 + 0x_6 = 0$, since it will be automatically satisfied by the solutions of the remaining equations.) Solving for the leading variables, we obtain

$$\begin{aligned} x_1 &= -3x_2 - 4x_4 - 2x_5 \\ x_3 &= -2x_4 \\ x_6 &= \frac{1}{3} \end{aligned}$$

If we assign $x_2$, $x_4$, and $x_5$ the arbitrary values $r$, $s$, and $t$ respectively, the solution set is given by the formulas

$$x_1 = -3r - 4s - 2t, \quad x_2 = r, \quad x_3 = -2s, \quad x_4 = s, \quad x_5 = t, \quad x_6 = \frac{1}{3}$$

**Example 7**

It is often more convenient to solve a system of linear equations by bringing the augmented matrix into row-echelon form without continuing all the way to the reduced row-echelon form. When this is done, the corresponding system of equations can be solved by a technique called ***back-substitution.*** We shall illustrate this method using the system of equations in Example 6.

From the computations in Example 6, a row-echelon form of the augmented matrix is

$$\begin{bmatrix} 1 & 3 & -2 & 0 & 2 & 0 & 0 \\ 0 & 0 & 1 & 2 & 0 & 3 & 1 \\ 0 & 0 & 0 & 0 & 0 & 1 & \frac{1}{3} \\ 0 & 0 & 0 & 0 & 0 & 0 & 0 \end{bmatrix}$$

To solve the corresponding system of equations

$$\begin{aligned} x_1 + 3x_2 - 2x_3 \quad\quad + 2x_5 \quad\quad &= 0 \\ x_3 + 2x_4 \quad\quad + 3x_6 &= 1 \\ x_6 &= \frac{1}{3} \end{aligned}$$

first, *solve these equations for the leading variables;* this yields

$$\begin{aligned} x_1 &= -3x_2 + 2x_3 - 2x_5 \\ x_3 &= \quad 1 \quad - 2x_4 - 3x_6 \\ x_6 &= \quad \frac{1}{3} \end{aligned}$$

Next, *beginning with the bottom equation and working upward, substitute each equation into all the equations above it.*

Substituting $x_6 = 1/3$ into the second equation yields

$$\begin{aligned} x_1 &= -3x_2 + 2x_3 - 2x_5 \\ x_3 &= -2x_4 \\ x_6 &= \quad \frac{1}{3} \end{aligned}$$

Substituting $x_3 = -2x_4$ into the first equation yields

$$\begin{aligned} x_1 &= -3x_2 - 4x_4 - 2x_5 \\ x_3 &= -2x_4 \\ x_6 &= \quad \frac{1}{3} \end{aligned}$$

Finally, *assign arbitrary values to the nonleading variables.*

If we assign $x_2$, $x_4$ and $x_5$ the arbitrary values $r$, $s$, and $t$ respectively, the solution set is given by the formulas

$$x_1 = -3r - 4s - 2t, \quad x_2 = r, \quad x_3 = -2s, \quad x_4 = s, \quad x_5 = t, \quad x_6 = \frac{1}{3}$$

This agrees with the solution obtained in Example 6.

The method of solving systems of linear equations by reducing the augmented matrix to row-echelon form is called **Gaussian elimination.**

## EXERCISE SET 1.2

**1.** Which of the following are in reduced row-echelon form?

(a) $\begin{bmatrix} 1 & 0 & 0 \\ 0 & 0 & 0 \\ 0 & 0 & 1 \end{bmatrix}$
(b) $\begin{bmatrix} 0 & 1 & 0 \\ 1 & 0 & 0 \\ 0 & 0 & 0 \end{bmatrix}$
(c) $\begin{bmatrix} 1 & 1 & 0 \\ 0 & 1 & 0 \\ 0 & 0 & 0 \end{bmatrix}$

(d) $\begin{bmatrix} 1 & 2 & 0 & 3 & 0 \\ 0 & 0 & 1 & 1 & 0 \\ 0 & 0 & 0 & 0 & 1 \\ 0 & 0 & 0 & 0 & 0 \end{bmatrix}$
(e) $\begin{bmatrix} 1 & 0 & 0 & 5 \\ 0 & 0 & 1 & 3 \\ 0 & 1 & 0 & 4 \end{bmatrix}$
(f) $\begin{bmatrix} 1 & 0 & 3 & 1 \\ 0 & 1 & 2 & 4 \end{bmatrix}$

**2.** Which of the following are in row-echelon form?

(a) $\begin{bmatrix} 1 & 2 & 3 \\ 0 & 0 & 0 \\ 0 & 0 & 1 \end{bmatrix}$
(b) $\begin{bmatrix} 1 & -7 & 5 & 5 \\ 0 & 1 & 3 & 2 \end{bmatrix}$
(c) $\begin{bmatrix} 1 & 1 & 0 \\ 0 & 1 & 0 \\ 0 & 0 & 0 \end{bmatrix}$

(d) $\begin{bmatrix} 1 & 3 & 0 & 2 & 0 \\ 1 & 0 & 2 & 2 & 0 \\ 0 & 0 & 0 & 0 & 1 \\ 0 & 0 & 0 & 0 & 0 \end{bmatrix}$
(e) $\begin{bmatrix} 2 & 3 & 4 \\ 0 & 1 & 2 \\ 0 & 0 & 3 \end{bmatrix}$
(f) $\begin{bmatrix} 0 & 0 & 0 \\ 0 & 0 & 0 \\ 0 & 0 & 0 \end{bmatrix}$

**3.** In each part suppose that the augmented matrix for a system of linear equations has been reduced by row operations to the given reduced row-echelon form. Solve the system.

(a) $\begin{bmatrix} 1 & 0 & 0 & 4 \\ 0 & 1 & 0 & 3 \\ 0 & 0 & 1 & 2 \end{bmatrix}$
(b) $\begin{bmatrix} 1 & 0 & 0 & 3 & 2 \\ 0 & 1 & 0 & -1 & 4 \\ 0 & 0 & 1 & 1 & 2 \end{bmatrix}$

(c) $\begin{bmatrix} 1 & 5 & 0 & 0 & 5 & -1 \\ 0 & 0 & 1 & 0 & 3 & 1 \\ 0 & 0 & 0 & 1 & 4 & 2 \\ 0 & 0 & 0 & 0 & 0 & 0 \end{bmatrix}$
(d) $\begin{bmatrix} 1 & 2 & 0 & 0 \\ 0 & 0 & 1 & 0 \\ 0 & 0 & 0 & 1 \end{bmatrix}$

**4.** In each part suppose that the augmented matrix for a system of linear equations has been reduced by row operations to the given row-echelon form. Solve the system.

(a) $\begin{bmatrix} 1 & 2 & -4 & 2 \\ 0 & 1 & -2 & -1 \\ 0 & 0 & 1 & 2 \end{bmatrix}$
(b) $\begin{bmatrix} 1 & 0 & 4 & 7 & 10 \\ 0 & 1 & -3 & -4 & -2 \\ 0 & 0 & 1 & 1 & 2 \end{bmatrix}$

(c) $\begin{bmatrix} 1 & 5 & -4 & 0 & -7 & -5 \\ 0 & 0 & 1 & 1 & 7 & 3 \\ 0 & 0 & 0 & 1 & 4 & 2 \\ 0 & 0 & 0 & 0 & 0 & 0 \end{bmatrix}$  (d) $\begin{bmatrix} 1 & 2 & 2 & 2 \\ 0 & 1 & 3 & 3 \\ 0 & 0 & 0 & 1 \end{bmatrix}$

5. Solve each of the following systems by Gauss-Jordan elimination.

(a) $\begin{aligned} 2x_1 + x_2 + x_3 &= 8 \\ 3x_1 - 2x_2 - x_3 &= 1 \\ 4x_1 - 7x_2 + 3x_3 &= 10 \end{aligned}$    (b) $\begin{aligned} x_1 + x_2 + x_3 &= 0 \\ -2x_1 + 5x_2 + 2x_3 &= 0 \\ -7x_1 + 7x_2 + x_3 &= 0 \end{aligned}$

(c) $\begin{aligned} x_1 + x_2 - 2x_3 + x_4 + 3x_5 &= 1 \\ 3x_1 + 2x_2 - 4x_3 - 3x_4 - 9x_5 &= 3 \\ 2x_1 - x_2 + 2x_3 + 2x_4 + 6x_5 &= 2 \\ 6x_1 + 2x_2 - 4x_3 &= 6 \\ 2x_2 - 4x_3 - 6x_4 - 18x_5 &= 0 \end{aligned}$

6. Solve each of the systems in Exercise 5 by Gaussian elimination.

7. Solve each of the following systems by Gauss-Jordan elimination.

(a) $\begin{aligned} 2x_1 - 3x_2 &= -2 \\ 2x_1 + x_2 &= 1 \\ 3x_1 + 2x_2 &= 1 \end{aligned}$    (b) $\begin{aligned} 3x_1 + 2x_2 - x_3 &= -15 \\ 5x_1 + 3x_2 + 2x_3 &= 0 \\ 3x_1 + x_2 + 3x_3 &= 11 \\ 11x_1 + 7x_2 &= -30 \end{aligned}$    (c) $\begin{aligned} 4x_1 - 8x_2 &= 12 \\ 3x_1 - 6x_2 &= 9 \\ -2x_1 + 4x_2 &= -6 \end{aligned}$

8. Solve each of the systems in Exercise 7 by Gaussian elimination.

9. Solve each of the following systems by Gauss-Jordan elimination.

(a) $\begin{aligned} 5x_1 + 2x_2 + 6x_3 &= 0 \\ -2x_1 + x_2 + 3x_3 &= 0 \end{aligned}$    (b) $\begin{aligned} x_1 - 2x_2 + x_3 - 4x_4 &= 1 \\ x_1 + 3x_2 + 7x_3 + 2x_4 &= 2 \\ x_1 - 12x_2 - 11x_3 - 16x_4 &= 5 \end{aligned}$

10. Solve each of the systems in Exercise 9 by Gaussian elimination.

11. Solve the following systems, where $a$, $b$, and $c$ are constants.

(a) $\begin{aligned} 2x + y &= a \\ 3x + 6y &= b \end{aligned}$    (b) $\begin{aligned} x_1 + x_2 + x_3 &= a \\ 2x_1 + 2x_3 &= b \\ 3x_2 + 3x_3 &= c \end{aligned}$

12. For which values of $a$ will the following system have no solutions? Exactly one solution? Infinitely many solutions?

$$\begin{aligned} x + 2y - 3z &= 4 \\ 3x - y + 5z &= 2 \\ 4x + y + (a^2 - 14)z &= a + 2 \end{aligned}$$

**13.** Solve the following system of nonlinear equations for the unknown angles $\alpha$, $\beta$, and $\gamma$, where $0 \leq \alpha \leq 2\pi$, $0 \leq \beta \leq 2\pi$ and $0 \leq \gamma < \pi$.

$$2 \sin \alpha - \cos \beta + 3 \tan \gamma = 3$$
$$4 \sin \alpha + 2 \cos \beta - 2 \tan \gamma = 2$$
$$6 \sin \alpha - 3 \cos \beta + \tan \gamma = 9$$

**14.** Describe the possible reduced row-echelon forms of

$$\begin{bmatrix} a & b & c \\ d & e & f \\ g & h & i \end{bmatrix}$$

**15.** Show that if $ad - bc \neq 0$, then the reduced row-echelon form of

$$\begin{bmatrix} a & b \\ c & d \end{bmatrix} \quad \text{is} \quad \begin{bmatrix} 1 & 0 \\ 0 & 1 \end{bmatrix}$$

**16.** Use Exercise 15 to show that if $ad - bc \neq 0$, then the system

$$ax + by = k$$
$$cx + dy = l$$

has exactly one solution.

## 1.3 HOMOGENEOUS SYSTEMS OF LINEAR EQUATIONS

As we have already pointed out, every system of linear equations has exactly one solution, infinitely many solutions, or no solutions at all. As we progress, there will be situations in which we will not be interested in finding solutions to a given system, but will rather be concerned with deciding how many solutions the system has. In this section we consider several cases in which it is possible to make statements about the number of solutions by inspection.

A system of linear equations is said to be *homogeneous* if all the constant terms are zero; that is, the system has the form

$$a_{11}x_1 + a_{12}x_2 + \cdots + a_{1n}x_n = 0$$
$$a_{21}x_1 + a_{22}x_2 + \cdots + a_{2n}x_n = 0$$
$$\vdots \qquad \vdots \qquad \qquad \vdots \qquad \vdots$$
$$a_{m1}x_1 + a_{m2}x_2 + \cdots + a_{mn}x_n = 0$$

Every homogeneous system of linear equations is consistent, since $x_1 = 0$, $x_2 = 0, \ldots, x_n = 0$ is always a solution. This solution is called the *trivial solution;* if there are other solutions, they are called *nontrivial solutions.*

Since a homogeneous system of linear equations must be consistent, there is either one solution or infinitely many solutions. Since one of these solutions is the trivial solution, we can make the following statement.

For a homogeneous system of linear equations, exactly one of the following is true.

1. *The system has only the trivial solution.*
2. *The system has infinitely many nontrivial solutions in addition to the trivial solution.*

There is one case in which a homogeneous system is assured of having nontrivial solutions; namely, whenever the system involves more unknowns than equations. To see why, consider the following example of four equations in five unknowns.

## Example 8

Solve the following homogeneous system of linear equations by Gauss-Jordan elimination.

$$\begin{aligned} 2x_1 + 2x_2 - x_3 \quad + x_5 &= 0 \\ -x_1 - x_2 + 2x_3 - 3x_4 + x_5 &= 0 \\ x_1 + x_2 - 2x_3 \quad - x_5 &= 0 \\ x_3 + x_4 + x_5 &= 0 \end{aligned} \tag{1.2}$$

The augmented matrix for the system is

$$\begin{bmatrix} 2 & 2 & -1 & 0 & 1 & 0 \\ -1 & -1 & 2 & -3 & 1 & 0 \\ 1 & 1 & -2 & 0 & -1 & 0 \\ 0 & 0 & 1 & 1 & 1 & 0 \end{bmatrix}$$

Reducing this matrix to reduced row-echelon form, we obtain

$$\begin{bmatrix} 1 & 1 & 0 & 0 & 1 & 0 \\ 0 & 0 & 1 & 0 & 1 & 0 \\ 0 & 0 & 0 & 1 & 0 & 0 \\ 0 & 0 & 0 & 0 & 0 & 0 \end{bmatrix}$$

The corresponding system of equations is

$$\begin{aligned} x_1 + x_2 \quad + x_5 &= 0 \\ x_3 \quad + x_5 &= 0 \\ x_4 \quad &= 0 \end{aligned} \tag{1.3}$$

Solving for the leading variables yields

$$\begin{aligned} x_1 &= -x_2 - x_5 \\ x_3 &= -x_5 \\ x_4 &= 0 \end{aligned}$$

The solution set is therefore given by

$$x_1 = -s - t, \qquad x_2 = s, \qquad x_3 = -t, \qquad x_4 = 0, \qquad x_5 = t$$

A moment's reflection will make it evident that if a matrix $A$ has a column of zeros, this column is left unchanged by performing a row operation on $A$. Thus, as illustrated in this example, if we begin with a homogeneous system of equations, then the system of equations corresponding to the reduced row-echelon form of the augmented matrix will also be homogeneous. Further, depending on whether the reduced row-echelon form of the augmented matrix has any zero rows, the number of equations in the reduced system is the same or less than the number of equations in the original system (compare equations 1.2 and 1.3 in Example 7). Suppose that a given homogeneous system of $m$ equations in $n$ unknowns has more unknowns than equations, and that after the augmented matrix has been reduced to reduced row-echelon form, the leading variables in the corresponding system of equations are $x_{k_1}, x_{k_2}, \ldots, x_{k_r}$, where $r < n$. The reduced system thus has the form

$$
\begin{aligned}
\cdots x_{k_1} \quad &+ \Sigma(\ \ ) = 0 \\
\cdots x_{k_2} \quad &+ \Sigma(\ \ ) = 0 \\
\ddots \quad &\quad \vdots \\
x_{k_r} + \Sigma(\ \ ) &= 0
\end{aligned}
\tag{1.4}
$$

where the symbol $\Sigma(\ \ )$ denotes sums that involve the $n - r$ nonleading variables. Solving for the leading variables gives

$$
\begin{aligned}
x_{k_1} &= -\Sigma(\ \ ) \\
x_{k_2} &= -\Sigma(\ \ ) \\
&\vdots \\
x_{k_r} &= -\Sigma(\ \ )
\end{aligned}
$$

As in Example 8, we can assign arbitrary values to the variables on the right-hand side and thus obtain infinitely many solutions to the system.

In summary, we have the following important theorem.

**Theorem 1.** *A homogeneous system of linear equations that has more unknowns than equations always has infinitely many solutions.*

## EXERCISE SET 1.3

1. Without using pencil and paper, determine which of the following homogeneous systems have nontrivial solutions.

(a) $\begin{aligned} x_1 + 3x_2 + 5x_3 + x_4 &= 0 \\ 4x_1 - 7x_2 - 3x_3 - x_4 &= 0 \\ 3x_1 + 2x_2 + 7x_3 + 8x_4 &= 0 \end{aligned}$

(b) $\begin{aligned} x_1 + 2x_2 + 3x_3 &= 0 \\ x_2 + 4x_3 &= 0 \\ 5x_3 &= 0 \end{aligned}$

(c) $\begin{aligned} a_{11}x_1 + a_{12}x_2 + a_{13}x_3 &= 0 \\ a_{21}x_1 + a_{22}x_2 + a_{23}x_3 &= 0 \end{aligned}$

(d) $\begin{aligned} x_1 + x_2 &= 0 \\ 2x_1 + 2x_2 &= 0 \end{aligned}$

In Exercises 2–5 solve the given homogeneous system of linear equations.

2.
$$2x_1 + x_2 + 3x_3 = 0$$
$$x_1 + 2x_2 \qquad = 0$$
$$x_2 + x_3 = 0$$

3.
$$3x_1 + x_2 + x_3 + x_4 = 0$$
$$5x_1 - x_2 + x_3 - x_4 = 0$$

4.
$$2x_1 - 4x_2 + x_3 + x_4 = 0$$
$$x_1 - 5x_2 + 2x_3 \qquad = 0$$
$$- 2x_2 - 2x_3 - x_4 = 0$$
$$x_1 + 3x_2 \qquad + x_4 = 0$$
$$x_1 - 2x_2 - x_3 + x_4 = 0$$

5.
$$x + 6y - 2z = 0$$
$$2x - 4y + z = 0$$

6. For which value(s) of $\lambda$ does the following system of equations have nontrivial solutions?

$$(\lambda - 3)x + \qquad y = 0$$
$$x + (\lambda - 3)y = 0$$

7. Consider the system of equations

$$ax + by = 0$$
$$cx + dy = 0$$
$$ex + fy = 0$$

Discuss the relative positions of the lines $ax + by = 0$, $cx + dy = 0$, and $ex + fy = 0$ when:
(a) the system has only the trivial solution
(b) the system has nontrivial solutions

8. Consider the system of equations

$$ax + by = 0$$
$$cx + dy = 0$$

(a) Show that if $x = x_0$, $y = y_0$ is any solution and $k$ is any constant, then $x = kx_0$, $y = ky_0$ is also a solution.
(b) Show that if $x = x_0$, $y = y_0$ and $x = x_1$, $y = y_1$ are any two solutions, then $x = x_0 + x_1$, $y = y_0 + y_1$ is also a solution.

9. Consider the systems of equations

$$(I) \quad \begin{matrix} ax + by = k \\ cx + dy = l \end{matrix} \qquad (II) \quad \begin{matrix} ax + by = 0 \\ cx + dy = 0 \end{matrix}$$

(a) Show that if $x = x_1$, $y = y_1$ and $x = x_2$, $y = y_2$ are both solutions of $I$, then $x = x_1 - x_2$, $y = y_1 - y_2$ is a solution of $II$.
(b) Show that if $x = x_1$, $y = y_1$ is a solution of $I$ and $x = x_0$, $y = y_0$ is a solution of $II$, then $x = x_1 + x_0$, $y = y_1 + y_0$ is a solution of $I$.

10. (a) In the system of equations numbered (1.4), explain why it would be incorrect to denote the leading variables by $x_1, x_2, \ldots, x_r$ rather than $x_{k_1}, x_{k_2}, \ldots, x_{k_r}$ as we have done.
(b) The system of equations numbered (1.3) is a specific case of (1.4). What value does $r$ have in this case? What are $x_{k_1}, x_{k_2}, \ldots, x_{k_r}$ in this case? Write out the sums denoted by $\Sigma(\ )$ in (1.4).

## 1.4 MATRICES AND MATRIX OPERATIONS

Rectangular arrays of real numbers arise in many contexts other than as augmented matrices for systems of linear equations. In this section we consider such arrays as objects in their own right and develop some of their properties for use in our later work.

**Definition.** A *matrix* is a rectangular array of numbers. The numbers in the array are called the *entries* in the matrix.

**Example 9**

The following are matrices.

$$\begin{bmatrix} 1 & 2 \\ 3 & 0 \\ -1 & 4 \end{bmatrix} \quad \begin{bmatrix} 2 & 1 & 0 & -3 \end{bmatrix} \quad \begin{bmatrix} -\sqrt{2} & \pi & e \\ 3 & \frac{1}{2} & 0 \\ 0 & 0 & 0 \end{bmatrix} \quad \begin{bmatrix} 1 \\ 3 \end{bmatrix} \quad [4]$$

As these examples indicate, matrices vary in size. The size of a matrix is described by specifying the number of rows (horizontal lines) and columns (vertical lines) that occur in the matrix. The first matrix in Example 9 has 3 rows and 2 columns so that its size is 3 by 2 (written $3 \times 2$). The first number always indicates the number of rows and the second indicates the number of columns. The remaining matrices in Example 9 thus have sizes $1 \times 4$, $3 \times 3$, $2 \times 1$, and $1 \times 1$ respectively.

We shall use capital letters to denote matrices and lower case letters to denote numerical quantities; thus, we might write

$$A = \begin{bmatrix} 2 & 1 & 7 \\ 3 & 4 & 2 \end{bmatrix} \quad \text{or} \quad C = \begin{bmatrix} a & b & c \\ d & e & f \end{bmatrix}$$

In accordance with common practice, numerical quantities will be referred to as *scalars.* In this text *all our scalars will be real numbers.*

If $A$ is a matrix, we will use $a_{ij}$ to denote the entry that occurs in row $i$ and column $j$ of $A$. A general $3 \times 4$ matrix can thus be written

$$A = \begin{bmatrix} a_{11} & a_{12} & a_{13} & a_{14} \\ a_{21} & a_{22} & a_{23} & a_{24} \\ a_{31} & a_{32} & a_{33} & a_{34} \end{bmatrix}$$

Naturally, if we use $B$ to denote the matrix, then we will use $b_{ij}$ for entry in row $i$ and column $j$. A general $m \times n$ matrix might thus be written

$$B = \begin{bmatrix} b_{11} & b_{12} & \cdots & b_{1n} \\ b_{21} & b_{22} & \cdots & b_{2n} \\ \vdots & \vdots & & \vdots \\ b_{m1} & b_{m2} & \cdots & b_{mn} \end{bmatrix}$$

A matrix, $A$, with $n$ rows and $n$ columns is called a **square matrix of order $n$,** and the entries $a_{11}, a_{22}, \ldots, a_{nn}$ are said to be on the **main diagonal** of $A$ (see Figure 1.2).

$$\begin{bmatrix} a_{11} & a_{12} & \cdots & a_{1n} \\ a_{21} & a_{22} & \cdots & a_{2n} \\ \vdots & \vdots & & \vdots \\ a_{n1} & a_{n2} & \cdots & a_{nn} \end{bmatrix}$$ Figure 1.2

So far we have used matrices to abbreviate the work in solving systems of linear equations. For other applications, however, it is desirable to develop an "arithmetic of matrices" in which matrices can be added and multiplied in a useful way. The remainder of this section will be devoted to developing this arithmetic.

Two matrices are said to be **equal** if they have the same size and the corresponding entries in the two matrices are equal.

**Example 10**

Consider the matrices

$$A = \begin{bmatrix} 2 & 1 \\ 3 & 4 \end{bmatrix} \qquad B = \begin{bmatrix} 2 & 1 \\ 3 & 5 \end{bmatrix} \qquad C = \begin{bmatrix} 2 & 1 & 0 \\ 3 & 4 & 0 \end{bmatrix}$$

Here $A \neq C$ since $A$ and $C$ do not have the same size. For the same reason $B \neq C$. Also, $A \neq B$ since not all the corresponding entries are equal.

**Definition.** If $A$ and $B$ are any two matrices of the same size, then the **sum $A + B$** is the matrix obtained by adding together the corresponding entries in the two matrices. Matrices with different sizes cannot be added.

**Example 11**

Consider the matrices

$$A = \begin{bmatrix} 2 & 1 & 0 & 3 \\ -1 & 0 & 2 & 4 \\ 4 & -2 & 7 & 0 \end{bmatrix} \qquad B = \begin{bmatrix} -4 & 3 & 5 & 1 \\ 2 & 2 & 0 & -1 \\ 3 & 2 & -4 & 5 \end{bmatrix} \qquad C = \begin{bmatrix} 1 & 1 \\ 2 & 2 \end{bmatrix}$$

Then

$$A + B = \begin{bmatrix} -2 & 4 & 5 & 4 \\ 1 & 2 & 2 & 3 \\ 7 & 0 & 3 & 5 \end{bmatrix}$$

while $A + C$ and $B + C$ are undefined.

**Definition.** If $A$ is any matrix and $c$ is any scalar, then the **product $cA$** is the matrix obtained by multiplying each entry of $A$ by $c$.

**Example 12**

If $A$ is the matrix

$$A = \begin{bmatrix} 4 & 2 \\ 1 & 3 \\ -1 & 0 \end{bmatrix}$$

then

$$2A = \begin{bmatrix} 8 & 4 \\ 2 & 6 \\ -2 & 0 \end{bmatrix} \quad \text{and} \quad (-1)A = \begin{bmatrix} -4 & -2 \\ -1 & -3 \\ 1 & 0 \end{bmatrix}$$

If $B$ is any matrix, then $-B$ will denote the product $(-1)B$. If $A$ and $B$ are two matrices of the same size, then $A - B$ is defined to be the sum $A + (-B) = A + (-1)B$.

**Example 13**

Consider the matrices

$$A = \begin{bmatrix} 2 & 3 & 4 \\ 1 & 2 & 1 \end{bmatrix} \quad \text{and} \quad B = \begin{bmatrix} 0 & 2 & 7 \\ 1 & -3 & 5 \end{bmatrix}$$

From the above definitions

$$-B = \begin{bmatrix} 0 & -2 & -7 \\ -1 & 3 & -5 \end{bmatrix}$$

and

$$A - B = \begin{bmatrix} 2 & 3 & 4 \\ 1 & 2 & 1 \end{bmatrix} + \begin{bmatrix} 0 & -2 & -7 \\ -1 & 3 & -5 \end{bmatrix} = \begin{bmatrix} 2 & 1 & -3 \\ 0 & 5 & -4 \end{bmatrix}$$

Observe that $A - B$ can be obtained directly by subtracting the entries of $B$ from the corresponding entries of $A$.

Above, we defined the multiplication of a matrix by a scalar. The next question, then, is how do we multiply two matrices together? Perhaps the most natural definition of matrix multiplication would seem to be "multiply corresponding entries together." Surprisingly, however, this definition would not be very useful for most problems. Experience has led mathematicians to the following less intuitive but more useful definition of matrix multiplication.

**Definition.** If $A$ is an $m \times r$ matrix and $B$ is an $r \times n$ matrix, then the **product** $AB$ is the $m \times n$ matrix whose entries are determined as follows. To find the entry in row $i$ and column $j$ of $AB$, single out row $i$ from the matrix $A$ and column $j$ from the matrix $B$. Multiply the corresponding entries from the row and column together and then add up the resulting products.

**Example 14**

Consider the matrices

$$A = \begin{bmatrix} 1 & 2 & 4 \\ 2 & 6 & 0 \end{bmatrix} \qquad B = \begin{bmatrix} 4 & 1 & 4 & 3 \\ 0 & -1 & 3 & 1 \\ 2 & 7 & 5 & 2 \end{bmatrix}$$

Since $A$ is a $2 \times 3$ matrix and $B$ is a $3 \times 4$ matrix, the product $AB$ is a $2 \times 4$ matrix. To determine, for example, the entry in row 2 and column 3 of $AB$, we single out row 2 from $A$ and column 3 from $B$. Then, as illustrated below, we multiply corresponding entries together and add up these products.

$$\begin{bmatrix} 1 & 2 & 4 \\ 2 & 6 & 0 \end{bmatrix} \begin{bmatrix} 4 & 1 & 4 & 3 \\ 0 & -1 & 3 & 1 \\ 2 & 7 & 5 & 2 \end{bmatrix} = \begin{bmatrix} \square & \square & \square & \square \\ \square & \square & 26 & \square \end{bmatrix}$$

$$(2 \cdot 4) + (6 \cdot 3) + (0 \cdot 5) = 26$$

The entry in row 1 and column 4 of $AB$ is computed as follows.

$$\begin{bmatrix} 1 & 2 & 4 \\ 2 & 6 & 0 \end{bmatrix} \begin{bmatrix} 4 & 1 & 4 & 3 \\ 0 & -1 & 3 & 1 \\ 2 & 7 & 5 & 2 \end{bmatrix} = \begin{bmatrix} \square & \square & \square & 13 \\ \square & \square & 26 & \square \end{bmatrix}$$

$$(1 \cdot 3) + (2 \cdot 1) + (4 \cdot 2) = 13$$

The computations for the remaining products are

$$\begin{array}{ll}
(1 \cdot 4) + (2 \cdot 0) + (4 \cdot 2) = & 12 \\
(1 \cdot 1) - (2 \cdot 1) + (4 \cdot 7) = & 27 \\
(1 \cdot 4) + (2 \cdot 3) + (4 \cdot 5) = & 30 \qquad AB = \begin{bmatrix} 12 & 27 & 30 & 13 \\ 8 & -4 & 26 & 12 \end{bmatrix} \\
(2 \cdot 4) + (6 \cdot 0) + (0 \cdot 2) = & 8 \\
(2 \cdot 1) - (6 \cdot 1) + (0 \cdot 7) = & -4 \\
(2 \cdot 3) + (6 \cdot 1) + (0 \cdot 2) = & 12
\end{array}$$

**Example 15**

If $A$ is a general $m \times r$ matrix and $B$ is a general $r \times n$ matrix, then, as suggested by the shading below, the entry in row $i$ and column $j$ of $AB$ is given by the formula

$$a_{i1}b_{1j} + a_{i2}b_{2j} + a_{i3}b_{3j} + \cdots + a_{ir}b_{rj}$$

$$AB = \begin{bmatrix} a_{11} & a_{12} & \cdots & a_{1r} \\ a_{21} & a_{22} & \cdots & a_{2r} \\ \vdots & \vdots & & \vdots \\ a_{i1} & a_{i2} & \cdots & a_{ir} \\ \vdots & \vdots & & \vdots \\ a_{m1} & a_{m2} & \cdots & a_{mr} \end{bmatrix} \begin{bmatrix} b_{11} & b_{12} & \cdots & b_{1j} & \cdots & b_{1n} \\ b_{21} & b_{22} & \cdots & b_{2j} & \cdots & b_{2n} \\ \vdots & \vdots & & \vdots & & \vdots \\ b_{r1} & b_{r2} & \cdots & b_{rj} & \cdots & b_{rn} \end{bmatrix}$$

The definition of matrix multiplication requires that the number of columns of the first factor $A$ be the same as the number of rows of the second factor $B$ in order to form the product $AB$. If this condition is not satisfied, the product is undefined. A convenient way to determine whether a product of two matrices is defined is to write down the size of the first factor and, to the right of it, write down the size of the second factor. If, as in Figure 1.3, the inside numbers are the same, then the product is defined. The outside numbers then give the size of the product.

**Example 16**

Suppose that $A$ is a $3 \times 4$ matrix, $B$ is a $4 \times 7$ matrix, and $C$ is a $7 \times 3$ matrix. Then $AB$ is defined and is a $3 \times 7$ matrix; $CA$ is defined and is a $7 \times 4$ matrix; $BC$ is defined and is a $4 \times 3$ matrix. The products $AC$, $CB$, and $BA$ are all undefined.

Matrix multiplication has an important application to systems of linear equations. Consider any system of $m$ linear equations in $n$ unknowns

$$\begin{aligned} a_{11}x_1 + a_{12}x_2 + \cdots + a_{1n}x_n &= b_1 \\ a_{21}x_1 + a_{22}x_2 + \cdots + a_{2n}x_n &= b_2 \\ \vdots \qquad\qquad \vdots \\ a_{m1}x_1 + a_{m2}x_2 + \cdots + a_{mn}x_n &= b_m \end{aligned}$$

Since two matrices are equal if and only if their corresponding entries are equal, we can replace the $m$ equations in this system by the single matrix equation

$$\begin{bmatrix} a_{11}x_1 + a_{12}x_2 + \cdots + a_{1n}x_n \\ a_{21}x_1 + a_{22}x_2 + \cdots + a_{2n}x_n \\ \vdots \qquad\qquad \vdots \\ a_{m1}x_1 + a_{m2}x_2 + \cdots + a_{mn}x_n \end{bmatrix} = \begin{bmatrix} b_1 \\ b_2 \\ \vdots \\ b_m \end{bmatrix}$$

Figure 1.3

The $m \times 1$ matrix on the left side of this equation can be written as a product to give

$$\begin{bmatrix} a_{11} & a_{12} & \cdots & a_{1n} \\ a_{21} & a_{22} & \cdots & a_{2n} \\ \vdots & \vdots & & \vdots \\ a_{m1} & a_{m2} & \cdots & a_{mn} \end{bmatrix} \begin{bmatrix} x_1 \\ x_2 \\ \vdots \\ x_n \end{bmatrix} = \begin{bmatrix} b_1 \\ b_2 \\ \vdots \\ b_m \end{bmatrix}$$

If we designate these matrices by $A$, $X$, and $B$, respectively, the original system of $m$ equations in $n$ unknowns has been replaced by the single matrix equation $AX = B$. Some of our later work will be devoted to solving matrix equations like this for the unknown matrix $X$. As a consequence of this matrix approach, we will obtain effective new methods for solving systems of linear equations.

**Example 17**

At times it is helpful to be able to find a particular row or column in a product $AB$ without computing the entire product. We leave it as an exercise to show that the entries in the $j$th column of $AB$ are the entries in the product $AB_j$, where $B_j$ is the matrix formed using only the $j$th column of $B$. Thus, if $A$ and $B$ are the matrices in Example 14, the second column of $AB$ can be obtained by the computation

$$\begin{bmatrix} 1 & 2 & 4 \\ 2 & 6 & 0 \end{bmatrix} \begin{bmatrix} 1 \\ -1 \\ 7 \end{bmatrix} = \begin{bmatrix} 27 \\ -4 \end{bmatrix}$$

$$\begin{array}{cc} \uparrow & \uparrow \\ \text{Second column} & \text{Second column} \\ \text{of } B & \text{of } AB \end{array}$$

Similarly, the entries in the $i$th row of $AB$ are the entries in the product $A_iB$, where $A_i$ is the matrix formed by using only the $i$th row of $A$. Thus the first row in the product $AB$ of Example 14 can be obtained by the computation

$$[1 \quad 2 \quad 4] \begin{bmatrix} 4 & 1 & 4 & 3 \\ 0 & -1 & 3 & 1 \\ 2 & 7 & 5 & 2 \end{bmatrix} = [12 \quad 27 \quad 30 \quad 13]$$

First row of $A$          First row of $AB$

## EXERCISE SET 1.4

1. Let $A$ and $B$ be $4 \times 5$ matrices and let $C$, $D$, and $E$ be $5 \times 2, 4 \times 2$, and $5 \times 4$ matrices, respectively. Determine which of the following matrix expressions are defined. For those which are defined, give the size of the resulting matrix.

(a) $BA$       (b) $AC + D$       (c) $AE + B$

(d) $AB + B$       (e) $E(A + B)$       (f) $E(AC)$

**2.** (a) Show that if $AB$ and $BA$ are both defined, then $AB$ and $BA$ are square matrices.
(b) Show that if $A$ is an $m \times n$ matrix and $A(BA)$ is defined, then $B$ is an $n \times m$ matrix.

**3.** Solve the following matrix equation for $a$, $b$, $c$, and $d$.

$$\begin{bmatrix} a-b & b+c \\ 3d+c & 2a-4d \end{bmatrix} = \begin{bmatrix} 8 & 1 \\ 7 & 6 \end{bmatrix}$$

*[handwritten: $a-b=8$, $3d+c=7$, $b+c=1$, $2a-4d=6$]*

**4.** Consider the matrices

$$A = \begin{bmatrix} 3 & 0 \\ -1 & 2 \\ 1 & 1 \end{bmatrix} \quad B = \begin{bmatrix} 4 & -1 \\ 0 & 2 \end{bmatrix} \quad C = \begin{bmatrix} 1 & 4 & 2 \\ 3 & 1 & 5 \end{bmatrix}$$

$$D = \begin{bmatrix} 1 & 5 & 2 \\ -1 & 0 & 1 \\ 3 & 2 & 4 \end{bmatrix} \quad E = \begin{bmatrix} 6 & 1 & 3 \\ -1 & 1 & 2 \\ 4 & 1 & 3 \end{bmatrix}$$

Compute

(a) $AB$    (b) $D + E$    (c) $D - E$
(d) $DE$    (e) $ED$    (f) $-7B$

**5.** Using the matrices from Exercise 4, compute (where possible)
(a) $3C - D$      (b) $(3E)D$
(c) $(AB)C$      (d) $A(BC)$
(e) $(4B)C + 2B$      (f) $D + E^2$(where $E^2 = EE$)

**6.** Let

$$A = \begin{bmatrix} 3 & -2 & 7 \\ 6 & 5 & 4 \\ 0 & 4 & 9 \end{bmatrix} \quad \text{and} \quad B = \begin{bmatrix} 6 & -2 & 4 \\ 0 & 1 & 3 \\ 7 & 7 & 5 \end{bmatrix}$$

Use the method of Example 17 to find
(a) the first row of $AB$      (b) the third row of $AB$
(c) the second column of $AB$      (d) the first column of $BA$
(e) the third row of $AA$      (f) the third column of $AA$

**7.** Let $C$, $D$, and $E$ be the matrices in Exercise 4. Using as few computations as possible, determine the entry in row 2 and column 3 of $C(DE)$.

**8.** (a) Show that if $A$ has a row of zeros and $B$ is any matrix for which $AB$ is defined, then $AB$ also has a row of zeros.
(b) Find a similar result involving a column of zeros.

**9.** Let $A$ be any $m \times n$ matrix and let $0$ be the $m \times n$ matrix each of whose entries is zero. Show that if $kA = 0$, then $k = 0$ or $A = 0$.

**10.** Let $I$ be the $n \times n$ matrix whose entry in row $i$ and column $j$ is

$$\begin{cases} 1 & \text{if} \quad i = j \\ 0 & \text{if} \quad i \neq j \end{cases}$$

Show that $AI = IA = A$ for every $n \times n$ matrix $A$.

11. A square matrix is called a ***diagonal matrix*** if all the entries off the main diagonal are zeros. Show that the product of diagonal matrices is again a diagonal matrix. State a rule for multiplying diagonal matrices.

12. (a) Show that the entries in the $j$th column of $AB$ are the entries in the product $AB_j$, where $B_j$ is the matrix formed from the $j$th column of $B$.
    (b) Show that the entries in the $i$th row of $AB$ are the entries in the product $A_i B$ where, $A_i$ is the matrix formed from the $i$th row of $A$.

## 1.5 RULES OF MATRIX ARITHMETIC

Although many of the rules of arithmetic for real numbers also hold for matrices, there are some exceptions. One of the most important exceptions occurs in the multiplication of matrices. For real numbers $a$ and $b$, we always have $ab = ba$. This property is often called the *commutative law for multiplication*. For matrices, however, $AB$ and $BA$ need not be equal. Equality can fail to hold for several reasons. It can happen, for example, that $AB$ is defined but $BA$ is undefined. This is the case if $A$ is a $2 \times 3$ matrix and $B$ is a $3 \times 4$ matrix. As a second possibility, it can happen that $AB$ and $BA$ are both defined but have different sizes. This is the situation, if $A$ is a $2 \times 3$ matrix and $B$ is a $3 \times 2$ matrix. Finally, as our next example shows, it is possible to have $AB \neq BA$ even if both $AB$ and $BA$ are defined and have the same size.

**Example 18**

Consider the matrices

$$A = \begin{bmatrix} -1 & 0 \\ 2 & 3 \end{bmatrix} \qquad B = \begin{bmatrix} 1 & 2 \\ 3 & 0 \end{bmatrix}$$

Multiplying gives

$$AB = \begin{bmatrix} -1 & -2 \\ 11 & 4 \end{bmatrix} \qquad BA = \begin{bmatrix} 3 & 6 \\ -3 & 0 \end{bmatrix}$$

Thus $AB \neq BA$.

Although the commutative law for multiplication is not valid in matrix arithmetic, many familiar laws of arithmetic are valid for matrices. Some of the most important ones and their names are summarized in the following theorem.

**Theorem 2.** *Assuming that the sizes of the matrices are such that the indicated operations can be performed, the following rules of matrix arithmetic are valid.*

(a) $A + B = B + A$          *(Commutative law for addition)*
(b) $A + (B + C) = (A + B) + C$    *(Associative law for addition)*

(c)  $A(BC) = (AB)C$      (*Associative law for multiplication*)
(d)  $A(B + C) = AB + AC$      (*Distributive law*)
(e)  $(B + C)A = BA + CA$      (*Distributive law*)
(f)  $A(B - C) = AB - AC$
(g)  $(B - C)A = BA - CA$
(h)  $a(B + C) = aB + aC$
(i)  $a(B - C) = aB - aC$
(j)  $(a + b)C = aC + bC$
(k)  $(a - b)C = aC - bC$
(l)  $(ab)C = a(bC)$
(m)  $a(BC) = (aB)C = B(aC)$

Each of the equations in this theorem asserts an equality between matrices. To verify one of these equalities, it is necessary to show that the matrix on the left side has the same size as the matrix on the right side, and that corresponding entries on the two sides are equal. To illustrate, we shall prove (*h*). Some of the remaining proofs are given as exercises.

*Proof of (h).* Since the left side involves the operation $B + C$, $B$ and $C$ must have the same size, say $m \times n$. It follows that $a(B + C)$ and $aB + aC$ are also $m \times n$ matrices, and thus have the same size.

Let $l_{ij}$ be any entry on the left side, and let $r_{ij}$ be the corresponding entry on the right side. To complete the proof, we must show $l_{ij} = r_{ij}$. If we let $a_{ij}$, $b_{ij}$, and $c_{ij}$ be the entries in the $i$th row and $j$th column of $A$, $B$, and $C$, respectively, then by direct computation, $l_{ij} = a(b_{ij} + c_{ij})$. Similarly on the right side, $r_{ij} = ab_{ij} + ac_{ij}$. Since $a(b_{ij} + c_{ij}) = ab_{ij} + ac_{ij}$, we have $l_{ij} = r_{ij}$, which completes the proof. ∎

Although the operations of matrix addition and matrix multiplication were defined for pairs of matrices, associative laws (*b*) and (*c*) enable us to denote sums and products of three matrices as $A + B + C$ and $ABC$ without inserting any parentheses. This is justified by the fact that no matter how parentheses are inserted, the associative laws guarantee that the same end result will be obtained. Without going into details, we observe that similar results are valid for sums or products involving four or more matrices. In general, *given any sum or any product of matrices, pairs of parentheses can be inserted or deleted anywhere within the expression without affecting the end result*.

**Example 19**

As an illustration of the associative law for matrix multiplication, consider

$$A = \begin{bmatrix} 1 & 2 \\ 3 & 4 \\ 0 & 1 \end{bmatrix} \quad B = \begin{bmatrix} 4 & 3 \\ 2 & 1 \end{bmatrix} \quad C = \begin{bmatrix} 1 & 0 \\ 2 & 3 \end{bmatrix}$$

Then

$$AB = \begin{bmatrix} 1 & 2 \\ 3 & 4 \\ 0 & 1 \end{bmatrix} \begin{bmatrix} 4 & 3 \\ 2 & 1 \end{bmatrix} = \begin{bmatrix} 8 & 5 \\ 20 & 13 \\ 2 & 1 \end{bmatrix}$$

and

$$(AB)C = \begin{bmatrix} 8 & 5 \\ 20 & 13 \\ 2 & 1 \end{bmatrix} \begin{bmatrix} 1 & 0 \\ 2 & 3 \end{bmatrix} = \begin{bmatrix} 18 & 15 \\ 46 & 39 \\ 4 & 3 \end{bmatrix}$$

On the other hand

$$BC = \begin{bmatrix} 4 & 3 \\ 2 & 1 \end{bmatrix} \begin{bmatrix} 1 & 0 \\ 2 & 3 \end{bmatrix} = \begin{bmatrix} 10 & 9 \\ 4 & 3 \end{bmatrix}$$

and

$$A(BC) = \begin{bmatrix} 1 & 2 \\ 3 & 4 \\ 0 & 1 \end{bmatrix} \begin{bmatrix} 10 & 9 \\ 4 & 3 \end{bmatrix} = \begin{bmatrix} 18 & 15 \\ 46 & 39 \\ 4 & 3 \end{bmatrix}$$

Thus $(AB)C = A(BC)$, as guaranteed by Theorem 2.

A matrix, all of whose entries are zero, such as

$$\begin{bmatrix} 0 & 0 \\ 0 & 0 \end{bmatrix}, \begin{bmatrix} 0 & 0 & 0 \\ 0 & 0 & 0 \\ 0 & 0 & 0 \end{bmatrix}, \begin{bmatrix} 0 & 0 & 0 & 0 \\ 0 & 0 & 0 & 0 \end{bmatrix}$$

is called a *zero matrix*. Zero matrices will be denoted by $0$; if it is important to emphasize the size, we shall write $0_{m \times n}$ for the $m \times n$ zero matrix.

If $A$ is any matrix and $0$ is the zero matrix with the same size, it is obvious that $A + 0 = A$. The matrix $0$ thus plays much the same role in this equation as the number 0 plays in the numerical equation $a + 0 = a$.

Since we already know that some of the rules of arithmetic for real numbers do not carry over to matrix arithmetic, it would be foolhardy to assume that all the properties of the real number zero carry over to zero matrices. To illustrate, consider the following two standard results in the arithmetic of real numbers.

 (i) If $ab = ac$ and $a \neq 0$, then $b = c$. (This is called the cancellation law.)
(ii) If $ad = 0$, then at least one of the factors on the left is 0.

As the next example shows, the corresponding results are false in matrix arithmetic.

**Example 20**

Consider the matrices

$$A = \begin{bmatrix} 0 & 1 \\ 0 & 2 \end{bmatrix} \quad B = \begin{bmatrix} 1 & 1 \\ 3 & 4 \end{bmatrix} \quad C = \begin{bmatrix} 2 & 5 \\ 3 & 4 \end{bmatrix} \quad D = \begin{bmatrix} 3 & 7 \\ 0 & 0 \end{bmatrix}$$

Here

$$AB = AC = \begin{bmatrix} 3 & 4 \\ 6 & 8 \end{bmatrix}$$

Although $A \neq 0$, it is *incorrect* to cancel the $A$ from both sides of the equation and write $B = C$. The cancellation law thus fails to hold for matrices.

Also, $AD = 0$; yet $A \neq 0$ and $D \neq 0$ so that result (ii) listed above does not carry over to matrix arithmetic.

In spite of these negative examples, a number of familiar properties of the real number 0 carry over to zero matrices. Some of the more important ones are summarized in the next theorem. The proofs are left as exercises.

**Theorem 3.** *Assuming that the sizes of the matrices are such that the indicated operations can be performed, the following rules of matrix arithmetic are valid.*

(a) $A + 0 = 0 + A = A$
(b) $A - A = 0$
(c) $0 - A = -A$
(d) $A0 = 0$;     $0A = 0$

As an application of our results on matrix arithmetic, we prove the following theorem, which was anticipated earlier in the text.

**Theorem 4.** *Every system of linear equations has no solutions, exactly one solution, or infinitely many solutions.*

*Proof.* If $AX = B$ is a system of linear equations, exactly one of the following is true: (a) the system has no solutions, (b) the system has exactly one solution, or (c) the system has more than one solution. The proof will be complete if we can show that the system has infinitely many solutions in case (c).

Assume that $AX = B$ has more than one solution; let $X_1$ and $X_2$ be two different solutions. Thus, $AX_1 = B$ and $AX_2 = B$. Subtracting these equations gives $AX_1 - AX_2 = 0$ or $A(X_1 - X_2) = 0$. If we let $X_0 = X_1 - X_2$ and let $k$ be any scalar, then

$$
\begin{aligned}
A(X_1 + kX_0) &= AX_1 + A(kX_0) \\
&= AX_1 + k(AX_0) \\
&= B + k0 \\
&= B + 0 \\
&= B
\end{aligned}
$$

But this says that $X_1 + kX_0$ is a solution of $AX = B$. Since there are infinitely many choices for $k$, $AX = B$ has infinitely many solutions. ∎

Of special interest are square matrices with 1's on the main diagonal and 0's off the main diagonal, such as

$$\begin{bmatrix} 1 & 0 \\ 0 & 1 \end{bmatrix}, \begin{bmatrix} 1 & 0 & 0 \\ 0 & 1 & 0 \\ 0 & 0 & 1 \end{bmatrix}, \begin{bmatrix} 1 & 0 & 0 & 0 \\ 0 & 1 & 0 & 0 \\ 0 & 0 & 1 & 0 \\ 0 & 0 & 0 & 1 \end{bmatrix}, \text{ etc.}$$

A matrix of this kind is called an **identity matrix** and is denoted by $I$. If it is important to emphasize the size, we shall write $I_n$ for the $n \times n$ identity matrix.

If $A$ is an $m \times n$ matrix, then, as illustrated in the next example, $AI_n = A$ and $I_mA = A$. Thus, an identity matrix plays much the same role in matrix arithmetic as the number 1 plays in the numerical relationships $a \cdot 1 = a$ and $1 \cdot a = a$.

**Example 21**

Consider the matrix

$$A = \begin{bmatrix} a_{11} & a_{12} & a_{13} \\ a_{21} & a_{22} & a_{23} \end{bmatrix}$$

Then

$$I_2A = \begin{bmatrix} 1 & 0 \\ 0 & 1 \end{bmatrix}\begin{bmatrix} a_{11} & a_{12} & a_{13} \\ a_{21} & a_{22} & a_{23} \end{bmatrix} = \begin{bmatrix} a_{11} & a_{12} & a_{13} \\ a_{21} & a_{22} & a_{23} \end{bmatrix} = A$$

and

$$AI_3 = \begin{bmatrix} a_{11} & a_{12} & a_{13} \\ a_{21} & a_{22} & a_{23} \end{bmatrix}\begin{bmatrix} 1 & 0 & 0 \\ 0 & 1 & 0 \\ 0 & 0 & 1 \end{bmatrix} = \begin{bmatrix} a_{11} & a_{12} & a_{13} \\ a_{21} & a_{22} & a_{23} \end{bmatrix} = A$$

**EXERCISE SET 1.5**

1. Let

$$A = \begin{bmatrix} 3 & 2 \\ -1 & 3 \end{bmatrix} \qquad B = \begin{bmatrix} 4 & 0 \\ 1 & 5 \end{bmatrix} \qquad C = \begin{bmatrix} 0 & -1 \\ 4 & 6 \end{bmatrix} \qquad a = -3 \qquad b = 2$$

Show
(a) $A + (B + C) = (A + B) + C$      (b) $(AB)C = A(BC)$
(c) $(a + b)C = aC + bC$      (d) $a(B - C) = aB - aC$

2. Using the matrices and scalars in Exercise 1, show
(a) $a(BC) = (aB)C = B(aC)$      (b) $A(B - C) = AB - AC$.

3. Find $2 \times 2$ matrices $A$ and $B$ such that

$$(A + B)^2 \neq A^2 + 2AB + B^2$$

4. Show that, if $A$ and $B$ are square matrices such that $AB = BA$, then

$$(A + B)^2 = A^2 + 2AB + B^2$$

5. Find an expansion of $(A + B)^2$ that is valid for all square matrices $A$ and $B$ having the same size.

6. Let $A$ and $B$ be square matrices of the same size. Is $(AB)^2 = A^2B^2$ a valid matrix identity? Justify your answer.

7. Let $A$, $B$, and $C$ be $n \times n$ matrices and suppose that $AC = CA = I$. In each part solve for $C$ in terms of $A$, $B$, and $I$.
   (a) $2C + 3B = A$
   (b) $C^2 + 3C = 2B$
   (c) $(A + C)^2 - C^2 = C + B$
   (d) $(BA)C - 4C^2 = 2A$

8. In Theorem 3 why didn't we write part (d) as $A0 = 0 = 0A$?

9. The real equation $a^2 = 1$ has exactly two solutions. Find at least eight different $3 \times 3$ matrices that satisfy the matrix equation $A^2 = I_3$. (**Hint.** Look for solutions in which all the entries off the main diagonal are zero.)

10. Let $AX = B$ be any consistent system of linear equations, and let $X_1$ be a fixed solution. Show that every solution to the system can be written in the form $X = X_1 + X_0$ where $X_0$ is a solution to $AX = 0$. Show also that every matrix of this form is a solution.

11. Apply parts (d) and (m) of Theorem 2 to the matrices $A$, $B$, and $(-1)C$ to derive the result in part (f).

12. Prove part (b) of Theorem 2.

13. Prove Theorem 3.

14. Prove part (c) of Theorem 2.

## 1.6 INVERSES OF MATRICES

It was shown in Section 1.4 that the problem of solving a system of linear equations can be reduced to solving a matrix equation $AX = B$ for the unknown matrix $X$. This matrix equation is analogous to the numerical equation $ax = b$, which can be solved, if $a \neq 0$, by multiplying both sides by the reciprocal (or multiplicative inverse) of $a$ to obtain $x = a^{-1}b$. In this section we shall develop the notion of the inverse of a matrix, and in later sections we shall use it to solve the matrix equation $AX = B$.

If $A$ is any square matrix, and if a matrix $B$ can be found such that $AB = BA = I$, then $A$ is said to be **invertible** and $B$ is called an **inverse** of $A$.

### Example 22

The matrix

$$B = \begin{bmatrix} 3 & 5 \\ 1 & 2 \end{bmatrix} \text{ is an inverse of } A = \begin{bmatrix} 2 & -5 \\ -1 & 3 \end{bmatrix}$$

since

$$AB = \begin{bmatrix} 2 & -5 \\ -1 & 3 \end{bmatrix}\begin{bmatrix} 3 & 5 \\ 1 & 2 \end{bmatrix} = \begin{bmatrix} 1 & 0 \\ 0 & 1 \end{bmatrix} = I$$

and

$$BA = \begin{bmatrix} 3 & 5 \\ 1 & 2 \end{bmatrix} \begin{bmatrix} 2 & -5 \\ -1 & 3 \end{bmatrix} = \begin{bmatrix} 1 & 0 \\ 0 & 1 \end{bmatrix} = I$$

## Example 23

The matrix

$$A = \begin{bmatrix} 1 & 4 & 0 \\ 2 & 5 & 0 \\ 3 & 6 & 0 \end{bmatrix}$$

is not invertible. To see why, let

$$B = \begin{bmatrix} b_{11} & b_{12} & b_{13} \\ b_{21} & b_{22} & b_{23} \\ b_{31} & b_{32} & b_{33} \end{bmatrix}$$

be any $3 \times 3$ matrix. From Example 17 the third column of $BA$ is

$$\begin{bmatrix} b_{11} & b_{12} & b_{13} \\ b_{21} & b_{22} & b_{23} \\ b_{31} & b_{32} & b_{33} \end{bmatrix} \begin{bmatrix} 0 \\ 0 \\ 0 \end{bmatrix} = \begin{bmatrix} 0 \\ 0 \\ 0 \end{bmatrix}$$

Thus

$$BA \neq I = \begin{bmatrix} 1 & 0 & 0 \\ 0 & 1 & 0 \\ 0 & 0 & 1 \end{bmatrix}$$

## Example 24

Consider the $2 \times 2$ matrix

$$A = \begin{bmatrix} a & b \\ c & d \end{bmatrix}$$

If $ad - bc \neq 0$, and if $B$ is the matrix

$$B = \frac{1}{ad - bc} \begin{bmatrix} d & -b \\ -c & a \end{bmatrix} = \begin{bmatrix} \dfrac{d}{ad - bc} & -\dfrac{b}{ad - bc} \\ -\dfrac{c}{ad - bc} & \dfrac{a}{ad - bc} \end{bmatrix} \tag{1.5}$$

then $AB = BA = I_2$. Thus, (1.5) gives a formula for an inverse of $A$. In the next section we shall show how to find inverses of invertible matrices whose sizes are greater than $2 \times 2$.

It is reasonable to ask whether an invertible matrix can have more than one inverse. The next theorem shows the answer is no; that is, an invertible matrix has exactly one inverse.

**Theorem 5.** *If B and C are both inverses of the matrix A, then $B = C$.*

*Proof.* Since $B$ is an inverse of $A$, $BA = I$. Multiplying both sides on the right by $C$ gives $(BA)C = IC = C$. But $(BA)C = B(AC) = BI = B$, so that $B = C$. ∎

As a consequence of this important result, we can now speak of "the" inverse of an invertible matrix. If $A$ is invertible, then its inverse will be denoted by the symbol $A^{-1}$.

**Theorem 6.** *If A and B are invertible matrices of the same size, then*

    *(a) AB is invertible*
    *(b) $(AB)^{-1} = B^{-1}A^{-1}$*

*Proof.* If we can show that $(AB)(B^{-1}A^{-1}) = (B^{-1}A^{-1})(AB) = I$, then we will have simultaneously established that $AB$ is invertible and that $(AB)^{-1} = B^{-1}A^{-1}$. But $(AB)(B^{-1}A^{-1}) = A(BB^{-1})A^{-1} = AIA^{-1} = AA^{-1} = I$. Similarly $(B^{-1}A^{-1})(AB) = I$. ∎

Although we will not prove it, this result can be extended to include three or more factors. We can thus state the following general rule. *A product of invertible matrices is always invertible, and the inverse of the product is the product of the inverses in the reverse order.*

**Example 25**

Consider the matrices

$$A = \begin{bmatrix} 1 & 2 \\ 1 & 3 \end{bmatrix} \quad B = \begin{bmatrix} 3 & 2 \\ 2 & 2 \end{bmatrix} \quad AB = \begin{bmatrix} 7 & 6 \\ 9 & 8 \end{bmatrix}$$

Applying the formula given in Example 24, we obtain

$$A^{-1} = \begin{bmatrix} 3 & -2 \\ -1 & 1 \end{bmatrix} \quad B^{-1} = \begin{bmatrix} 1 & -1 \\ -1 & \frac{3}{2} \end{bmatrix} \quad (AB)^{-1} = \begin{bmatrix} 4 & -3 \\ -\frac{9}{2} & \frac{7}{2} \end{bmatrix}$$

Also

$$B^{-1}A^{-1} = \begin{bmatrix} 1 & -1 \\ -1 & \frac{3}{2} \end{bmatrix} \begin{bmatrix} 3 & -2 \\ -1 & 1 \end{bmatrix} = \begin{bmatrix} 4 & -3 \\ -\frac{9}{2} & \frac{7}{2} \end{bmatrix}$$

Therefore $(AB)^{-1} = B^{-1}A^{-1}$.

If $A$ is a square matrix and $n$ is a positive integer, we define

$$A^n = \underbrace{AA \ldots A}_{n\text{-factors}}$$

If, in addition, $A$ is invertible, we define

$$A^{-n} = (A^{-1})^n = \underbrace{A^{-1}A^{-1} \ldots A^{-1}}_{n\text{-factors}}$$

As might be expected, we define $A^0 = I$.

**Theorem 7.** *If $A$ is an invertible matrix, then:*

(a) *For any nonzero scalar $k$, $kA$ is invertible and $(kA)^{-1} = \dfrac{1}{k} A^{-1}$.*

(b) *$A^{-1}$ is invertible and $(A^{-1})^{-1} = A$.*

(c) *$A^n$ is invertible and $(A^n)^{-1} = (A^{-1})^n$ for $n = 0, 1, 2, \ldots$.*

*Proof.* (a) If $k$ is any nonzero scalar, results ($l$) and ($m$) of Theorem 2 enable us to write

$$(kA)\left(\frac{1}{k} A^{-1}\right) = \frac{1}{k}(kA)A^{-1} = \left(\frac{1}{k} k\right)AA^{-1} = I$$

Similarly $\left(\dfrac{1}{k} A^{-1}\right)(kA) = I$ so that $kA$ is invertible and $(kA)^{-1} = \dfrac{1}{k}A^{-1}$.

(b) Since $AA^{-1} = A^{-1}A = I$, $A^{-1}$ is invertible and $(A^{-1})^{-1} = A$.

(c) This part is left as an exercise. ▪

We conclude this section by observing that if $A$ is a square matrix and $r$ and $s$ are integers, then the following familiar laws of exponents are valid:

$$A^r A^s = A^{r+s} \qquad (A^r)^s = A^{rs}$$

The proofs are left as exercises.

## EXERCISE SET 1.6

**1.** Use the formula given in Example 24 to compute the inverses of the following matrices.

$$A = \begin{bmatrix} 3 & 1 \\ 5 & 2 \end{bmatrix} \qquad B = \begin{bmatrix} 2 & -3 \\ 4 & 4 \end{bmatrix} \qquad C = \begin{bmatrix} 2 & 0 \\ 0 & 3 \end{bmatrix}$$

**2.** Verify that the matrices $A$ and $B$ in Exercise 1 satisfy the relationship $(AB)^{-1} = B^{-1}A^{-1}$.

**3.** Let $A$ be an invertible matrix whose inverse is

$$\begin{bmatrix} 3 & 4 \\ 5 & 6 \end{bmatrix}$$

Find the matrix $A$.

**4.** Let $A$ be an invertible matrix, and suppose that the inverse of $7A$ is

$$\begin{bmatrix} -1 & 2 \\ 4 & -7 \end{bmatrix}$$

Find the matrix $A$.

**5.** Let $A$ be the matrix

$$\begin{bmatrix} 1 & 0 \\ 2 & 3 \end{bmatrix}$$

Compute $A^3$, $A^{-3}$, and $3A^2 - 2A + I$.

**6.** Let $A$ be the matrix

$$\begin{bmatrix} 1 & 1 & 0 \\ 0 & 1 & 1 \\ 1 & 0 & 1 \end{bmatrix}$$

Determine if $A$ is invertible, and if so, find its inverse. (**Hint.** Solve $AX = I$ by equating corresponding entries on the two sides.)

**7.** Show that for every $\theta$

$$\begin{bmatrix} \cos \theta & -\sin \theta \\ -\sin \theta & -\cos \theta \end{bmatrix}^{-1} = \begin{bmatrix} \cos \theta & -\sin \theta \\ -\sin \theta & -\cos \theta \end{bmatrix}$$

**8.** Consider the matrix

$$A = \begin{bmatrix} a_{11} & 0 & 0 & \cdots & 0 \\ 0 & a_{22} & 0 & \cdots & 0 \\ \vdots & \vdots & \vdots & & \vdots \\ 0 & 0 & 0 & \cdots & a_{nn} \end{bmatrix}$$

where $a_{11}a_{22} \cdots a_{nn} \neq 0$. Show that $A$ is invertible and find its inverse.

**9.** Assume that $A$ is a square matrix which satisfies $A^2 - 3A + I = 0$. Show that $A^{-1} = 3I - A$.

**10.** (a) Show that a matrix with a row of zeros cannot have an inverse.
(b) Show that a matrix with a column of zeros cannot have an inverse.

**11.** Is the sum of two invertible matrices necessarily invertible?

**12.** Let $A$ and $B$ be square matrices such that $AB = 0$. Show that $A$ cannot be invertible unless $B = 0$.

**13.** Prove part (c) of Theorem 7.

**14.** Consider the laws of exponents $A^r A^s = A^{r+s}$ and $(A^r)^s = A^{rs}$.
(a) Show that if $A$ is any square matrix, these laws are valid for all non-negative integer values of $r$ and $s$.

(b) Show that if $A$ is invertible, then these laws are valid for all integer values of $r$ and $s$.

**15.** Show that if $A$ is invertible and $k$ is any nonzero scalar, then $(kA)^n = k^n A^n$ for all integer values of $n$.

**16.** (a) Show that if $A$ is invertible and $AB = AC$ then $B = C$.
(b) Explain why part (a) and Example 20 do not contradict one another.

## 1.7 ELEMENTARY MATRICES AND A METHOD FOR FINDING A$^{-1}$

In this section we shall develop a simple scheme or algorithm for finding the inverse of an invertible matrix.

In Section 1.2 we introduced three row operations on matrices: multiply a row through by a nonzero constant; interchange two rows; and add a multiple of one row to another. We shall call these *elementary row operations;* if a matrix $B$ can be obtained from a matrix $A$ by performing a finite sequence of elementary row operations, then we shall say that $B$ is *row equivalent* to $A$.

**Definition.** An $n \times n$ matrix is called an *elementary matrix* if it can be obtained from the $n \times n$ identity matrix $I_n$ by performing a single elementary row operation.

**Example 26**

Listed below are four elementary matrices and the operations that produce them.

$$\text{(i) } \begin{bmatrix} 1 & 0 \\ 0 & -3 \end{bmatrix} \quad \text{(ii) } \begin{bmatrix} 1 & 0 & 0 & 0 \\ 0 & 0 & 0 & 1 \\ 0 & 0 & 1 & 0 \\ 0 & 1 & 0 & 0 \end{bmatrix} \quad \text{(iii) } \begin{bmatrix} 1 & 0 & 3 \\ 0 & 1 & 0 \\ 0 & 0 & 1 \end{bmatrix} \quad \text{(iv) } \begin{bmatrix} 1 & 0 & 0 \\ 0 & 1 & 0 \\ 0 & 0 & 1 \end{bmatrix}$$

$$\begin{pmatrix} \text{Multiply the} \\ \text{second row of} \\ I_2 \text{ by } -3 \end{pmatrix} \begin{pmatrix} \text{Interchange the second} \\ \text{and fourth rows of} \\ I_4 \end{pmatrix} \begin{pmatrix} \text{Add three times} \\ \text{the third row of} \\ I_3 \text{ to the first row} \end{pmatrix} \begin{pmatrix} \text{Multiply the} \\ \text{first row of} \\ I_3 \text{ by } 1 \end{pmatrix}$$

We now observe without proof that *an elementary row operation can be performed indirectly on an $m \times n$ matrix $A$ by first performing the row operation on the identity matrix $I_m$, and then multiplying $A$ on the left by the resulting elementary matrix.* The following example will illustrate.

**Example 27**

Consider the matrix

$$A = \begin{bmatrix} 1 & 0 & 2 & 3 \\ 2 & -1 & 3 & 6 \\ 1 & 4 & 4 & 0 \end{bmatrix}$$

If we add 3 times the first row of $A$ to the third, we obtain the matrix

$$B = \begin{bmatrix} 1 & 0 & 2 & 3 \\ 2 & -1 & 3 & 6 \\ 4 & 4 & 10 & 9 \end{bmatrix}$$

On the other hand, if we perform the same row operation on the $3 \times 3$ identity matrix, the resulting elementary matrix is

$$E = \begin{bmatrix} 1 & 0 & 0 \\ 0 & 1 & 0 \\ 3 & 0 & 1 \end{bmatrix}$$

A direct computation will show that $EA = B$.

This result is primarily of theoretical interest and will be used for developing some results about matrices and systems of linear equations. From a computational viewpoint, it is preferable to perform row operations directly rather than use this indirect procedure. The following theorem formalizes the observations we made above.

**Theorem 8.** *Let $A$ be an $m \times n$ matrix and let $B$ be a matrix obtained by performing a single row operation on $A$. If $E$ is the elementary matrix that results when the same row operation is performed on $I_m$, then $B = EA$.*

If an elementary row operation is applied to an identity matrix $I$ to produce an elementary matrix $E$, then there is a second row operation which, when applied to $E$, produces $I$ back again. For example, if $E$ is obtained by multiplying the $i$th row of $I$ by a nonzero constant $c$, then $I$ can be recovered if the $i$th row of $E$ is multiplied by $1/c$. The various possibilities are listed in Figure 1.4.

The operations on the right side of this table are called the ***inverse operations*** of the corresponding operations on the left.

| Row operation on I that produces E | Row operation on E that reproduces I |
|---|---|
| Multiply row $i$ by $c \neq 0$. | Multiply row $i$ by $1/c$. |
| Interchange rows $i$ and $j$. | Interchange rows $i$ and $j$. |
| Add $c$ times row $i$ to row $j$. | Add $-c$ times row $i$ to row $j$ |

**Figure 1.4**

**Example 28**

Using the results in Figure 1.4, the first three elementary matrices given in Example 26 can be restored to identity matrices by applying the following row operations: multiply the second row in (i) by $-(1/3)$; interchange the second and fourth rows in (ii); add $-3$ times the third row of (iii) to the first row.

The next theorem gives an important property of elementary matrices.

**Theorem 9.** *Every elementary matrix is invertible, and its inverse is also an elementary matrix.*

*Proof.* If $E$ is any elementary matrix, then by the observations above, $I$ can be obtained from $E$ by performing a single elementary row operation. Let $E_0$ be the elementary matrix that results when this row operation is performed on $I$. Applying Theorem 8, we find that

$$E_0 E = I \qquad (1.6)$$

To complete the proof, we shall show that

$$EE_0 = I$$

Since $E_0$ is an elementary matrix, there is an elementary row operation which when applied to $E_0$, produces $I$. Let $E_1$ be the elementary matrix that results when this row operation is performed on $I$. Applying Theorem 8 again we obtain

$$E_1 E_0 = I \qquad (1.7)$$

Multiplying both sides of equation (1.6) on the left by $E_1$ gives $E_1 E_0 E = E_1$ or $IE = E_1$ or $E = E_1$. Substituting $E$ for $E_1$ in equation (1.7) gives $EE_0 = I$, which completes the proof. ▊

The next theorem establishes some fundamental relationships between $n \times n$ matrices and systems of $n$ linear equations in $n$ unknowns. These results are extremely important and will be used many times in later sections.

**Theorem 10.** *If $A$ is an $n \times n$ matrix, then the following statements are equivalent, that is, are all true or all false.*

(a) *$A$ is invertible.*
(b) *$AX = 0$ has only the trivial solution.*
(c) *$A$ is row equivalent to $I_n$.*

*Proof.* We shall prove the equivalence by establishing the following chain of implications: $(a) \Longrightarrow (b) \Longrightarrow (c) \Longrightarrow (a)$.

$(a) \Rightarrow (b)$: Assume $A$ is invertible and let $X_0$ be any solution to $AX = 0$. Thus $AX_0 = 0$. Multiplying both sides of this equation by $A^{-1}$ gives $A^{-1}(AX_0) = A^{-1}0$, or $(A^{-1}A)X_0 = 0$, or $IX_0 = 0$, or $X_0 = 0$. Thus $AX = 0$ has only the trivial solution.

$(b) \Rightarrow (c)$: Let $AX = 0$ be the matrix form of the system

$$\begin{aligned}
a_{11}x_1 + a_{12}x_2 + \cdots + a_{1n}x_n &= 0 \\
a_{21}x_1 + a_{22}x_2 + \cdots + a_{2n}x_n &= 0 \\
\vdots \qquad \vdots \qquad\qquad \vdots \qquad \vdots \\
a_{n1}x_1 + a_{n2}x_2 + \cdots + a_{nn}x_n &= 0
\end{aligned} \tag{1.8}$$

and assume the system has only the trivial solution. If we solve by Gauss-Jordan elimination, then the system of equations corresponding to the reduced row-echelon form of the augmented matrix will be

$$\begin{aligned}
x_1 \qquad\qquad\quad &= 0 \\
x_2 \qquad\quad &= 0 \\
\ddots \\
x_n &= 0
\end{aligned} \tag{1.9}$$

Thus the augmented matrix

$$\begin{bmatrix}
a_{11} & a_{12} & \cdots & a_{1n} & 0 \\
a_{21} & a_{22} & \cdots & a_{2n} & 0 \\
\vdots & \vdots & & \vdots & \vdots \\
a_{n1} & a_{n2} & \cdots & a_{nn} & 0
\end{bmatrix}$$

for (1.8) can be reduced to the augmented matrix

$$\begin{bmatrix}
1 & 0 & 0 & \cdots & 0 & 0 \\
0 & 1 & 0 & \cdots & 0 & 0 \\
0 & 0 & 1 & \cdots & 0 & 0 \\
\vdots & \vdots & \vdots & & \vdots & \vdots \\
0 & 0 & 0 & \cdots & 1 & 0
\end{bmatrix}$$

for (1.9) by a sequence of elementary row operations. If we disregard the last column (of zeros) in each of these matrices, we can conclude that $A$ can be reduced to $I_n$ by a sequence of elementary row operations; that is, $A$ is row equivalent to $I_n$.

$(c) \Rightarrow (a)$: Assume $A$ is row equivalent to $I_n$, so that $A$ can be reduced to $I_n$ by a finite sequence of elementary row operations. By Theorem 8 each of these operations can be accomplished by multiplying on the left by an appropriate elementary matrix. Thus, we can find elementary matrices $E_1, E_2, \ldots, E_k$ such that

$$E_k \ldots E_2 E_1 A = I_n \tag{1.10}$$

By Theorem 9, $E_1, E_2, \ldots, E_k$ are invertible. Multiplying both sides of equation (1.10) on the left successively by $E_k^{-1}, \ldots, E_2^{-1}, E_1^{-1}$ we obtain

$$A = E_1^{-1}E_2^{-1} \ldots E_k^{-1}I_n = E_1^{-1}E_2^{-1} \ldots E_k^{-1} \tag{1.11}$$

Since (1.11) expresses $A$ as a product of invertible matrices, we can conclude that $A$ is invertible. ∎

As our first application of this theorem, we shall establish a method for determining the inverse of an invertible matrix.

Inverting the left and right sides of (1.11) yields $A^{-1} = E_k \ldots E_2E_1$, or equivalently

$$A^{-1} = E_k \ldots E_2E_1I_n \tag{1.12}$$

which tells us that $A^{-1}$ can be obtained by multiplying $I_n$ successively on the left by the elementary matrices $E_1, E_2, \ldots, E_k$. Since each multiplication on the left by one of these elementary matrices performs a row operation, it follows, by comparing equations (1.10) and (1.12), that *the sequence of row operations that reduces $A$ to $I_n$ will reduce $I_n$ to $A^{-1}$*. Thus, to find the inverse of an invertible matrix $A$, we must find a sequence of elementary row operations that reduces $A$ to the identity and then perform this same sequence of operations on $I_n$ to obtain $A^{-1}$. A simple method for carrying out this procedure is given in the following example.

**Example 29**

Find the inverse of

$$A = \begin{bmatrix} 1 & 2 & 3 \\ 2 & 5 & 3 \\ 1 & 0 & 8 \end{bmatrix}$$

We wish to reduce $A$ to the identity matrix by row operations and simultaneously apply these operations to $I$ to produce $A^{-1}$. This can be accomplished by adjoining the identity matrix to the right of $A$ and applying row operations to both sides until the left side is reduced to $I$. The final matrix will then have the form $[I | A^{-1}]$. The computations can be carried out as follows.

$$\left[ \begin{array}{ccc|ccc} 1 & 2 & 3 & 1 & 0 & 0 \\ 2 & 5 & 3 & 0 & 1 & 0 \\ 1 & 0 & 8 & 0 & 0 & 1 \end{array} \right]$$

$$\left[ \begin{array}{ccc|ccc} 1 & 2 & 3 & 1 & 0 & 0 \\ 0 & 1 & -3 & -2 & 1 & 0 \\ 0 & -2 & 5 & -1 & 0 & 1 \end{array} \right] \quad \left( \begin{array}{l} \text{We added } -2 \text{ times the first} \\ \text{row to the second and } -1 \text{ times} \\ \text{the first row to the third.} \end{array} \right)$$

$$\left[ \begin{array}{ccc|ccc} 1 & 2 & 3 & 1 & 0 & 0 \\ 0 & 1 & -3 & -2 & 1 & 0 \\ 0 & 0 & -1 & -5 & 2 & 1 \end{array} \right] \quad \left( \begin{array}{l} \text{We added 2 times the} \\ \text{second row to the third.} \end{array} \right)$$

$$\begin{bmatrix} 1 & 2 & 3 & \vdots & 1 & 0 & 0 \\ 0 & 1 & -3 & \vdots & -2 & 1 & 0 \\ 0 & 0 & 1 & \vdots & 5 & -2 & -1 \end{bmatrix}$$ $\left(\begin{array}{l}\text{We multiplied the} \\ \text{third row by } -1.\end{array}\right)$

$$\begin{bmatrix} 1 & 2 & 0 & \vdots & -14 & 6 & 3 \\ 0 & 1 & 0 & \vdots & 13 & -5 & -3 \\ 0 & 0 & 1 & \vdots & 5 & -2 & -1 \end{bmatrix}$$ $\left(\begin{array}{l}\text{We added 3 times the third} \\ \text{row to the second and } -3 \text{ times} \\ \text{the third row to the first.}\end{array}\right)$

$$\begin{bmatrix} 1 & 0 & 0 & \vdots & -40 & 16 & 9 \\ 0 & 1 & 0 & \vdots & 13 & -5 & -3 \\ 0 & 0 & 1 & \vdots & 5 & -2 & -1 \end{bmatrix}$$ $\left(\begin{array}{l}\text{We added } -2 \text{ times the} \\ \text{second row to the first.}\end{array}\right)$

Thus

$$A^{-1} = \begin{bmatrix} -40 & 16 & 9 \\ 13 & -5 & -3 \\ 5 & -2 & -1 \end{bmatrix}$$

Often it will not be known in advance whether a given matrix is invertible. If the procedure used in this example is attempted on a matrix that is not invertible, then by part (c) of Theorem 10, it is impossible to reduce the left side to $I$ by row operations. At some point in the computation a row of zeros will occur on the left side. It can then be concluded that the given matrix is not invertible and the computations can be stopped.

**Example 30**

Consider the matrix

$$A = \begin{bmatrix} 1 & 6 & 4 \\ 2 & 4 & -1 \\ -1 & 2 & 5 \end{bmatrix}$$

Applying the procedure of Example 29 yields

$$\begin{bmatrix} 1 & 6 & 4 & \vdots & 1 & 0 & 0 \\ 2 & 4 & -1 & \vdots & 0 & 1 & 0 \\ -1 & 2 & 5 & \vdots & 0 & 0 & 1 \end{bmatrix}$$

$$\begin{bmatrix} 1 & 6 & 4 & \vdots & 1 & 0 & 0 \\ 0 & -8 & -9 & \vdots & -2 & 1 & 0 \\ 0 & 8 & 9 & \vdots & 1 & 0 & 1 \end{bmatrix}$$ $\left(\begin{array}{l}\text{We added } -2 \text{ times the first} \\ \text{row to the second and added} \\ \text{the first row to the third.}\end{array}\right)$

$$\begin{bmatrix} 1 & 6 & 4 & \vdots & 1 & 0 & 0 \\ 0 & -8 & -9 & \vdots & -2 & 1 & 0 \\ 0 & 0 & 0 & \vdots & -1 & 1 & 1 \end{bmatrix}$$ $\left(\begin{array}{l}\text{We added the second} \\ \text{row to the third.}\end{array}\right)$

Since we have obtained a row of zeros on the left side, $A$ is not invertible.

**Example 31**

In Example 29 we showed that

$$A = \begin{bmatrix} 1 & 2 & 3 \\ 2 & 5 & 3 \\ 1 & 0 & 8 \end{bmatrix}$$

is an invertible matrix. From Theorem 10 we can now conclude that the system of equations

$$\begin{array}{rcl} x_1 + 2x_2 + 3x_3 & = & 0 \\ 2x_1 + 5x_2 + 3x_3 & = & 0 \\ x_1 \qquad\quad + 8x_3 & = & 0 \end{array}$$

has only the trivial solution.

**EXERCISE SET 1.7**

1. Which of the following are elementary matrices?

(a) $\begin{bmatrix} 2 & 0 \\ 0 & 1 \end{bmatrix}$  (b) $\begin{bmatrix} 1 & 0 \\ 3 & 1 \end{bmatrix}$  (c) $\begin{bmatrix} 2 & 0 \\ 0 & 2 \end{bmatrix}$

(d) $\begin{bmatrix} 0 & 1 & 0 \\ 1 & 0 & 0 \\ 0 & 0 & 1 \end{bmatrix}$  (e) $\begin{bmatrix} 0 & 1 & 0 \\ 0 & 0 & 1 \\ 1 & 0 & 0 \end{bmatrix}$

(f) $\begin{bmatrix} 1 & 0 & 0 \\ 0 & 1 & -3 \\ 0 & 0 & 1 \end{bmatrix}$  (g) $\begin{bmatrix} 1 & 0 & 0 & 0 \\ 0 & 1 & 0 & 0 \\ 0 & 1 & 1 & 0 \\ 0 & 0 & 0 & 1 \end{bmatrix}$

2. Determine the row operation that will restore the given elementary matrix to an identity matrix.

(a) $\begin{bmatrix} 1 & 0 \\ 5 & 1 \end{bmatrix}$  (b) $\begin{bmatrix} 0 & 0 & 1 \\ 0 & 1 & 0 \\ 1 & 0 & 0 \end{bmatrix}$  (c) $\begin{bmatrix} 1 & 0 & 0 & 0 \\ 0 & 8 & 0 & 0 \\ 0 & 0 & 1 & 0 \\ 0 & 0 & 0 & 1 \end{bmatrix}$

3. Consider the matrices

$$A = \begin{bmatrix} 1 & 2 & 3 \\ 4 & 5 & 6 \\ 7 & 8 & 9 \end{bmatrix} \quad B = \begin{bmatrix} 7 & 8 & 9 \\ 4 & 5 & 6 \\ 1 & 2 & 3 \end{bmatrix} \quad C = \begin{bmatrix} 1 & 2 & 3 \\ 4 & 5 & 6 \\ 9 & 12 & 15 \end{bmatrix}$$

Find elementary matrices $E_1$, $E_2$, $E_3$, and $E_4$ such that
(a) $E_1A = B$  (b) $E_2B = A$  (c) $E_3A = C$  (d) $E_4C = A$

**4.** In Exercise 3 is it possible to find an elementary matrix $E$ such that $EB = C$? Justify your answer.

In Exercises 5–7 use the method shown in Examples 29 and 30 to find the inverse of the given matrix if the matrix is invertible.

**5.** (a) $\begin{bmatrix} 1 & 2 \\ 3 & 5 \end{bmatrix}$  (b) $\begin{bmatrix} -2 & 3 \\ 3 & -5 \end{bmatrix}$  (c) $\begin{bmatrix} 8 & -6 \\ -4 & 3 \end{bmatrix}$

**6.** (a) $\begin{bmatrix} 3 & 4 & -1 \\ 1 & 0 & 3 \\ 2 & 5 & -4 \end{bmatrix}$  (b) $\begin{bmatrix} 3 & 1 & 5 \\ 2 & 4 & 1 \\ -4 & 2 & -9 \end{bmatrix}$  (c) $\begin{bmatrix} 1 & 0 & 1 \\ 0 & 1 & 1 \\ 1 & 1 & 0 \end{bmatrix}$

(d) $\begin{bmatrix} 2 & 6 & 6 \\ 2 & 7 & 6 \\ 2 & 7 & 7 \end{bmatrix}$  (e) $\begin{bmatrix} 1 & 0 & 1 \\ -1 & 1 & 1 \\ 0 & 1 & 0 \end{bmatrix}$  (f) $\begin{bmatrix} \dfrac{1}{5} & \dfrac{1}{5} & \dfrac{1}{5} \\ \dfrac{1}{5} & \dfrac{1}{5} & -\dfrac{4}{5} \\ -\dfrac{2}{5} & \dfrac{1}{10} & \dfrac{1}{10} \end{bmatrix}$

**7.** (a) $\begin{bmatrix} \dfrac{1}{\sqrt{2}} & \dfrac{1}{\sqrt{2}} & 0 \\ -\dfrac{1}{\sqrt{2}} & \dfrac{1}{\sqrt{2}} & 0 \\ 0 & 0 & 1 \end{bmatrix}$  (b) $\begin{bmatrix} 1 & 0 & 0 & 0 \\ 1 & 2 & 0 & 0 \\ 1 & 2 & 4 & 0 \\ 1 & 2 & 4 & 8 \end{bmatrix}$  (c) $\begin{bmatrix} 5 & 11 & 7 & 3 \\ 2 & 1 & 4 & -5 \\ 3 & -2 & 8 & 7 \\ 0 & 0 & 0 & 0 \end{bmatrix}$

**8.** Show that the matrix

$$A = \begin{bmatrix} \cos\theta & \sin\theta & 0 \\ -\sin\theta & \cos\theta & 0 \\ 0 & 0 & 1 \end{bmatrix}$$

is invertible for all values of $\theta$ and find $A^{-1}$.

**9.** Consider the matrix

$$A = \begin{bmatrix} 1 & 0 \\ 3 & 4 \end{bmatrix}$$

(a) Find elementary matrices $E_1$ and $E_2$ such that $E_2 E_1 A = I$.
(b) Write $A^{-1}$ as a product of two elementary matrices.
(c) Write $A$ as a product of two elementary matrices.

**10.** Perform the following row operations on

$$A = \begin{bmatrix} 3 & 1 & 0 \\ -2 & 1 & 4 \\ 3 & 5 & 5 \end{bmatrix}$$

by multiplying $A$ on the left by a suitable elementary matrix. Check your answer in each case by performing the row operation directly on $A$.
(a) Interchange the first and third rows.

(b) Multiply the second row by $1/3$.

(c) Add twice the second row to the first row.

**11.** Express the matrix

$$A = \begin{bmatrix} 1 & 3 & 3 & 8 \\ -2 & -5 & 1 & -8 \\ 0 & 1 & 7 & 8 \end{bmatrix}$$

in the form $A = EFR$, where $E$, $F$, and $R$ are elementary matrices, and $R$ is in row-echelon form.

**12.** Show that if

$$A = \begin{bmatrix} 1 & 0 & 0 \\ 0 & 1 & 0 \\ a & b & c \end{bmatrix}$$

is an elementary matrix, then at least one entry in the third row must be a zero.

**13.** Find the inverse of each of the following $4 \times 4$ matrices, where $k_1$, $k_2$, $k_3$, $k_4$ and $k$ are all nonzero.

(a) $\begin{bmatrix} k_1 & 0 & 0 & 0 \\ 0 & k_2 & 0 & 0 \\ 0 & 0 & k_3 & 0 \\ 0 & 0 & 0 & k_4 \end{bmatrix}$  (b) $\begin{bmatrix} 0 & 0 & 0 & k_1 \\ 0 & 0 & k_2 & 0 \\ 0 & k_3 & 0 & 0 \\ k_4 & 0 & 0 & 0 \end{bmatrix}$  (c) $\begin{bmatrix} k & 0 & 0 & 0 \\ 1 & k & 0 & 0 \\ 0 & 1 & k & 0 \\ 0 & 0 & 1 & k \end{bmatrix}$

**14.** Prove that if $A$ is an $m \times n$ matrix, there is an invertible matrix $C$ such that $CA$ is in reduced row-echelon form.

**15.** Prove that if $A$ is an invertible matrix and $B$ is row equivalent to $A$, then $B$ is also invertible.

## 1.8 FURTHER RESULTS ON SYSTEMS OF EQUATIONS AND INVERTIBILITY

In this section we shall establish more results about systems of linear equations and invertibility of matrices. Our work will lead to a method for solving $n$ equations in $n$ unknowns that is more efficient than Gaussian elimination for certain kinds of problems.

**Theorem 11.** *If $A$ is an invertible $n \times n$ matrix, then for each $n \times 1$ matrix $B$, the system of equations $AX = B$ has exactly one solution, namely, $X = A^{-1}B$.*

*Proof.* Since $A(A^{-1}B) = B$, $X = A^{-1}B$ is a solution of $AX = B$. To show that this is the only solution, we will assume that $X_0$ is an arbitrary solution, and then show that $X_0$ must be the solution $A^{-1}B$.

If $X_0$ is any solution, then $AX_0 = B$. Multiplying both sides by $A^{-1}$, we obtain $X_0 = A^{-1}B$. ▮

**Example 32**

Consider the system of linear equations

$$\begin{aligned} x_1 + 2x_2 + 3x_3 &= 5 \\ 2x_1 + 5x_2 + 3x_3 &= 3 \\ x_1 \qquad\quad + 8x_3 &= 17 \end{aligned}$$

In matrix form this system can be written as $AX = B$, where

$$A = \begin{bmatrix} 1 & 2 & 3 \\ 2 & 5 & 3 \\ 1 & 0 & 8 \end{bmatrix} \quad X = \begin{bmatrix} x_1 \\ x_2 \\ x_3 \end{bmatrix} \quad B = \begin{bmatrix} 5 \\ 3 \\ 17 \end{bmatrix}$$

In Example 29 we showed that $A$ is invertible and

$$A^{-1} = \begin{bmatrix} -40 & 16 & 9 \\ 13 & -5 & -3 \\ 5 & -2 & -1 \end{bmatrix}$$

By Theorem 11 the solution of the system is

$$X = A^{-1}B = \begin{bmatrix} -40 & 16 & 9 \\ 13 & -5 & -3 \\ 5 & -2 & -1 \end{bmatrix} \begin{bmatrix} 5 \\ 3 \\ 17 \end{bmatrix} = \begin{bmatrix} 1 \\ -1 \\ 2 \end{bmatrix}$$

or $x_1 = 1$, $x_2 = -1$, $x_3 = 2$.

**Example 33**

In Example 24 we showed that a $2 \times 2$ matrix

$$A = \begin{bmatrix} a & b \\ c & d \end{bmatrix}$$

is invertible if $ad - bc \neq 0$. If this condition is satisfied, then by Theorem 11 the system of equations

$$\begin{aligned} ax_1 + bx_2 &= k_1 \\ cx_1 + dx_2 &= k_2 \end{aligned}$$

has a unique solution for all values of $k_1$ and $k_2$. From the formula for $A^{-1}$ given in Example 24, this solution is

$$\begin{bmatrix} x_1 \\ x_2 \end{bmatrix} = \frac{1}{ad - bc} \begin{bmatrix} d & -b \\ -c & a \end{bmatrix} \begin{bmatrix} k_1 \\ k_2 \end{bmatrix} = \begin{bmatrix} \dfrac{k_1 d - bk_2}{ad - bc} \\ \dfrac{ak_2 - k_1 c}{ad - bc} \end{bmatrix}$$

or
$$x_1 = \frac{k_1 d - b k_2}{ad - bc} \qquad x_2 = \frac{a k_2 - k_1 c}{ad - bc}$$

We now consider an important class of problems for which the method of solution given in the last two examples is more efficient than Gaussian elimination. In certain applied problems, physical systems are considered that can be described as *black boxes*. This term indicates that the system has been stripped to its bare essentials. As illustrated in Figure 1.5, one imagines simply that if a certain input is applied to the system, then a certain output for the system will result. The internal workings of the system are either unknown or unimportant for the problem—hence, the term *black box*. For many important black-box systems, both the input and the output can be described mathematically as matrices having a single column. For example, if the black box consists of certain electronic circuitry, then the input might be an $n \times 1$ matrix whose entries are $n$ voltages read across certain input terminals, and the output might be an $m \times 1$ matrix whose entries are the resulting currents in $m$ output wires. Mathematically speaking, such a system does nothing more than transform an $n \times 1$ input matrix into an $m \times 1$ output matrix. For a large class of black-box systems the input and output matrices are related in the following simple way. Associated with the system is a certain fixed $m \times n$ matrix $A$ whose entries are physical parameters determined by the system. The $m \times 1$ output matrix $B$ that results from an $n \times 1$ input matrix $C$ can be obtained by forming the product $AC$; that is, $B$ and $C$ are related by the equation

$$AC = B$$

A system of this kind is an example of what is called a *linear physical system*. In applications it is often of importance to determine what input must be applied to the system to achieve a certain desired output. For a linear physical system of the type we have just discussed, this amounts to solving the equation $AX = B$ for the unknown input $X$, given the desired output $B$.

Suppose now that $A$ is a square matrix and we have a succession of different output matrices $B_1, \ldots, B_k$. If we want to determine the input matrices that produce these given outputs, then we must successively solve the $k$ systems of linear equations

$$AX = B_j \qquad j = 1, 2, \ldots, k$$

**Figure 1.5**

The thing to notice about these systems is that they all have the same coefficient matrix $A$. If we follow the method of Examples 32 and 33, then these $k$ systems can be solved by first computing $A^{-1}$ and then successively forming the products $A^{-1}B_1, A^{-1}B_2, \ldots, A^{-1}B_k$. The solution of all $k$ systems thus involves one matrix inversion and $k$ matrix multiplications. This procedure is more efficient than separately applying Gaussian elimination to each of the $k$ systems.

In our later work the following fundamental problem will occur over and over again in various contexts.

**A Fundamental Problem.** Let $A$ be a fixed $m \times n$ matrix. Find all $m \times 1$ matrices $B$ such that the system of equations $AX = B$ is consistent.

If $A$ is an invertible matrix, Theorem 11 completely solves this problem by asserting that for every $m \times 1$ matrix $B$, $AX = B$ has the unique solution $X = A^{-1}B$. If $A$ is not square, or if $A$ is square but not invertible, then Theorem 11 does not apply. In these cases we would like to determine what conditions, if any, the matrix $B$ must satisfy in order for $AX = B$ to be consistent. The following example illustrates how Gaussian elimination can be used to determine such conditions.

**Example 34**

What conditions must $b_1$, $b_2$, and $b_3$ satisfy in order for the system of equations

$$\begin{aligned}
x_1 + x_2 + 2x_3 &= b_1 \\
x_1 \phantom{{}+ x_2} + x_3 &= b_2 \\
2x_1 + x_2 + 3x_3 &= b_3
\end{aligned}$$

to be consistent?

*Solution.* The augmented matrix is

$$\begin{bmatrix} 1 & 1 & 2 & b_1 \\ 1 & 0 & 1 & b_2 \\ 2 & 1 & 3 & b_3 \end{bmatrix}$$

which can be reduced to row-echelon form as follows.

$$\begin{bmatrix} 1 & 1 & 2 & b_1 \\ 0 & -1 & -1 & b_2 - b_1 \\ 0 & -1 & -1 & b_3 - 2b_1 \end{bmatrix}$$
$\left( \begin{array}{l} -1 \text{ times the first row was added} \\ \text{to the second and } -2 \text{ times the} \\ \text{first row was added to the third.} \end{array} \right)$

$$\begin{bmatrix} 1 & 1 & 2 & b_1 \\ 0 & 1 & 1 & b_1 - b_2 \\ 0 & -1 & -1 & b_3 - 2b_1 \end{bmatrix}$$
$\left( \begin{array}{l} \text{The second row was} \\ \text{multiplied by } -1. \end{array} \right)$

$$\begin{bmatrix} 1 & 1 & 1 & b_2 \\ 0 & 1 & 1 & b_1 - b_2 \\ 0 & 0 & 0 & b_3 - b_2 - b_1 \end{bmatrix}$$
$\left( \begin{array}{l} \text{The second row was added} \\ \text{to the third.} \end{array} \right)$

It is now evident from the third row in the matrix that the system has a solution if and only if $b_1$, $b_2$ and $b_3$ satisfy the condition

$$b_3 - b_2 - b_1 = 0 \quad \text{or} \quad b_3 = b_1 + b_2$$

To express this condition another way, $AX = B$ is consistent if and only if $B$ is a matrix of the form

$$B = \begin{bmatrix} b_1 \\ b_2 \\ b_1 + b_2 \end{bmatrix}$$

where $b_1$ and $b_2$ are arbitrary.

The next theorem simplifies the problem of showing a matrix is invertible. Up to now, to show an $n \times n$ matrix $A$ is invertible, it was necessary to find an $n \times n$ matrix $B$ such that

$$AB = I \quad \text{and} \quad BA = I$$

The next theorem shows that if we produce an $n \times n$ matrix $B$ satisfying *either* condition, then the other condition holds automatically.

**Theorem 12.** *Let $A$ be a square matrix.*

(a) *If $B$ is a square matrix satisfying $BA = I$, then $B = A^{-1}$.*
(b) *If $B$ is a square matrix satisfying $AB = I$, then $B = A^{-1}$.*

*Proof.* We shall prove (a) and leave (b) as an exercise.
(a) Assume $BA = I$. If we can show that $A$ is invertible, the proof can be completed by multiplying $BA = I$ on both sides by $A^{-1}$ to obtain

$$BAA^{-1} = IA^{-1} \quad \text{or} \quad BI = IA^{-1} \quad \text{or} \quad B = A^{-1}$$

We shall show that the system $AX = 0$ has only the trivial solution, and thus conclude from Theorem 10 that $A$ is invertible. Multiplying both sides of $AX = 0$ on the left by $B$, we obtain $BAX = B0$ or $IX = 0$ or $X = 0$. The system of equations $AX = 0$ therefore has only the trivial solution. ▮

The following corollary adds a fourth statement equivalent to the three given in Theorem 10.

**Corollary.** *If $A$ is an $n \times n$ matrix, then the following statements are equivalent.*

(a) *$A$ is invertible.*
(b) *$AX = 0$ has only the trivial solution.*
(c) *$A$ is row equivalent to $I_n$.*
(d) *$AX = B$ is consistent for every $n \times 1$ matrix $B$.*

*Proof.* See Exercise 14.

**EXERCISE SET 1.8**

In Exercises 1–6 solve the system using the method of Example 32.

1. $\begin{aligned} x_1 + 2x_2 &= 7 \\ 2x_1 + 5x_2 &= -3 \end{aligned}$

2. $\begin{aligned} 3x_1 - 6x_2 &= 8 \\ 2x_1 + 5x_2 &= 1 \end{aligned}$

3. $\begin{aligned} x_1 + 2x_2 + 2x_3 &= -1 \\ x_1 + 3x_2 + x_3 &= 4 \\ x_1 + 3x_2 + 2x_3 &= 3 \end{aligned}$

4. $\begin{aligned} 2x_1 + x_2 + x_3 &= 7 \\ 3x_1 + 2x_2 + x_3 &= -3 \\ x_2 + x_3 &= 5 \end{aligned}$

5. $\begin{aligned} \frac{1}{5}x + \frac{1}{5}y + \frac{1}{5}z &= 1 \\ \frac{1}{5}x + \frac{1}{5}y - \frac{4}{5}z &= 2 \\ -\frac{2}{5}x + \frac{1}{10}y + \frac{1}{10}z &= 0 \end{aligned}$

6. $\begin{aligned} 3w + x + 7y + 9z &= 4 \\ w + x + 4y + 4z &= 7 \\ -w - 2y - 3z &= 0 \\ -2w - x - 4y - 6z &= 6 \end{aligned}$

7. Solve the system

$$\begin{aligned} x_1 + 2x_2 + x_3 &= b_1 \\ x_1 - x_2 + x_3 &= b_2 \\ x_1 + x_2 &= b_3 \end{aligned}$$

when

(a) $b_1 = -1, b_2 = 3, b_3 = 4$    (b) $b_1 = 5, b_2 = 0, b_3 = 0$

(c) $b_1 = -1, b_2 = -1, b_3 = 3$   (d) $b_1 = \frac{1}{2}, b_2 = 3, b_3 = \frac{1}{7}$

8. What conditions must the $b$'s satisfy in order for the given system to be consistent?

(a) $\begin{aligned} x_1 - x_2 + 3x_3 &= b_1 \\ 3x_1 - 3x_2 + 9x_3 &= b_2 \\ -2x_1 + 2x_2 - 6x_3 &= b_3 \end{aligned}$

(b) $\begin{aligned} 2x_1 + 3x_2 - x_3 + x_4 &= b_1 \\ x_1 + 5x_2 + x_3 - 2x_4 &= b_2 \\ -x_1 + 2x_2 + 2x_3 - 3x_4 &= b_3 \\ 3x_1 + x_2 - 3x_3 + 4x_4 &= b_4 \end{aligned}$

9. Consider the matrices

$$A = \begin{bmatrix} 2 & 2 & 3 \\ 1 & 2 & 1 \\ 2 & -2 & 1 \end{bmatrix} \quad \text{and} \quad X = \begin{bmatrix} x_1 \\ x_2 \\ x_3 \end{bmatrix}$$

(a) Show that the equation $AX = X$ can be rewritten as $(A - I)X = 0$ and use this result to solve $AX = X$ for $X$.

(b) Solve $AX = 4X$.

10. Without using pencil and paper, determine if the following matrices are invertible.

(a) $\begin{bmatrix} 2 & 1 & -3 & 1 \\ 0 & 5 & 4 & 3 \\ 0 & 0 & 1 & 2 \\ 0 & 0 & 0 & 3 \end{bmatrix}$   (b) $\begin{bmatrix} 5 & 1 & 4 & 1 \\ 0 & 0 & 2 & -1 \\ 0 & 0 & 1 & 1 \\ 0 & 0 & 0 & 7 \end{bmatrix}$

*Hint.* Consider the associated homogeneous systems

$$2x_1 + x_2 - 3x_3 + x_4 = 0$$
$$5x_2 + 4x_3 + 3x_4 = 0$$
$$\text{and}$$
$$x_3 + 2x_4 = 0$$
$$3x_4 = 0$$

$$5x_1 + x_2 + 4x_3 + x_4 = 0$$
$$2x_3 - x_4 = 0$$
$$x_3 + x_4 = 0$$
$$7x_4 = 0$$

11. Let $AX = 0$ be a homogeneous system of $n$ linear equations in $n$ unknowns that has only the trivial solution. Show that if $k$ is any positive integer, then the system $A^k X = 0$ also has only the trivial solution.

12. Let $AX = 0$ be a homogeneous system of $n$ linear equations in $n$ unknowns, and let $Q$ be an invertible matrix. Show that $AX = 0$ has just the trivial solution if and only if $(QA)X = 0$ has just the trivial solution.

13. Show that an $n \times n$ matrix $A$ is invertible if and only if it can be written as a product of elementary matrices.

14. Prove the corollary to Theorem 12.

15. Use part (a) of Theorem 12 to prove part (b). (**Hint.** Interchange the matrices $A$ and $B$ in part *a*.)

# Determinants

## 2.1 THE DETERMINANT FUNCTION

We are all familiar with functions like $f(x) = \sin x$ and $f(x) = x^2$, which associate a real number $f(x)$ with a real value of the variable $x$. Since both $x$ and $f(x)$ assume only real values, such functions can be described as real-valued functions of a real variable. In this section we initiate the study of real-valued functions of a matrix variable, that is, functions that associate a real number $f(X)$ with a matrix $X$. Our primary effort will be devoted to the study of one such function called the determinant function. Our work on the determinant function will have important applications to the theory of systems of linear equations and will also lead us to an explicit formula for the inverse of an invertible matrix.

Before we shall be able to define the determinant function, it will be necessary to establish some results concerning permutations.

**Definition.** A *permutation* of the set of integers $\{1, 2, \ldots, n\}$ is an arrangement of these integers in some order without repetitions.

**Example 1**

There are six different permutations of the set of integers $\{1, 2, 3\}$. These are

$$(1, 2, 3) \quad (2, 1, 3) \quad (3, 1, 2)$$
$$(1, 3, 2) \quad (2, 3, 1) \quad (3, 2, 1)$$

One convenient method for systematically listing permutations is to use a *permutation tree*. This method will be illustrated in our next example.

**Example 2**

List all permutations of the set of integers $\{1, 2, 3, 4\}$.

Consider the following figure.

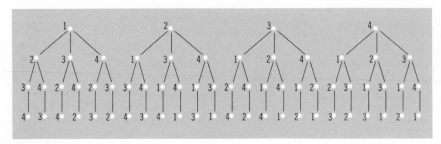

**Figure 2.1**

The four dots labeled $1, 2, 3, 4$ at the top of Figure 2.1 represent the possible choices for the first number in the permutation. The three branches emanating from these dots represent the possible choices for the second position in the permutation. Thus, if the permutation begins $(2, -, -, -)$, the three possibilities for the second position are 1, 3, and 4. The two branches emanating from each dot in the second position represent the possible choices for the third position. Thus, if the permutation begins $(2, 3, -, -,)$, the two possible choices for the third position are 1 and 4. Finally, the single branch emanating from each dot in the third position represents the only possible choice for the fourth position. Thus, if the permutation begins $(2, 3, 4, -)$, the only choice for the fourth position is 1. The different permutations can now be listed by tracing out all the possible paths through the "tree" from the first position to the last position. We obtain the following list by this process.

| | | | |
|---|---|---|---|
| $(1, 2, 3, 4)$ | $(2, 1, 3, 4)$ | $(3, 1, 2, 4)$ | $(4, 1, 2, 3)$ |
| $(1, 2, 4, 3)$ | $(2, 1, 4, 3)$ | $(3, 1, 4, 2)$ | $(4, 1, 3, 2)$ |
| $(1, 3, 2, 4)$ | $(2, 3, 1, 4)$ | $(3, 2, 1, 4)$ | $(4, 2, 1, 3)$ |
| $(1, 3, 4, 2)$ | $(2, 3, 4, 1)$ | $(3, 2, 4, 1)$ | $(4, 2, 3, 1)$ |
| $(1, 4, 2, 3)$ | $(2, 4, 1, 3)$ | $(3, 4, 1, 2)$ | $(4, 3, 1, 2)$ |
| $(1, 4, 3, 2)$ | $(2, 4, 3, 1)$ | $(3, 4, 2, 1)$ | $(4, 3, 2, 1)$ |

From this example we see that there are 24 permutations of $\{1, 2, 3, 4\}$. This result could have been anticipated without actually listing the permutations by arguing as follows. Since the first position can be filled in 4 ways and then the second position in 3 ways, there are $4 \cdot 3$ ways of filling the first two positions. Since the third position can then be filled in 2 ways, there are $4 \cdot 3 \cdot 2$ ways of filling the first 3 positions. Finally, since the last position can then be filled in only 1 way, there are $4 \cdot 3 \cdot 2 \cdot 1 = 24$ ways of filling all four positions. In general, the set $\{1, 2, \ldots, n\}$ will have $n(n-1)(n-2) \cdots 2 \cdot 1 = n!$ different permutations.

To denote a general permutation of the set $\{1, 2, \ldots, n\}$, we shall write $(j_1, j_2, \ldots, j_n)$. Here, $j_1$ is the first integer in the permutation, $j_2$ is the second, etc. An *inversion* is said to occur in a permutation $(j_1, j_2, \ldots, j_n)$ whenever a larger integer precedes a smaller one. A moment's reflection will make it evident that the total number of inversions occurring in a permutation can be obtained as follows. (1) Count the number of integers that are less than $j_1$ and that follow $j_1$ in the permutation. (2) Count the number of integers that are less than $j_2$ and that follow $j_2$ in the permutation. Continue this counting process for $j_3, \ldots, j_{n-1}$. The sum of these numbers will be the number of inversions in the permutation.

**Example 3**

Determine the number of inversions in the following permutations:

$$\text{(i) } (6, 1, 3, 4, 5, 2) \quad \text{(ii) } (2, 4, 1, 3) \quad \text{(iii) } (1, 2, 3, 4)$$

(i) The number of inversions is $5 + 0 + 1 + 1 + 1 = 8$.
(ii) The number of inversions is $1 + 2 + 0 = 3$.
(iii) There are no inversions in this permutation.

**Definition.** A permutation is called *even* if the total number of inversions is an even integer and is called *odd* if the total number of inversions is an odd integer.

**Example 4**

The following table classifies the various permutations of $\{1, 2, 3\}$ as even or odd.

| Permutation | Number of inversions | Classification |
|:---:|:---:|:---:|
| $(1, 2, 3)$ | 0 | even |
| $(1, 3, 2)$ | 1 | odd |
| $(2, 1, 3)$ | 1 | odd |
| $(2, 3, 1)$ | 2 | even |
| $(3, 1, 2)$ | 2 | even |
| $(3, 2, 1)$ | 3 | odd |

Consider the $n \times n$ matrix

$$A = \begin{bmatrix} a_{11} & a_{12} & \cdots & a_{1n} \\ a_{21} & a_{22} & \cdots & a_{2n} \\ \vdots & \vdots & & \vdots \\ a_{n1} & a_{n2} & \cdots & a_{nn} \end{bmatrix}$$

By an *elementary product from A* we shall mean any product of $n$ entries from $A$, no two of which come from the same row or same column.

### Example 5

List all elementary products from the matrices

$$(i) \begin{bmatrix} a_{11} & a_{12} \\ a_{21} & a_{22} \end{bmatrix} \quad (ii) \begin{bmatrix} a_{11} & a_{12} & a_{13} \\ a_{21} & a_{22} & a_{23} \\ a_{31} & a_{32} & a_{33} \end{bmatrix}$$

(i) Since each elementary product has two factors, and since each factor comes from a different row, an elementary product can be written in the form

$$a_{1\_}a_{2\_}$$

where the blanks designate column numbers. Since no two factors in the product come from the same column, the column numbers must be 1 2 or 2 1. The only elementary products are therefore $a_{11}a_{22}$ and $a_{12}a_{21}$.

(ii) Since each elementary product has three factors, each of which comes from a different row, an elementary product can be written in the form

$$a_{1\_}a_{2\_}a_{3\_}$$

Since no two factors in the product come from the same column, the column numbers have no repetitions; consequently, they must form a permutation of the set $\{1, 2, 3\}$. These $3! = 6$ permutations yield the following list of elementary products.

$$\begin{array}{ccc} a_{11}a_{22}a_{33} & a_{12}a_{21}a_{33} & a_{13}a_{21}a_{32} \\ a_{11}a_{23}a_{32} & a_{12}a_{23}a_{31} & a_{13}a_{22}a_{31} \end{array}$$

As this example points out, an $n \times n$ matrix $A$ has $n!$ elementary products. They are the products of the form $a_{1j_1}a_{2j_2} \ldots a_{nj_n}$, where $(j_1, j_2, \ldots, j_n)$ is a permutation of the set $\{1, 2, \ldots, n\}$. By a *signed elementary product from A* we shall mean an elementary product $a_{1j_1}a_{2j_2} \ldots a_{nj_n}$ multiplied by $+1$ or $-1$. We use the $+$ if $(j_1, j_2, \ldots, j_n)$ is an even permutation and the $-$ if $(j_1, j_2, \ldots, j_n)$ is an odd permutation.

### Example 6

List all signed elementary products from the matrices

$$(i) \begin{bmatrix} a_{11} & a_{12} \\ a_{21} & a_{22} \end{bmatrix} \quad (ii) \begin{bmatrix} a_{11} & a_{12} & a_{13} \\ a_{21} & a_{22} & a_{23} \\ a_{31} & a_{32} & a_{33} \end{bmatrix}$$

(i)

| *Elementary product* | *Associated permutation* | *Even or odd* | *Signed elementary product* |
|---|---|---|---|
| $a_{11}a_{22}$ | $(1, 2)$ | even | $a_{11}a_{22}$ |
| $a_{12}a_{21}$ | $(2, 1)$ | odd | $-a_{12}a_{21}$ |

| (ii) Elementary product | Associated permutation | Even or odd | Signed elementary product |
|---|---|---|---|
| $a_{11}a_{22}a_{33}$ | $(1,2,3)$ | even | $a_{11}a_{22}a_{33}$ |
| $a_{11}a_{23}a_{32}$ | $(1,3,2)$ | odd | $-a_{11}a_{23}a_{32}$ |
| $a_{12}a_{21}a_{33}$ | $(2,1,3)$ | odd | $-a_{12}a_{21}a_{33}$ |
| $a_{12}a_{23}a_{31}$ | $(2,3,1)$ | even | $a_{12}a_{23}a_{31}$ |
| $a_{13}a_{21}a_{32}$ | $(3,1,2)$ | even | $a_{13}a_{21}a_{32}$ |
| $a_{13}a_{22}a_{31}$ | $(3,2,1)$ | odd | $-a_{13}a_{22}a_{31}$ |

We are now in a position to define the determinant function.

**Definition.** Let $A$ be a square matrix. The **determinant function** is denoted by **det**, and we define $\det(A)$ to be the sum of all signed elementary products from $A$.

**Example 7**

Referring to Example 6, we obtain

(i) $\det\begin{bmatrix} a_{11} & a_{12} \\ a_{21} & a_{22} \end{bmatrix} = a_{11}a_{22} - a_{12}a_{21}$

(ii) $\det\begin{bmatrix} a_{11} & a_{12} & a_{13} \\ a_{21} & a_{22} & a_{23} \\ a_{31} & a_{32} & a_{33} \end{bmatrix} = a_{11}a_{22}a_{33} + a_{12}a_{23}a_{31} + a_{13}a_{21}a_{32}$
$$- a_{13}a_{22}a_{31} - a_{12}a_{21}a_{33} - a_{11}a_{23}a_{32}$$

It is useful to have the two formulas in this example available for ready reference. To avoid memorizing these unwieldy expressions, however, we suggest using the mnemonic devices illustrated in Figure 2.2.

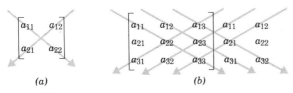

(a)                    (b)

**Figure 2.2**

The first formula in Example 7 is obtained from Figure 2.2(a) by multiplying the entries on the rightward arrow and subtracting the product of the entries on the leftward arrow. The second formula in Example 7 is obtained by repeating the first and second columns as shown in Figure 2.2(b). The determinant is then computed by summing up the products on the rightward arrows and subtracting off the products on the leftward arrows.

**Example 8**

Evaluate the determinants of

$$A = \begin{bmatrix} 3 & 1 \\ 4 & -2 \end{bmatrix} \quad \text{and} \quad B = \begin{bmatrix} 1 & 2 & 3 \\ -4 & 5 & 6 \\ 7 & -8 & 9 \end{bmatrix}$$

Using the method of Figure 2.2(a) gives

$$\det(A) = (3)(-2) - (1)(4) = -10$$

Using the method of Figure 2.2(b) gives

$$\det(B) = (45) + (84) + (96) - (105) - (-48) - (-72) = 240$$

We emphasize that the methods described in Figure 2.2 *do not* work for determinants of $4 \times 4$ matrices or higher.

Directly applying the definition to evaluate determinants leads to computational difficulties. Indeed, the direct evaluation of a $4 \times 4$ determinant would involve computing $4! = 24$ signed elementary products and a $10 \times 10$ determinant would involve $10! = 3{,}628{,}800$ signed elementary products. Even the fastest of our current digital computers cannot handle the computation of a $25 \times 25$ determinant by this method in a practical amount of time. Much of the remainder of this chapter is devoted, therefore, to developing properties of determinants that will simplify their evaluation.

We conclude this section by noting that the determinant of $A$ is often written symbolically as

$$\det(A) = \Sigma \pm a_{1j_1} a_{2j_2} \dots a_{nj_n}$$

where $\Sigma$ indicates that the terms are to be summed over all permutations $(j_1, j_2, \dots, j_n)$ and the $+$ or $-$ is selected in each term according to whether the permutation is even or odd. We also observe that historically the original notation for the determinant of a matrix $A$ was $|A|$ rather than $\det(A)$. Since this notation is still in common use, we mention it here. With this notation, the determinant of the matrix $A$ in Example 8 can be written

$$\begin{vmatrix} 3 & 1 \\ 4 & -2 \end{vmatrix} = -10$$

**EXERCISE SET 2.1**

1. Find the number of inversions in each of the following permutations of $\{1, 2, 3, 4, 5\}$.
   (a) (3 4 1 5 2)    (b) (4 2 5 3 1)    (c) (5 4 3 2 1)
   (d) (1 2 3 4 5)    (e) (1 3 5 4 2)    (f) (2 3 5 4 1)

2. Classify each of the permutations in Exercise 1 as even or odd.

   In Exercises 3–10 evaluate the determinant.

3. $\begin{vmatrix} 1 & 2 \\ -1 & 3 \end{vmatrix}$    4. $\begin{vmatrix} 6 & 4 \\ 3 & 2 \end{vmatrix}$    5. $\begin{vmatrix} -1 & 7 \\ -8 & -3 \end{vmatrix}$    6. $\begin{vmatrix} k-1 & 2 \\ 4 & k-3 \end{vmatrix}$

7. $\begin{vmatrix} 1 & -2 & 7 \\ 3 & 5 & 1 \\ 4 & 3 & 8 \end{vmatrix}$    8. $\begin{vmatrix} 8 & 2 & -1 \\ -3 & 4 & -6 \\ 1 & 7 & 2 \end{vmatrix}$    9. $\begin{vmatrix} 1 & 0 & 3 \\ 4 & 0 & -1 \\ 2 & 8 & 6 \end{vmatrix}$    10. $\begin{vmatrix} k & -3 & 9 \\ 2 & 4 & k+1 \\ 1 & k^2 & 3 \end{vmatrix}$

11. Find all values of $\lambda$ for which $\det(A) = 0$.

    (a) $A = \begin{bmatrix} \lambda - 1 & -2 \\ 1 & \lambda - 4 \end{bmatrix}$    (b) $A = \begin{bmatrix} \lambda - 6 & 0 & 0 \\ 0 & \lambda & -1 \\ 0 & 4 & \lambda - 4 \end{bmatrix}$

12. Construct a table for a $4 \times 4$ matrix like those given in Example 6 for $2 \times 2$ and $3 \times 3$ matrices.

13. Use the table in Exercise 12 to construct a formula for the determinant of a $4 \times 4$ matrix.

14. Use the formula obtained in Exercise 13 to evaluate

$$\begin{vmatrix} 1 & 4 & -3 & 1 \\ 2 & 0 & 6 & 3 \\ 4 & -1 & 2 & 5 \\ 1 & 0 & -2 & 4 \end{vmatrix}$$

15. Use the determinant definition to evaluate

    (a) $\begin{vmatrix} 0 & 0 & 0 & 0 & 1 \\ 0 & 0 & 0 & 2 & 0 \\ 0 & 0 & 3 & 0 & 0 \\ 0 & 4 & 0 & 0 & 0 \\ 5 & 0 & 0 & 0 & 0 \end{vmatrix}$    (b) $\begin{vmatrix} 0 & 4 & 0 & 0 & 0 \\ 0 & 0 & 0 & 2 & 0 \\ 0 & 0 & 3 & 0 & 0 \\ 0 & 0 & 0 & 0 & 1 \\ 5 & 0 & 0 & 0 & 0 \end{vmatrix}$

16. Prove that if a square matrix $A$ has a column of zeros, then $\det(A) = 0$.

## 2.2 EVALUATING DETERMINANTS BY ROW REDUCTION

In this section we show that the determinant of a matrix can be evaluated by reducing the matrix to row-echelon form. This method is of importance since it avoids the lengthy computations involved in directly applying the determinant definition.

We first consider two classes of matrices whose determinants can be easily evaluated, regardless of the size of the matrix.

**Theorem 1.** *If A is any square matrix that contains a row of zeros, then det(A) = 0.*

*Proof.* Since a signed elementary product from $A$ contains one factor from each row of $A$, every signed elementary product contains a factor from the row of zeros and consequently has value zero. Since $\det(A)$ is the sum of all signed elementary products, we obtain $\det(A) = 0$. ▊

A square matrix is called ***upper triangular*** if all the entries below the main diagonal are zeros. Similarly, a square matrix is called ***lower triangular*** if all the entries above the main diagonal are zeros. A matrix that is either upper or lower triangular is called ***triangular.***

**Example 9**

A general $4 \times 4$ upper triangular matrix has the form

$$\begin{bmatrix} a_{11} & a_{12} & a_{13} & a_{14} \\ 0 & a_{22} & a_{23} & a_{24} \\ 0 & 0 & a_{33} & a_{34} \\ 0 & 0 & 0 & a_{44} \end{bmatrix}$$

A general $4 \times 4$ lower triangular matrix has the form

$$\begin{bmatrix} a_{11} & 0 & 0 & 0 \\ a_{21} & a_{22} & 0 & 0 \\ a_{31} & a_{32} & a_{33} & 0 \\ a_{41} & a_{42} & a_{43} & a_{44} \end{bmatrix}$$

**Example 10**

Evaluate $\det(A)$, where

$$A = \begin{bmatrix} a_{11} & 0 & 0 & 0 \\ a_{21} & a_{22} & 0 & 0 \\ a_{31} & a_{32} & a_{33} & 0 \\ a_{41} & a_{42} & a_{43} & a_{44} \end{bmatrix}$$

The only elementary product from $A$ that can be nonzero is $a_{11}a_{22}a_{33}a_{44}$. To see this, consider a typical elementary product $a_{1j_1}a_{2j_2}a_{3j_3}a_{4j_4}$. Since $a_{12} = a_{13} = a_{14} = 0$, we must have $j_1 = 1$ in order to have a nonzero elementary product. If $j_1 = 1$, we must have $j_2 \neq 1$ since no two factors come from the same column. Further, since $a_{23} = a_{24} = 0$, we must have $j_2 = 2$ in order to have a nonzero product. Continuing in this way, we obtain $j_3 = 3$ and $j_4 = 4$. Since $a_{11}a_{22}a_{33}a_{44}$ is multiplied by $+1$ in forming the signed elementary product, we obtain

$$\det(A) = a_{11}a_{22}a_{33}a_{44}$$

An argument similar to the one just presented can be applied to any triangular matrix to yield the following general result.

**Theorem 2.** *If A is an n × n triangular matrix, then det(A) is the product of the entries on the main diagonal; that is, $\det(A) = a_{11}a_{22} \ldots a_{nn}$.*

**Example 11**

$$\det \begin{bmatrix} 2 & 7 & -3 & 8 & 3 \\ 0 & -3 & 7 & 5 & 1 \\ 0 & 0 & 6 & 7 & 6 \\ 0 & 0 & 0 & 9 & 8 \\ 0 & 0 & 0 & 0 & 4 \end{bmatrix} = (2)(-3)(6)(9)(4) = -1296$$

The next theorem shows how an elementary row operation on a matrix affects the value of its determinant.

**Theorem 3.** *Let A be any n × n matrix.*

(a) *If A′ is the matrix that results when a single row of A is multiplied by a constant k, then $\det(A') = k \det(A)$.*

(b) *If A′ is the matrix that results when two rows of A are interchanged, then $\det(A') = -\det(A)$.*

(c) *If A′ is the matrix that results when a multiple of one row of A is added to another row, then $\det(A') = \det(A)$.*

We shall omit the proof (see Exercise 15).

**Example 12**

Consider the matrices

$$A = \begin{bmatrix} 1 & 2 & 3 \\ 0 & 1 & 4 \\ 1 & 2 & 1 \end{bmatrix} \quad A_1 = \begin{bmatrix} 4 & 8 & 12 \\ 0 & 1 & 4 \\ 1 & 2 & 1 \end{bmatrix} \quad A_2 = \begin{bmatrix} 0 & 1 & 4 \\ 1 & 2 & 3 \\ 1 & 2 & 1 \end{bmatrix}$$

$$A_3 = \begin{bmatrix} 1 & 2 & 3 \\ -2 & -3 & 2 \\ 1 & 2 & 1 \end{bmatrix}$$

If we evaluate the determinants of these matrices by the method used in Example 8 we obtain

$$\det(A) = -2, \ \det(A_1) = -8, \ \det(A_2) = 2, \ \det(A_3) = -2$$

Observe now that $A_1$ is the matrix that results when the first row of A is multiplied by 4; $A_2$ results when the first and second rows of A are interchanged; and $A_3$ results when $-2$ times the third row of $A$ is added to the second row. As predicted by Theorem 3, we have the relationships

$$\det(A_1) = 4\det(A) \quad \det(A_2) = -\det(A) \quad \text{and} \quad \det(A_3) = \det(A)$$

**Example 13**

Statement ($a$) in Theorem 3 has an alternate interpretation that is sometimes useful. This result allows us to bring a "common factor" from any row of a square matrix through the det sign. To illustrate, consider the matrices

$$A = \begin{bmatrix} a_{11} & a_{12} & a_{13} \\ a_{21} & a_{22} & a_{23} \\ a_{31} & a_{32} & a_{33} \end{bmatrix} \qquad B = \begin{bmatrix} a_{11} & a_{12} & a_{13} \\ ka_{21} & ka_{22} & ka_{23} \\ a_{31} & a_{32} & a_{33} \end{bmatrix}$$

where the second row of $B$ has a common factor of $k$. Since $B$ is the matrix that results when the second row of $A$ is multiplied by $k$, statement ($a$) in Theorem 3 asserts that $\det(B) = k\det(A)$; that is,

$$\det \begin{bmatrix} a_{11} & a_{12} & a_{13} \\ ka_{21} & ka_{22} & ka_{23} \\ a_{31} & a_{32} & a_{33} \end{bmatrix} = k\det \begin{bmatrix} a_{11} & a_{12} & a_{13} \\ a_{21} & a_{22} & a_{23} \\ a_{31} & a_{32} & a_{33} \end{bmatrix}$$

We shall now formulate an alternative method for evaluating determinants that will avoid the large amount of computation involved in directly applying the determinant definition. The basic idea of this method is to apply elementary row operations to reduce the given matrix $A$ to a matrix $R$ that is in row-echelon form. Since a row-echelon form of a square matrix is upper triangular (Exercise 14), $\det(R)$ can be evaluated using Theorem 2. The value of $\det(A)$ can then be obtained by using Theorem 3 to relate the unknown value of $\det(A)$ to the known value of $\det(R)$. The following example illustrates this method.

**Example 14**

Evaluate $\det(A)$ where

$$A = \begin{bmatrix} 0 & 1 & 5 \\ 3 & -6 & 9 \\ 2 & 6 & 1 \end{bmatrix}$$

*Solution.* Reducing $A$ to row-echelon form and applying Theorem 3, we obtain

$$\det(A) = -\det \begin{bmatrix} 3 & -6 & 9 \\ 0 & 1 & 5 \\ 2 & 6 & 1 \end{bmatrix} \quad \begin{pmatrix} \text{The first and second rows of } A \\ \text{were interchanged.} \end{pmatrix}$$

$$= -3 \det \begin{bmatrix} 1 & -2 & 3 \\ 0 & 1 & 5 \\ 2 & 6 & 1 \end{bmatrix} \quad \begin{pmatrix} \text{A common factor of 3 from the first} \\ \text{row of the preceding matrix was} \\ \text{taken through the det sign} \\ \text{(see Example 13).} \end{pmatrix}$$

$$= -3 \det \begin{bmatrix} 1 & -2 & 3 \\ 0 & 1 & 5 \\ 0 & 10 & -5 \end{bmatrix} \quad \begin{pmatrix} -2 \text{ times the first row of} \\ \text{the preceding matrix was added to} \\ \text{the third row.} \end{pmatrix}$$

$$= -3 \det \begin{bmatrix} 1 & -2 & 3 \\ 0 & 1 & 5 \\ 0 & 0 & -55 \end{bmatrix} \quad \begin{pmatrix} -10 \text{ times the second row of} \\ \text{the preceding matrix was added to} \\ \text{the third row.} \end{pmatrix}$$

$$= (-3)(-55) \det \begin{bmatrix} 1 & -2 & 3 \\ 0 & 1 & 5 \\ 0 & 0 & 1 \end{bmatrix} \quad \begin{pmatrix} \text{A common factor of } -55 \text{ from the} \\ \text{last row of the preceding matrix} \\ \text{was taken through the det sign.} \end{pmatrix}$$

$$= (-3)(-55)(1) = 165$$

**Example 15**

Evaluate $\det(A)$, where

$$A = \begin{bmatrix} 1 & 3 & -2 & 4 \\ 2 & 6 & -4 & 8 \\ 3 & 9 & 1 & 5 \\ 1 & 1 & 4 & 8 \end{bmatrix}$$

$$\det(A) = \det \begin{bmatrix} 1 & 3 & -2 & 4 \\ 0 & 0 & 0 & 0 \\ 3 & 9 & 1 & 5 \\ 1 & 1 & 4 & 8 \end{bmatrix} \quad \begin{pmatrix} -2 \text{ times the first} \\ \text{row of } A \text{ was added} \\ \text{to the second row.} \end{pmatrix}$$

No further reduction is needed since it follows from Theorem 1 that $\det(A) = 0$.

It should be evident from this example that whenever a square matrix has two proportional rows (like the first and second rows of $A$), it is possible to introduce a row of zeros by adding a suitable multiple of one of these rows to the other. Thus, *if a square matrix has two proportional rows, its determinant is zero.*

**Example 16**

Each of the following matrices has two proportional rows; thus, by inspection, each has a zero determinant.

$$\begin{bmatrix} -1 & 4 \\ -2 & 8 \end{bmatrix} \qquad \begin{bmatrix} 2 & 7 & 8 \\ 3 & 2 & 4 \\ 2 & 7 & 8 \end{bmatrix} \qquad \begin{bmatrix} 3 & -1 & 4 & -5 \\ 6 & -2 & 5 & 2 \\ 5 & 8 & 1 & 4 \\ -9 & 3 & -12 & 15 \end{bmatrix}$$

## EXERCISE SET 2.2

**1.** Evaluate the following by inspection.

(a) $\begin{vmatrix} 2 & -40 & 17 \\ 0 & 1 & 11 \\ 0 & 0 & 3 \end{vmatrix}$    (b) $\begin{vmatrix} 1 & 0 & 0 & 0 \\ -9 & -1 & 0 & 0 \\ 12 & 7 & 8 & 0 \\ 4 & 5 & 7 & 2 \end{vmatrix}$

(c) $\begin{vmatrix} 1 & 2 & 3 \\ 3 & 7 & 6 \\ 1 & 2 & 3 \end{vmatrix}$    (d) $\begin{vmatrix} 3 & -1 & 2 \\ 6 & -2 & 4 \\ 1 & 7 & 3 \end{vmatrix}$

In exercises 2 to 9 evaluate the determinants of the given matrices by reducing the matrix to row-echelon form.

**2.** $\begin{bmatrix} 2 & 3 & 4 \\ 0 & 0 & -3 \\ -1 & 2 & 7 \end{bmatrix}$    **3.** $\begin{bmatrix} 2 & 1 & 1 \\ 4 & 2 & 3 \\ 1 & 3 & 0 \end{bmatrix}$

**4.** $\begin{bmatrix} 6 & 6 & 2 \\ -3 & 3 & 1 \\ 3 & 9 & 2 \end{bmatrix}$    **5.** $\begin{bmatrix} 3 & 1 & 4 \\ 1 & 7 & 3 \\ 5 & -12 & 5 \end{bmatrix}$

**6.** $\begin{bmatrix} -1 & 2 & 1 & 2 \\ 1 & 2 & 4 & 1 \\ 2 & 0 & -1 & 3 \\ 3 & 2 & -1 & 0 \end{bmatrix}$    **7.** $\begin{bmatrix} 4 & 6 & 8 & -6 \\ 0 & -3 & 0 & -1 \\ 3 & 3 & -4 & -2 \\ -2 & 3 & 4 & 2 \end{bmatrix}$

**8.** $\begin{bmatrix} \dfrac{1}{2} & \dfrac{1}{2} & 1 & \dfrac{1}{2} \\ -\dfrac{1}{2} & \dfrac{1}{2} & 0 & \dfrac{1}{2} \\ \dfrac{2}{3} & \dfrac{1}{3} & \dfrac{1}{3} & 0 \\ \dfrac{1}{3} & 1 & \dfrac{1}{3} & 0 \end{bmatrix}$    **9.** $\begin{bmatrix} 0 & 0 & 2 & 2 & 2 \\ 3 & -3 & 3 & 3 & 3 \\ -4 & 4 & 4 & 4 & 4 \\ 4 & 3 & 1 & 9 & 2 \\ 1 & 2 & 1 & 3 & 1 \end{bmatrix}$

**10.** Assume $\det \begin{bmatrix} a & b & c \\ d & e & f \\ g & h & i \end{bmatrix} = 5$. Find

(a) $\det \begin{bmatrix} d & e & f \\ g & h & i \\ a & b & c \end{bmatrix}$    (b) $\det \begin{bmatrix} -a & -b & -c \\ 2d & 2e & 2f \\ -g & -h & -i \end{bmatrix}$

(c) $\det \begin{bmatrix} a+d & b+e & c+f \\ d & e & f \\ g & h & i \end{bmatrix}$    (d) $\det \begin{bmatrix} a & b & c \\ d-3a & e-3b & f-3c \\ 2g & 2h & 2i \end{bmatrix}$

**11.** Show that

$$\begin{vmatrix} 1 & 1 & 1 \\ a & b & c \\ a^2 & b^2 & c^2 \end{vmatrix} = (b - a)(c - a)(c - b)$$

**12.** Use an argument like that given in Example 10 to show

(a) $\det \begin{bmatrix} 0 & 0 & a_{13} \\ 0 & a_{22} & a_{23} \\ a_{31} & a_{32} & a_{33} \end{bmatrix} = -a_{13}a_{22}a_{31}$

(b) $\det \begin{bmatrix} 0 & 0 & 0 & a_{14} \\ 0 & 0 & a_{23} & a_{24} \\ 0 & a_{32} & a_{33} & a_{34} \\ a_{41} & a_{42} & a_{43} & a_{44} \end{bmatrix} = a_{14}a_{23}a_{32}a_{41}$

**13.** Prove that Theorem 1 is true when the word "row" is replaced by "column."

**14.** Prove that a row-echelon form of a square matrix is upper triangular.

**15.** Prove the following special cases of Theorem 3.

(a) $\begin{vmatrix} ka_{11} & ka_{12} & ka_{13} \\ a_{21} & a_{22} & a_{23} \\ a_{31} & a_{32} & a_{33} \end{vmatrix} = k \begin{vmatrix} a_{11} & a_{12} & a_{13} \\ a_{21} & a_{22} & a_{23} \\ a_{31} & a_{32} & a_{33} \end{vmatrix}$

(b) $\begin{vmatrix} a_{21} & a_{22} & a_{23} \\ a_{11} & a_{12} & a_{13} \\ a_{31} & a_{32} & a_{33} \end{vmatrix} = - \begin{vmatrix} a_{11} & a_{12} & a_{13} \\ a_{21} & a_{22} & a_{23} \\ a_{31} & a_{32} & a_{33} \end{vmatrix}$

(c) $\begin{vmatrix} a_{11} + ka_{21} & a_{12} + ka_{22} & a_{13} + ka_{23} \\ a_{21} & a_{22} & a_{23} \\ a_{31} & a_{32} & a_{33} \end{vmatrix} = \begin{vmatrix} a_{11} & a_{12} & a_{13} \\ a_{21} & a_{22} & a_{23} \\ a_{31} & a_{32} & a_{33} \end{vmatrix}$

## 2.3 PROPERTIES OF THE DETERMINANT FUNCTION

In this section we develop some of the fundamental properties of the determinant function. Our work here will give us some further insight into the relationship between a square matrix and its determinant. One of the immediate consequences of this material will be an important test for the invertibility of a matrix that utilizes the determinant function.

If $A$ is any $m \times n$ matrix, then the **transpose of $A$** is denoted by $A^t$ and is defined to be the $n \times m$ matrix whose first column is the first row of $A$, whose second column is the second row of $A$, whose third column is the third row of $A$, etc.

**Example 17**

The transposes of the matrices

$$A = \begin{bmatrix} a_{11} & a_{12} & a_{13} & a_{14} \\ a_{21} & a_{22} & a_{23} & a_{24} \\ a_{31} & a_{32} & a_{33} & a_{34} \end{bmatrix}$$

$$B = \begin{bmatrix} 2 & 3 \\ 1 & 4 \\ 5 & 6 \end{bmatrix} \qquad C = [1 \ 3 \ 5] \qquad D = \begin{bmatrix} 3 & 5 & -2 \\ 5 & 4 & 1 \\ -2 & 1 & 7 \end{bmatrix}$$

are

$$A^t = \begin{bmatrix} a_{11} & a_{21} & a_{31} \\ a_{12} & a_{22} & a_{32} \\ a_{13} & a_{23} & a_{33} \\ a_{14} & a_{24} & a_{34} \end{bmatrix}$$

$$B^t = \begin{bmatrix} 2 & 1 & 5 \\ 3 & 4 & 6 \end{bmatrix} \qquad C^t = \begin{bmatrix} 1 \\ 3 \\ 5 \end{bmatrix} \qquad D^t = \begin{bmatrix} 3 & 5 & -2 \\ 5 & 4 & 1 \\ -2 & 1 & 7 \end{bmatrix}$$

Assuming that the sizes of the matrices are such that the operations can be performed, the transpose operation has the following properties (see Exercise 13):

(i) $(A^t)^t = A$

(ii) $(A + B)^t = A^t + B^t$

(iii) $(kA)^t = kA^t$, *where k is a scalar*

(iv) $(AB)^t = B^t A^t$

Recall that the determinant of an $n \times n$ matrix $A$ is defined to be the sum of all signed elementary products from $A$. Since an elementary product has one factor from each row and one factor from each column, it is obvious that $A$ and $A^t$ have precisely the same set of elementary products. Although we shall omit the details, it can be shown that $A$ and $A^t$ actually have the same *signed* elementary products; this yields the following theorem.

**Theorem 4.** *If A is any square matrix, then* $det(A) = det(A^t)$.

Because of this result, nearly every theorem about determinants that contains the word "row" in its statement is also true when the word "column" is substituted for "row." In particular, this applies to Theorem 3 of the previous section. To illustrate, suppose that $A'$ is the matrix that results when column $r$ and column $s$ of $A$ are interchanged. Thus $(A')^t$ is the matrix that results when *row r* and *row s* of $A^t$ are interchanged. From part (b) of Theorem 3

$$det(A')^t = -det(A^t)$$

But $\det(A')^t = \det(A')$ and $\det(A^t) = \det(A)$; thus

$$\det(A') = -\det(A)$$

This establishes that part (b) of Theorem 3 is true when the word "row" is replaced by "column." The basic idea of this proof was to first transpose the matrix in question in order to convert the column statements to row statements. The proof was then completed by applying the known result for rows.

The following examples illustrate several points about determinants that depend on column properties of the matrix.

**Example 18**

By inspection, the matrix

$$\begin{bmatrix} 1 & -2 & 7 \\ -4 & 8 & 5 \\ 2 & -4 & 3 \end{bmatrix}$$

has a zero determinant since the first and second columns are proportional.

**Example 19**

Compute the determinant of

$$A = \begin{bmatrix} 1 & 0 & 0 & 3 \\ 2 & 7 & 0 & 6 \\ 0 & 6 & 3 & 0 \\ 7 & 3 & 1 & -5 \end{bmatrix}$$

This determinant could be computed as before by using elementary row operations to reduce $A$ to upper triangular form. On the other hand, we can put $A$ in lower triangular form in one step by adding $-3$ times the first column to the fourth to obtain

$$\det(A) = \det \begin{bmatrix} 1 & 0 & 0 & 0 \\ 2 & 7 & 0 & 0 \\ 0 & 6 & 3 & 0 \\ 7 & 3 & 1 & -26 \end{bmatrix} = (1)(7)(3)(-26) = -546$$

This example points out that it is always wise to keep an eye open for a clever column operation that will shorten the computations.

Suppose that $A$ and $B$ are $n \times n$ matrices and $k$ is any scalar. We now consider possible relationships between $\det(A)$, $\det(B)$, and the values of

$$\det(kA), \det(A + B), \text{ and } \det(AB)$$

Since a common factor of any row of a matrix can be moved through the det sign, and since each of the $n$ rows in $kA$ has a common factor of $k$, we obtain

$$\det(kA) = k^n \det(A) \tag{2.1}$$

**Example 20**

Consider the matrices

$$A = \begin{bmatrix} 3 & 1 \\ 2 & 2 \end{bmatrix} \quad \text{and} \quad 5A = \begin{bmatrix} 15 & 5 \\ 10 & 10 \end{bmatrix}$$

By direct calculation $\det(A) = 4$ and $\det(5A) = 100$. This agrees with relationship (2.1), which asserts that $\det(5A) = 5^2\det(A)$.

Unfortunately, no simple relationship exists between $\det(A)$, $\det(B)$, and $\det(A + B)$ in general. In particular, we emphasize that $\det(A + B)$ is usually *not* equal to $\det(A) + \det(B)$. The following example illustrates this point.

**Example 21**

Consider

$$A = \begin{bmatrix} 1 & 2 \\ 2 & 5 \end{bmatrix} \quad B = \begin{bmatrix} 3 & 1 \\ 1 & 3 \end{bmatrix} \quad A + B = \begin{bmatrix} 4 & 3 \\ 3 & 8 \end{bmatrix}$$

We have $\det(A) = 1$, $\det(B) = 8$, and $\det(A + B) = 23$; thus $\det(A + B) \neq \det(A) + \det(B)$.

In spite of this negative result, there is one important relationship concerning sums of determinants that is often useful. To illustrate, consider two $2 \times 2$ matrices

$$A = \begin{bmatrix} a_{11} & a_{12} \\ a_{21} & a_{22} \end{bmatrix} \quad \text{and} \quad A' = \begin{bmatrix} a_{11} & a_{12} \\ a'_{21} & a'_{22} \end{bmatrix}$$

that differ only in the second row. From the formula in Example 7, we obtain

$$\det(A) + \det(A') = (a_{11}a_{22} - a_{12}a_{21}) + (a_{11}a'_{22} - a_{12}a'_{21})$$
$$= a_{11}(a_{22} + a'_{22}) - a_{12}(a_{21} + a'_{21})$$

But

$$a_{11}(a_{22} + a'_{22}) - a_{12}(a_{21} + a'_{21}) = \det \begin{bmatrix} a_{11} & a_{12} \\ a_{21} + a'_{21} & a_{22} + a'_{22} \end{bmatrix}$$

Thus

$$\det \begin{bmatrix} a_{11} & a_{12} \\ a_{21} & a_{22} \end{bmatrix} + \det \begin{bmatrix} a_{11} & a_{12} \\ a'_{21} & a'_{22} \end{bmatrix} = \det \begin{bmatrix} a_{11} & a_{12} \\ a_{21} + a'_{21} & a_{22} + a'_{22} \end{bmatrix}$$

This example is a special case of the following general result:

*Let A, A', and A'' be n × n matrices that differ only in a single row, say the rth, and assume that the rth row of A'' can be obtained by adding corresponding entries in the rth rows of A and A'. Then*

$$\det(A'') = \det(A) + \det(A')$$

A similar result holds for columns.

**Example 22**

By evaluating the determinants, the reader can check that

$$\det \begin{bmatrix} 1 & 7 & 5 \\ 2 & 0 & 3 \\ 1+0 & 4+1 & 7+(-1) \end{bmatrix} = \det \begin{bmatrix} 1 & 7 & 5 \\ 2 & 0 & 3 \\ 1 & 4 & 7 \end{bmatrix} + \det \begin{bmatrix} 1 & 7 & 5 \\ 2 & 0 & 3 \\ 0 & 1 & -1 \end{bmatrix}$$

**Theorem 5.** *If A and B are square matrices of the same size, then $\det(AB) = \det(A)\det(B)$.*

The elegant simplicity of this result contrasted with the complex nature of both matrix multiplication and the determinant definition is both refreshing and surprising. We shall omit the proof.

**Example 23**

Consider the matrices

$$A = \begin{bmatrix} 3 & 1 \\ 2 & 1 \end{bmatrix} \qquad B = \begin{bmatrix} -1 & 3 \\ 5 & 8 \end{bmatrix} \qquad AB = \begin{bmatrix} 2 & 17 \\ 3 & 14 \end{bmatrix}$$

We have $\det(A)\det(B) = (1)(-23) = -23$. On the other hand, by direct computation $\det(AB) = -23$, so that $\det(AB) = \det(A)\det(B)$.

In the corollary to Theorem 12 of Chapter 1, we listed three important statements that are equivalent to the invertibility of a matrix. The next example will help us to add another result to that list.

**Example 24**

The purpose of this example is to show that if the reduced row-echelon form $R$ of a *square* matrix has no rows consisting entirely of zeros, then $R$ must be the identity matrix. This can be illustrated by considering the following $3 \times 3$ matrix, which we assume to be in reduced row-echelon form

$$R = \begin{bmatrix} r_{11} & r_{12} & r_{13} \\ r_{21} & r_{22} & r_{23} \\ r_{31} & r_{32} & r_{33} \end{bmatrix}$$

Either the last row in this matrix consists entirely of zeros or it does not. If not, the matrix contains no zero rows, and consequently each of the three rows has a leftmost nonzero entry of 1. Since these leading 1's occur progressively further to the right as we move down the matrix, each of these 1's must occur on the main diagonal. Since the other entries in the same column as one of these 1's are zero, $R$ must be $I$. Thus either $R$ has a row of zeros or $R = I$.

**Theorem 6.** *A square matrix $A$ is invertible if and only if $\det(A) \neq 0$.*

*Proof.* If $A$ is invertible, then $I = AA^{-1}$ so that $1 = \det(I) = \det(A)\det(A^{-1})$. Thus $\det(A) \neq 0$. Conversely, assume that $\det(A) \neq 0$. We shall show that $A$ is row equivalent to $I$, and thus conclude from Theorem 10 in Chapter 1 that $A$ is invertible. Let $R$ be the reduced row-echelon form of $A$. Since $R$ can be obtained from $A$ by a finite sequence of elementary row operations, we can find elementary matrices $E_1, E_2, \ldots, E_k$ such that $E_k \ldots E_2 E_1 A = R$ or $A = E_1^{-1} E_2^{-1} \ldots E_k^{-1} R$. Thus

$$\det(A) = \det(E_1^{-1})\det(E_2^{-1}) \ldots \det(E_k^{-1})\det(R)$$

It now follows from our hypothesis $\det(A) \neq 0$ that $\det(R) \neq 0$. Therefore $R$ does not have any zero rows, so that $R = I$ (see Example 24). ▮

**Corollary.** If $A$ is invertible then

$$\det(A^{-1}) = \frac{1}{\det(A)}$$

*Proof.* Since $A^{-1}A = I$, $\det(A^{-1}A) = \det(I)$; that is, $\det(A^{-1})\det(A) = 1$. Since $\det(A) \neq 0$, the proof can be completed by dividing through by $\det(A)$. ▮

**Example 25**

Since the first and third rows of

$$A = \begin{bmatrix} 1 & 2 & 3 \\ 1 & 0 & 1 \\ 2 & 4 & 6 \end{bmatrix}$$

are proportional, $\det(A) = 0$. Thus $A$ is not invertible.

**EXERCISE SET 2.3**

1. Find the transposes of

(a) $\begin{bmatrix} 2 & 1 \\ -3 & 1 \\ 0 & 2 \end{bmatrix}$ (b) $\begin{bmatrix} 6 & 1 & 1 \\ -8 & 4 & 3 \\ 0 & 1 & 3 \end{bmatrix}$ (c) $[7 \quad 0 \quad 2]$ (d) $\begin{bmatrix} a_{11} & a_{12} & a_{13} \\ a_{21} & a_{22} & a_{23} \end{bmatrix}$

2. Verify Theorem 4 for

$$A = \begin{bmatrix} 1 & 2 & 7 \\ -1 & 0 & 6 \\ 3 & 2 & 8 \end{bmatrix}$$

3. Verify that $\det(AB) = \det(A)\det(B)$ when

$$A = \begin{bmatrix} 2 & 1 & 0 \\ 3 & 4 & 0 \\ 0 & 0 & 2 \end{bmatrix} \quad \text{and} \quad B = \begin{bmatrix} 1 & -1 & 3 \\ 7 & 1 & 2 \\ 5 & 0 & 1 \end{bmatrix}$$

4. Use Theorem 6 to determine which of the following matrices are invertible.

(a) $\begin{bmatrix} 1 & 0 & 0 \\ 3 & 6 & 7 \\ 0 & 8 & -1 \end{bmatrix}$ (b) $\begin{bmatrix} -2 & 1 & -4 \\ 1 & 1 & 2 \\ 3 & 1 & 6 \end{bmatrix}$

(c) $\begin{bmatrix} 7 & 2 & 1 \\ 7 & 2 & 1 \\ 3 & 6 & 6 \end{bmatrix}$ (d) $\begin{bmatrix} 0 & 7 & 5 \\ 0 & 1 & -1 \\ 0 & 3 & 2 \end{bmatrix}$

5. Assume $\det(A) = 5$, where

$$A = \begin{bmatrix} a & b & c \\ d & e & f \\ g & h & i \end{bmatrix}$$

Find
(a) $\det(3A)$ (b) $\det(2A^{-1})$ (c) $\det((2A)^{-1})$

(d) $\det \begin{bmatrix} a & g & d \\ b & h & e \\ c & i & f \end{bmatrix}$

6. Without directly evaluating, show that

$$\det \begin{bmatrix} 1 & 1 & 1 \\ a & b & c \\ bc & ca & ab \end{bmatrix} = \det \begin{bmatrix} 1 & 1 & 1 \\ a & b & c \\ a^2 & b^2 & c^2 \end{bmatrix}$$

7. Without directly evaluating, show that

$$\det \begin{bmatrix} b+c & c+a & b+a \\ a & b & c \\ 1 & 1 & 1 \end{bmatrix} = 0$$

**8.** For which value(s) of $k$ does $A$ fail to be invertible?

(a) $A = \begin{bmatrix} k-3 & -2 \\ -2 & k-2 \end{bmatrix}$    (b) $A = \begin{bmatrix} 1 & 2 & 4 \\ 3 & 1 & 6 \\ k & 3 & 2 \end{bmatrix}$

**9.** Let $A$ and $B$ be $n \times n$ matrices. Show that if $A$ is invertible, then $\det(B) = \det(A^{-1}BA)$.

**10.** (a) Find a nonzero $3 \times 3$ matrix $A$ such that $A = A^t$.
(b) Find a nonzero $3 \times 3$ matrix $A$ such that $A = -A^t$.

**11.** Let $a_{ij}$ be the entry in the $i$th row and $j$th column of $A$. In which row and column of $A^t$ will $a_{ij}$ appear?

**12.** Let $AX = 0$ be a system of $n$ linear equations in $n$ unknowns. Show that the system has a nontrivial solution if and only if $\det(A) = 0$.

**13.** Let

$$A = \begin{bmatrix} a_{11} & a_{12} & a_{13} \\ a_{21} & a_{22} & a_{23} \\ a_{31} & a_{32} & a_{33} \end{bmatrix} \quad \text{and} \quad B = \begin{bmatrix} b_{11} & b_{12} & b_{13} \\ b_{21} & b_{22} & b_{23} \\ b_{31} & b_{32} & b_{33} \end{bmatrix}$$

Show that

(a) $(A^t)^t = A$    (b) $(A + B)^t = A^t + B^t$    (c) $(AB)^t = B^t A^t$    (d) $(kA)^t = kA^t$

**14.** Prove that $(A^t B^t)^t = BA$.

**15.** A square matrix $A$ is called *symmetric* if $A^t = A$ and *skew-symmetric* if $A^t = -A$. Show that if $B$ is a square matrix, then
(a) $BB^t$ and $B + B^t$ are symmetric    (b) $B - B^t$ is skew-symmetric

## 2.4 COFACTOR EXPANSION; CRAMER'S RULE

In this section we consider another method for evaluating determinants. As a consequence of our work here, we shall obtain a formula for the inverse of an invertible matrix as well as a formula for the solution to certain systems of linear equations in terms of determinants.

**Definition.** If $A$ is a square matrix, then the *minor of entry* $a_{ij}$ is denoted by $M_{ij}$ and is defined to be the determinant of the submatrix that remains after the $i$th row and $j$th column are deleted from $A$. The number $(-1)^{i+j}M_{ij}$ is denoted by $C_{ij}$ and is called the *cofactor of entry* $a_{ij}$.

**Example 26**

Let

$$A = \begin{bmatrix} 3 & 1 & -4 \\ 2 & 5 & 6 \\ 1 & 4 & 8 \end{bmatrix}$$

The minor of entry $a_{11}$ is

$$M_{11} = \det \begin{bmatrix} 3 & 1 & -4 \\ 2 & 5 & 6 \\ 1 & 4 & 8 \end{bmatrix} = \det \begin{bmatrix} 5 & 6 \\ 4 & 8 \end{bmatrix} = 16$$

The cofactor of $a_{11}$ is

$$C_{11} = (-1)^{1+1} M_{11} = M_{11} = 16$$

Similarly, the minor of entry $a_{32}$ is

$$M_{32} = \det \begin{bmatrix} 3 & 1 & -4 \\ 2 & 5 & 6 \\ 1 & 4 & 8 \end{bmatrix} = \det \begin{bmatrix} 3 & -4 \\ 2 & 6 \end{bmatrix} = 26$$

The cofactor of $a_{32}$ is

$$C_{32} = (-1)^{3+2} M_{32} = -M_{32} = -26$$

Notice that the cofactor and the minor of an element $a_{ij}$ differ only in sign, that is, $C_{ij} = \pm M_{ij}$. A quick way for determining whether to use the $+$ or $-$ is to use the fact that the sign relating $C_{ij}$ and $M_{ij}$ is in the $i$th row and $j$th column of the array

$$\begin{bmatrix} + & - & + & - & + & \cdots \\ - & + & - & + & - & \cdots \\ + & - & + & - & + & \cdots \\ - & + & - & + & - & \cdots \\ \vdots & \vdots & \vdots & \vdots & \vdots & \end{bmatrix}$$

For example, $C_{11} = M_{11}$, $C_{21} = -M_{21}$, $C_{12} = -M_{12}$, $C_{22} = M_{22}$, etc.

Consider the general $3 \times 3$ matrix

$$A = \begin{bmatrix} a_{11} & a_{12} & a_{13} \\ a_{21} & a_{22} & a_{23} \\ a_{31} & a_{32} & a_{33} \end{bmatrix}$$

In Example 7 we showed

$$\det(A) = a_{11}a_{22}a_{33} + a_{12}a_{23}a_{31} + a_{13}a_{21}a_{32}$$
$$- a_{13}a_{22}a_{31} - a_{12}a_{21}a_{33} - a_{11}a_{23}a_{32} \quad (2.2)$$

This equation can be rewritten as

$$\det(A) = a_{11}(a_{22}a_{33} - a_{23}a_{32}) + a_{21}(a_{13}a_{32} - a_{12}a_{33}) + a_{31}(a_{12}a_{23} - a_{13}a_{22})$$

or equivalently

$$\det(A) = a_{11}C_{11} + a_{21}C_{21} + a_{31}C_{31} \qquad (2.3)$$

Equation (2.3) shows that the determinant of $A$ can be computed by multiplying the entries in the first column of $A$ by their cofactors and adding the resulting products. This method of evaluating $\det(A)$ is called **cofactor expansion** along the first column of $A$.

**Example 27**

Let

$$A = \begin{bmatrix} 3 & 1 & 0 \\ -2 & -4 & 3 \\ 5 & 4 & -2 \end{bmatrix}$$

Evaluate $\det(A)$ by cofactor expansion along the first column of $A$.

*Solution.* From (2.3)

$$\det(A) = 3 \det \begin{bmatrix} -4 & 3 \\ 4 & -2 \end{bmatrix} - (-2) \det \begin{bmatrix} 1 & 0 \\ 4 & -2 \end{bmatrix} + 5 \det \begin{bmatrix} 1 & 0 \\ -4 & 3 \end{bmatrix}$$

$$= 3(-4) - (-2)(-2) + 5(3) = -1$$

By rearranging the terms in (2.2) in various ways, it is possible to obtain other formulas like (2.3). There should be no trouble checking that all of the following are correct (see Exercise 23):

$$\begin{aligned}
\det(A) &= a_{11}C_{11} + a_{12}C_{12} + a_{13}C_{13} \\
&= a_{11}C_{11} + a_{21}C_{21} + a_{31}C_{31} \\
&= a_{21}C_{21} + a_{22}C_{22} + a_{23}C_{23} \\
&= a_{12}C_{12} + a_{22}C_{22} + a_{32}C_{32} \\
&= a_{31}C_{31} + a_{32}C_{32} + a_{33}C_{33} \\
&= a_{13}C_{13} + a_{23}C_{23} + a_{33}C_{33}
\end{aligned} \qquad (2.4)$$

Notice that in each equation the entries and cofactors all come from the same row or column. These equations are called the **cofactor expansions** of $\det(A)$.

The results we have just given for $3 \times 3$ matrices form a special case of the following general theorem, which we state without proof.

**Theorem 7.** *The determinant of an $n \times n$ matrix $A$ can be computed by multiplying the entries in any row (or column) by their cofactors and adding the resulting products; that is, for each $1 \leq i \leq n$ and $1 \leq j \leq n$,*

$$\det(A) = a_{1j}C_{1j} + a_{2j}C_{2j} + \cdots + a_{nj}C_{nj}$$
*(cofactor expansion along the jth column)*

*and*

$$det(A) = a_{i1}C_{i1} + a_{i2}C_{i2} + \cdots + a_{in}C_{in}$$
*(cofactor expansion along the ith row)*

**Example 28**

Let $A$ be the matrix in Example 27. Evaluate det($A$) by cofactor expansion along the first row.

*Solution.*

$$det(A) = 3 \det \begin{bmatrix} -4 & 3 \\ 4 & -2 \end{bmatrix} - (1) \det \begin{bmatrix} -2 & 3 \\ 5 & -2 \end{bmatrix} + 0 \det \begin{bmatrix} -2 & -4 \\ 5 & 4 \end{bmatrix}$$

$$= 3(-4) - (1)(-11) = -1$$

This agrees with the result obtained in Example 27.

Note that in this example it was unnecessary to compute the last cofactor, since it was multiplied by zero. In general, the best strategy for evaluating a determinant by cofactor expansion is to expand along a row or column having the largest number of zeros.

Although cofactor expansion is not usually as efficient as reduction to triangular form for the evaluation of determinants, the two methods can sometimes be combined in particular situations to yield an effective computational technique. The following example illustrates this idea.

**Example 29**

Evaluate det($A$) where

$$A = \begin{bmatrix} 3 & 5 & -2 & 6 \\ 1 & 2 & -1 & 1 \\ 2 & 4 & 1 & 5 \\ 3 & 7 & 5 & 3 \end{bmatrix}$$

*Solution.* By adding suitable multiples of the second row to the remaining rows, we obtain

$$det(A) = \det \begin{bmatrix} 0 & -1 & 1 & 3 \\ 1 & 2 & -1 & 1 \\ 0 & 0 & 3 & 3 \\ 0 & 1 & 8 & 0 \end{bmatrix}$$

Expanding by cofactors along the first column gives

$$det(A) = -\det \begin{bmatrix} -1 & 1 & 3 \\ 0 & 3 & 3 \\ 1 & 8 & 0 \end{bmatrix}$$

The determinant of this $3 \times 3$ matrix can now be evaluated using any of the methods we have studied to obtain

$$\det(A) = 18$$

In a cofactor expansion we compute $\det(A)$ by multiplying the entries in a row or column by their cofactors and adding the resulting products. It turns out that if one multiplies the entries in any row by the corresponding cofactors from a *different* row, the sum of these products is always zero. (This result holds also for columns.) Although we omit the general proof, the next example illustrates the idea of the proof in a special case.

**Example 30**

Let

$$A = \begin{bmatrix} a_{11} & a_{12} & a_{13} \\ a_{21} & a_{22} & a_{23} \\ a_{31} & a_{32} & a_{33} \end{bmatrix}$$

Consider the quantity

$$a_{11}C_{31} + a_{12}C_{32} + a_{13}C_{33}$$

which is formed by multiplying the entries in the first row by the cofactors of the corresponding entries in the third row and adding the resulting products. We now show this quantity is equal to zero by the following trick. Construct a new matrix $A'$ by replacing the third row of $A$ with another copy of the first row. Thus

$$A' = \begin{bmatrix} a_{11} & a_{12} & a_{13} \\ a_{21} & a_{22} & a_{23} \\ a_{11} & a_{12} & a_{13} \end{bmatrix}$$

Let $C'_{31}$, $C'_{32}$ and $C'_{33}$ be the cofactors of the entries in the third row of $A'$. Since the first two rows of $A$ and $A'$ are the same, and since the computation of $C_{31}$, $C_{32}$, $C_{33}$, $C'_{31}$, $C'_{32}$, and $C'_{33}$ involve only entries from the first two rows of $A$ and $A'$, it follows that

$$C_{31} = C'_{31}, \ C_{32} = C'_{32}, \ C_{33} = C'_{33}$$

Since $A'$ has two identical rows

$$\det(A') = 0 \qquad (2.5)$$

On the other hand, evaluating $\det(A')$ by cofactor expansion along the third row gives

$$\begin{aligned} \det(A') &= a_{11}C'_{31} + a_{12}C'_{32} + a_{13}C'_{33} \\ &= a_{11}C_{31} + a_{12}C_{32} + a_{13}C_{33} \end{aligned} \qquad (2.6)$$

From (2.5) and (2.6) we obtain

$$a_{11}C_{31} + a_{12}C_{32} + a_{13}C_{33} = 0$$

**Definition.** If $A$ is any $n \times n$ matrix and $C_{ij}$ is the cofactor of $a_{ij}$, then the matrix

$$\begin{bmatrix} C_{11} & C_{12} & \cdots & C_{1n} \\ C_{21} & C_{22} & \cdots & C_{2n} \\ \vdots & \vdots & & \vdots \\ C_{n1} & C_{n2} & \cdots & C_{nn} \end{bmatrix}$$

is called the ***matrix of cofactors from A.*** The transpose of this matrix is called the ***adjoint*** of $A$ and is denoted by adj($A$).

**Example 31**

Let

$$A = \begin{bmatrix} 3 & 2 & -1 \\ 1 & 6 & 3 \\ 2 & -4 & 0 \end{bmatrix}$$

The cofactors of $A$ are

$$\begin{array}{lll} C_{11} = 12 & C_{12} = 6 & C_{13} = -16 \\ C_{21} = -4 & C_{22} = 2 & C_{23} = 16 \\ C_{31} = 12 & C_{32} = -10 & C_{33} = 16 \end{array}$$

so that the matrix of cofactors is

$$\begin{bmatrix} 12 & 6 & -16 \\ -4 & 2 & 16 \\ 12 & -10 & 16 \end{bmatrix}$$

and the adjoint of $A$ is

$$\text{adj}(A) = \begin{bmatrix} 12 & -4 & 12 \\ 6 & 2 & -10 \\ -16 & 16 & 16 \end{bmatrix}$$

We are now in a position to establish a formula for the inverse of an invertible matrix.

**Theorem 8.** *If $A$ is an invertible matrix, then*

$$A^{-1} = \frac{1}{\det(A)} \, \text{adj}(A)$$

*Proof.* We show first that

$$A \text{ adj}(A) = \det(A)I$$

Consider the product

$$
A \text{ adj}(A) =
\begin{bmatrix}
a_{11} & a_{12} & \cdots & a_{1n} \\
a_{21} & a_{22} & \cdots & a_{2n} \\
\vdots & \vdots & & \vdots \\
a_{i1} & a_{i2} & \cdots & a_{in} \\
\vdots & \vdots & & \vdots \\
a_{n1} & a_{n2} & \cdots & a_{nn}
\end{bmatrix}
\begin{bmatrix}
C_{11} & C_{21} & \cdots & C_{j1} & \cdots & C_{n1} \\
C_{12} & C_{22} & \cdots & C_{j2} & \cdots & C_{n2} \\
\vdots & \vdots & & \vdots & & \vdots \\
C_{1n} & C_{2n} & \cdots & C_{jn} & \cdots & C_{nn}
\end{bmatrix}
$$

The entry in the $i$th row and $j$th column of $A \text{ adj}(A)$ is

$$a_{i1}C_{j1} + a_{i2}C_{j2} + \cdots + a_{in}C_{jn} \tag{2.7}$$

(see the shaded lines above).

If $i = j$, then (2.7) is the cofactor expansion of $\det(A)$ along the $i$th row of $A$ (Theorem 7). On the other hand, if $i \neq j$, then the $a$'s and the cofactors come from different rows of $A$, so the value of (2.7) is zero. Therefore

$$
A \text{ adj}(A) =
\begin{bmatrix}
\det(A) & 0 & \cdots & 0 \\
0 & \det(A) & \cdots & 0 \\
\vdots & \vdots & & \vdots \\
0 & 0 & \cdots & \det(A)
\end{bmatrix}
= \det(A)I \tag{2.8}
$$

Since $A$ is invertible, $\det(A) \neq 0$. Therefore, equation (2.8) can be rewritten as

$$\frac{1}{\det(A)}[A \text{ adj}(A)] = I$$

or

$$A\left[\frac{1}{\det(A)} \text{ adj}(A)\right] = I$$

By Theorem 12 in Chapter 1

$$A^{-1} = \frac{1}{\det(A)} \text{ adj}(A) \quad \blacksquare \tag{2.9}$$

**Example 32**

Use (2.9) to find the inverse of the matrix $A$ in Example 31.

*Solution.* The reader can check that $\det(A) = 64$. Thus

$$A^{-1} = \frac{1}{\det(A)} \operatorname{adj}(A) = \frac{1}{64} \begin{bmatrix} 12 & -4 & 12 \\ 6 & 2 & -10 \\ -16 & 16 & 16 \end{bmatrix}$$

$$= \begin{bmatrix} \dfrac{12}{64} & -\dfrac{4}{64} & \dfrac{12}{64} \\[2mm] \dfrac{6}{64} & \dfrac{2}{64} & -\dfrac{10}{64} \\[2mm] -\dfrac{16}{64} & \dfrac{16}{64} & \dfrac{16}{64} \end{bmatrix}$$

We note that for matrices larger than $3 \times 3$ the matrix inversion method in this example is computationally inferior to the technique given in Section 1.7.

On the other hand, the inversion method in Section 1.7 is just a computational procedure or algorithm for computing $A^{-1}$ and is not very useful for studying properties of the inverse. Formula (2.9) can often be used to obtain properties of the inverse without actually computing it (Exercise 20).

In a similar vein, it is often useful to have a formula for the solution of a system of equations that can be used to study properties of the solution without solving the system. The next theorem establishes such a formula for systems of $n$ equations in $n$ unknowns. The formula is known as **Cramer's rule.**

**Theorem 9** (*Cramer's Rule*). *If $AX = B$ is a system of $n$ linear equations in $n$ unknowns such that $\det(A) \neq 0$, then the system has a unique solution. This solution is*

$$x_1 = \frac{\det(A_1)}{\det(A)}, \ x_2 = \frac{\det(A_2)}{\det(A)}, \ldots, x_n = \frac{\det(A_n)}{\det(A)}$$

*where $A_j$ is the matrix obtained by replacing the entries of jth column of A by the entries in the matrix*

$$B = \begin{bmatrix} b_1 \\ b_2 \\ \vdots \\ b_n \end{bmatrix}$$

*Proof.* If $\det(A) \neq 0$, then $A$ is invertible and, by Theorem 11 in Section 1.8, $X = A^{-1}B$ is the unique solution of $AX = B$. Therefore, by Theorem 8 we have

$$X = A^{-1}B = \frac{1}{\det(A)} \operatorname{adj}(A)B = \frac{1}{\det(A)} \begin{bmatrix} C_{11} & C_{21} & \cdots & C_{n1} \\ C_{12} & C_{22} & \cdots & C_{n2} \\ \vdots & \vdots & & \vdots \\ C_{1n} & C_{2n} & \cdots & C_{nn} \end{bmatrix} \begin{bmatrix} b_1 \\ b_2 \\ \vdots \\ b_n \end{bmatrix}$$

Multiplying the matrices out gives

$$X = \frac{1}{\det(A)} \begin{bmatrix} b_1C_{11} + b_2C_{21} + \cdots + b_nC_{n1} \\ b_1C_{12} + b_2C_{22} + \cdots + b_nC_{n2} \\ \vdots \qquad \vdots \qquad \qquad \vdots \\ b_1C_{1n} + b_2C_{2n} + \cdots + b_nC_{nn} \end{bmatrix}$$

The entry in the $j$th row of $X$ is therefore

$$x_j = \frac{b_1C_{1j} + b_2C_{2j} + \cdots + b_nC_{nj}}{\det(A)} \tag{2.10}$$

Now let

$$A_j = \begin{bmatrix} a_{11} & a_{12} & \cdots & a_{1j-1} & b_1 & a_{1j+1} & \cdots & a_{1n} \\ a_{21} & a_{22} & \cdots & a_{2j-1} & b_2 & a_{2j+1} & \cdots & a_{2n} \\ \vdots & \vdots & & \vdots & \vdots & \vdots & & \vdots \\ a_{n1} & a_{n2} & \cdots & a_{nj-1} & b_n & a_{nj+1} & \cdots & a_{nn} \end{bmatrix}$$

Since $A_j$ differs from $A$ only in the $j$th column, the cofactors of entries $b_1, b_2, \ldots, b_n$ in $A_j$ are the same as the cofactors of the corresponding entries in the $j$th column of $A$. The cofactor expansion of $\det(A_j)$ along the $j$th column is therefore

$$\det(A_j) = b_1C_{1j} + b_2C_{2j} + \cdots + b_nC_{nj}$$

Substituting this result in (2.10) gives

$$x_j = \frac{\det(A_j)}{\det(A)}$$

## Example 33

Use Cramer's Rule to solve

$$\begin{aligned} x_1 + \phantom{4x_2} + 2x_3 &= 6 \\ -3x_1 + 4x_2 + 6x_3 &= 30 \\ -x_1 - 2x_2 + 3x_3 &= 8 \end{aligned}$$

*Solution.*

$$A = \begin{bmatrix} 1 & 0 & 2 \\ -3 & 4 & 6 \\ -1 & -2 & 3 \end{bmatrix} \quad A_1 = \begin{bmatrix} 6 & 0 & 2 \\ 30 & 4 & 6 \\ 8 & -2 & 3 \end{bmatrix}$$

$$A_2 = \begin{bmatrix} 1 & 6 & 2 \\ -3 & 30 & 6 \\ -1 & 8 & 3 \end{bmatrix} \quad A_3 = \begin{bmatrix} 1 & 0 & 6 \\ -3 & 4 & 30 \\ -1 & -2 & 8 \end{bmatrix}$$

Therefore

$$x_1 = \frac{\det(A_1)}{\det(A)} = \frac{-40}{44} = \frac{-10}{11}, \qquad x_2 = \frac{\det(A_2)}{\det(A)} = \frac{72}{44} = \frac{18}{11},$$

$$x_3 = \frac{\det(A_3)}{\det(A)} = \frac{152}{44} = \frac{38}{11}$$

To solve a system of $n$ equations in $n$ unknowns by Cramer's Rule, it is necessary to evaluate $n + 1$ determinants of $n \times n$ matrices. For systems with more than three equations, Gaussian elimination is superior computationally since it is only necessary to reduce one $n$ by $n + 1$ augmented matrix. Cramer's rule, however, gives a formula for the solution.

## EXERCISE SET 2.4

**1.** Let

$$A = \begin{bmatrix} 1 & 6 & -3 \\ -2 & 7 & 1 \\ 3 & -1 & 4 \end{bmatrix}$$

(a) Find all the minors.
(b) Find all the cofactors.

**2.** Let

$$A = \begin{bmatrix} 4 & 0 & 4 & 4 \\ -1 & 0 & 1 & 1 \\ 1 & -3 & 0 & 3 \\ 6 & 3 & 14 & 2 \end{bmatrix}$$

Find
(a) $M_{13}$ and $C_{13}$    (b) $M_{23}$ and $C_{23}$
(c) $M_{22}$ and $C_{22}$    (d) $M_{21}$ and $C_{21}$

**3.** Evaluate the determinant of the matrix in Exercise 1 by a cofactor expansion along
(a) the first row      (b) the first column
(c) the second row    (d) the second column
(e) the third row      (f) the third column

**4.** For the matrix in Exercise 1, find
(a) adj($A$)
(b) $A^{-1}$ using the method of Example 32.

In Exercises **5–10** evaluate det($A$) by a cofactor expansion along a row or column of your choice.

**5.** $A = \begin{bmatrix} 0 & 6 & 0 \\ 8 & 6 & 8 \\ 3 & 2 & 2 \end{bmatrix}$    **6.** $A = \begin{bmatrix} 1 & 3 & 7 \\ 2 & 0 & -8 \\ -1 & -3 & 4 \end{bmatrix}$

**7.** $\begin{bmatrix} 1 & 1 & 1 \\ k & k & k \\ k^2 & k^2 & k^2 \end{bmatrix}$  **8.** $\begin{bmatrix} k-1 & 2 & 3 \\ 2 & k-3 & 4 \\ 3 & 4 & k-4 \end{bmatrix}$

**9.** $A = \begin{bmatrix} 4 & 4 & 0 & 4 \\ 1 & 1 & 0 & -1 \\ 3 & 0 & -3 & 1 \\ 6 & 14 & 3 & 6 \end{bmatrix}$  **10.** $A = \begin{bmatrix} 4 & 3 & 1 & 9 & 2 \\ 0 & 3 & 2 & 4 & 2 \\ 0 & 3 & 4 & 6 & 4 \\ 1 & -1 & 2 & 2 & 2 \\ 0 & 0 & 3 & 3 & 3 \end{bmatrix}$

**11.** Let

$$A = \begin{bmatrix} 1 & 3 & 1 & 1 \\ 2 & 5 & 2 & 2 \\ 1 & 3 & 8 & 9 \\ 1 & 3 & 2 & 2 \end{bmatrix}$$

(a) Evaluate $A^{-1}$ using the method of Example 32.

(b) Evaluate $A^{-1}$ using the method of Example 29 in Section 1.7.

(c) Which method involves less computation?

In Exercises 12–17 solve by Cramer's rule, where it applies.

**12.** $\begin{aligned} 3x_1 - 4x_2 &= -5 \\ 2x_1 + x_2 &= 4 \end{aligned}$

**13.** $\begin{aligned} 4x + 5y &= 2 \\ 11x + y + 2z &= 3 \\ x + 5y + 2z &= 1 \end{aligned}$

**14.** $\begin{aligned} x + y - 2z &= 1 \\ 2x - y + z &= 2 \\ x - 2y - 4z &= -4 \end{aligned}$

**15.** $\begin{aligned} x_1 - 3x_2 + x_3 &= 4 \\ 2x_1 - x_2 &= -2 \\ 4x_1 - 3x_3 &= 0 \end{aligned}$

**16.** $\begin{aligned} 2x_1 - x_2 + x_3 - 4x_4 &= -32 \\ 7x_1 + 2x_2 + 9x_3 - x_4 &= 14 \\ 3x_1 - x_2 + x_3 + x_4 &= 11 \\ x_1 + x_2 - 4x_3 - 2x_4 &= -4 \end{aligned}$

**17.** $\begin{aligned} 2x_1 - x_2 + x_3 &= 8 \\ 4x_1 + 3x_2 + x_3 &= 7 \\ 6x_1 + 2x_2 + 2x_3 &= 15 \end{aligned}$

**18.** Use Cramer's rule to solve for $z$ without solving for $x$, $y$, and $w$.

$$\begin{aligned} 4x + y + z + w &= 6 \\ 3x + 7y - z + w &= 1 \\ 7x + 3y - 5z + 8w &= -3 \\ x + y + z + 2w &= 3 \end{aligned}$$

**19.** Let $AX = B$ be a system of $n$ linear equations in $n$ unknowns.

(a) List three different methods for solving the system.

(b) Solve the system in Exercise 18 by each method.

(c) Which method involves the least amount of computation?

**20.** Prove that if $\det(A) = 1$ and all the entries in $A$ are integers, then all the entries in $A^{-1}$ are integers.

**21.** Let $AX = B$ be a system of $n$ linear equations in $n$ unknowns with integer coefficients and integer constants. Prove that if $\det(A) = 1$, then the solution $X$ has integer entries.

**22.** Prove that if $A$ is an invertible upper triangular matrix, then $A^{-1}$ is upper triangular.

**23.** Derive the first and last cofactor expansions listed in (2.4).

# Chapter 3 | *Vectors in 2-Space and 3-Space*

## 3.1 INTRODUCTION TO VECTORS (GEOMETRIC)

In this section vectors in 2-space and 3-space are introduced geometrically. Arithmetic operations on vectors are defined and some basic properties of these operations are established.

Many physical quantities like area, length, and mass are completely described once a real number representing the magnitude of the quantity is given. Other physical quantities, called *vectors,* are not completely determined until both a magnitude and a direction are specified. Force, displacement, and velocity are examples of vectors.

Vectors can be represented geometrically as directed line segments or arrows in 2-space or 3-space; the direction of the arrow specifies the direction of the vector and the length of the arrow describes its magnitude. The tail of the arrow is called the *initial point* of the vector, and the tip of the arrow the *terminal point.* We shall denote vectors by lower case boldface type like **a**, **k**, **v**, **w** and **x**. When discussing vectors, we shall refer to numbers as *scalars.* All our scalars will be real numbers and will be denoted by ordinary lower-case type like $a$, $k$, $v$, $w$ and $x$.

If as in Figure 3.1($a$) the initial point of a vector **v** is $A$ and the terminal point is $B$, we write

$$\mathbf{v} = \overrightarrow{AB}$$

Vectors having the same length and same direction, like those in Figure 3.1($b$), are called *equivalent.* Since we want a vector to be determined solely by its length and direction, equivalent vectors are regarded as *equal* even though they may be located in different positions. If **v** and **w** are equivalent we write

$$\mathbf{v} = \mathbf{w}$$

**Figure 3.1** (*a*) The vector $\overrightarrow{AB}$. (*b*) Equivalent vectors.

**Definition.** If **v** and **w** are any two vectors, then the ***sum*** **v** + **w** is the vector determined as follows. Position the vector **w** so that its initial point coincides with the terminal point of **v**. The vector **v** + **w** is represented by the arrow from the initial point of **v** to the terminal point of **w** (Figure 3.2).

In Figure 3.3 we have constructed two sums, **v** + **w** (black arrows) and **w** + **v** (white arrows). It is evident that

$$\mathbf{v} + \mathbf{w} = \mathbf{w} + \mathbf{v}$$

and that the sum coincides with the diagonal of the parallelogram determined by **v** and **w** when these vectors are located so they have the same initial point.

A vector of length zero is called a ***zero vector*** and is denoted by **0**. We define

$$\mathbf{0} + \mathbf{v} = \mathbf{v} + \mathbf{0} = \mathbf{v}$$

for every vector **v**. Since there is no natural direction for a zero vector, we shall agree that it can be assigned any direction that is convenient for the problem being considered. If **v** is any nonzero vector, it is obvious that the only vector **w** satisfying **v** + **w** = **0** is the vector having the same magnitude as **v** but oppositely directed. This vector is called the ***negative*** (or ***additive inverse***) of **v** and we write

$$\mathbf{w} = -\mathbf{v}$$

If **v** = **0**, then we define **0** = −**0**.

**Figure 3.2**

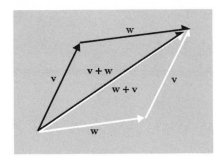

**Figure 3.3**

**Definition.** If **v** and **w** are any two vectors, then subtraction is defined by

$$\mathbf{v} - \mathbf{w} = \mathbf{v} + (-\mathbf{w})$$

(Figure 3.4*a*)

To obtain the difference **v** − **w** without constructing −**w**, position **v** and **w** so their initial points coincide; the vector from the terminal point of **w** to the terminal point of **v** is then the vector **v** − **w** (Figure 3.4*b*).

**Definition.** If **v** is a vector and *k* is a real number (scalar), then the ***product*** *k***v** is defined to be the vector whose length is |*k*| times the length of **v** and whose direction is the same as that of **v** if *k* > 0 and opposite to that of **v** if *k* < 0. If **v** = **0**, we define *k***0** = **0**.

Figure 3.5 illustrates the relation between a vector **v** and the vectors $\frac{1}{2}\mathbf{v}$, (−1)**v**, 2**v**, and (−3)**v**.

Note that the vector (−1)**v** has the same length as **v** but is oppositely directed. Thus (−1)**v** is just the negative of **v**; that is,

$$(-1)\mathbf{v} = -\mathbf{v}$$

Problems involving vectors can often be simplified by introducing a rectangular coordinate system. For the moment we shall restrict the discussion to vectors in

**Figure 3.4**

**Figure 3.5**

2-space (the plane). Let **v** be any vector in the plane, and assume, as in Figure 3.6, that **v** has been positioned so its initial point is at the origin of a rectangular coordinate system. The coordinates $(v_1, v_2)$ of the terminal point of **v** are called the *components* of **v**, and we write

$$\mathbf{v} = (v_1, v_2)$$

If equivalent vectors, **v** and **w**, are located so their initial points fall at the origin, then it is obvious that their terminal points must coincide (since the vectors have the same length and direction). The vectors thus have the same components. It is equally obvious that vectors with the same components must have the same length and same direction, and consequently be equivalent. In summary, two vectors

$$\mathbf{v} = (v_1, v_2) \quad \text{and} \quad \mathbf{w} = (w_1, w_2)$$

are equivalent if and only if

$$v_1 = w_1 \quad \text{and} \quad v_2 = w_2$$

The operations of vector addition and multiplication by scalars are very easy to carry out in terms of components. As illustrated in Figure 3.7, if

$$\mathbf{v} = (v_1, v_2) \quad \text{and} \quad \mathbf{w} = (w_1, w_2)$$

then

$$\mathbf{v} + \mathbf{w} = (v_1 + w_1, v_2 + w_2)$$

**Figure 3.6**

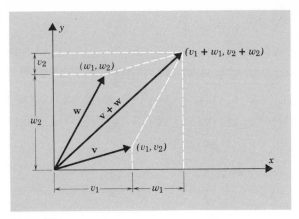

**Figure 3.7**

If $\mathbf{v} = (v_1, v_2)$ and $k$ is any scalar, then by using a geometric argument involving similar triangles, it can be shown (Exercise 14) that

$$\mathbf{kv} = (kv_1, kv_2)$$

(Figure 3.8).

Thus, for example, if $\mathbf{v} = (1, -2)$ and $\mathbf{w} = (7, 6)$, then

$$\mathbf{v} + \mathbf{w} = (1, -2) + (7, 6) = (1 + 7, -2 + 6) = (8, 4)$$

and

$$4\mathbf{v} = 4(1, -2) = (4(1), 4(-2)) = (4, -8)$$

Just as vectors in the plane can be described by pairs of real numbers, vectors in 3-space can be described by triples of real numbers, by introducing a ***rectangular coordinate*** system. To construct such a coordinate system, select a point $O$, called

**Figure 3.8**

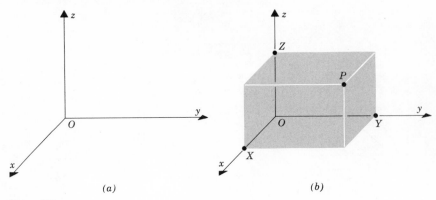

(a)                                    (b)

**Figure 3.9**

the *origin,* and choose three mutually perpendicular lines, called *coordinate axes,* passing through the origin. Label these axes $x$, $y$, and $z$ and select a positive direction for each coordinate axis as well as a unit of length for measuring distances. (Figure 3.9a). Each pair of coordinate axes determines a plane called a *coordinate plane.* These are referred to as the *xy-plane,* the *xz-plane* and the *yz-plane.* To each point $P$ in 3-space we assign a triple of numbers $(x, y, z)$ called the *coordinates of P* as follows. Pass three planes through $P$ parallel to the coordinate planes, and denote the points of intersections of these planes with the three coordinate axes by $X$, $Y$, and $Z$ (Figure 3.9b). The coordinates of $P$ are defined to be the signed lengths

$$x = OX \qquad y = OY \qquad z = OZ$$

In Figure 3.10 we have constructed the points whose coordinates are $(4, 5, 6)$ and $(-3, 2, -4)$.

**Figure 3.10**

**Figure 3.11** (*a*) Right-handed. (*b*) Left-handed.

Rectangular coordinate systems in 3-space fall into two categories, **left-handed** and **right-handed.** A right-handed system has the property that an ordinary screw pointed in the positive direction on the $z$-axis would be advanced if the positive $x$-axis is rotated 90° toward the positive $y$-axis (Figure 3.11*a*). The system is left-handed if the screw would retract (Figure 3.11*b*).

In this text we shall use only right-hand coordinate systems.

If, as in Figure 3.12, a vector **v** in 3-space is located so its initial point is at the origin of a rectangular coordinate system, then the coordinates of the terminal point are called the **components** of **v** and we write

$$\mathbf{v} = (v_1, v_2, v_3)$$

If $\mathbf{v} = (v_1, v_2, v_3)$ and $\mathbf{w} = (w_1, w_2, w_3)$ are two vectors in 3-space, then arguments similar to those used for vectors in a plane can be used to establish the following results.

(i) **v** and **w** are equivalent if and only if $v_1 = w_1$, $v_2 = w_2$ and $v_3 = w_3$
(ii) $\mathbf{v} + \mathbf{w} = (v_1 + w_1, v_2 + w_2, v_3 + w_3)$
(iii) $k\mathbf{v} = (kv_1, kv_2, kv_3)$ where $k$ is any scalar

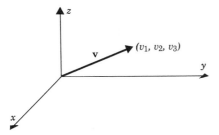

**Figure 3.12**

**Example 1**

If $\mathbf{v} = (1, -3, 2)$ and $\mathbf{w} = (4, 2, 1)$, then

$$\mathbf{v} + \mathbf{w} = (5, -1, 3),\ 2\mathbf{v} = (2, -6, 4),\ -\mathbf{w} = (-4, -2, -1),\ \text{and}$$
$$\mathbf{v} - \mathbf{w} = \mathbf{v} + (-\mathbf{w}) = (-3, -5, 1).$$

Sometimes vectors arise that do not have their initial points at the origin. To find the components of a vector $\mathbf{v}$ having initial point $P_1(x_1, y_1, z_1)$ and terminal point $P_2(x_2, y_2, z_2)$, we construct an equivalent vector whose initial point is at the origin. In Figure 3.13 $\overrightarrow{OQ}$ is such a vector. Therefore, the components of $\mathbf{v} = \overrightarrow{P_1P_2}$ are the coordinates $(a, b, c)$ of the point $Q$.

From Figure 3.13, $\overrightarrow{OQ} + \overrightarrow{OP_1} = \overrightarrow{OP_2}$; in terms of components

$$(a, b, c) + (x_1, y_1, z_1) = (x_2, y_2, z_2)$$

Equivalently

$$(a + x_1, b + y_1, c + z_1) = (x_2, y_2, z_2)$$

By equating corresponding components and solving for $a$, $b$, and $c$, we find that the components $(a, b, c)$ of $\mathbf{v} = \overrightarrow{P_1P_2}$ are given by

$$a = x_2 - x_1 \qquad b = y_2 - y_1 \qquad c = z_2 - z_1$$

**Example 2**

The components of the vector $\mathbf{v} = \overrightarrow{P_1P_2}$ with initial point $P_1(2, -1, 4)$ and terminal point $P_2(7, 5, -8)$ are

$$\mathbf{v} = (7 - 2, 5 - (-1), (-8) - 4) = (5, 6, -12)$$

Analogously, in 2-space the vector $\mathbf{v}$ with initial point $P_1(x_1, x_2)$ and terminal point $P_2(x_2, y_2)$ is $\mathbf{v} = (x_2 - x_1, y_2 - y_1)$.

**Figure 3.13**

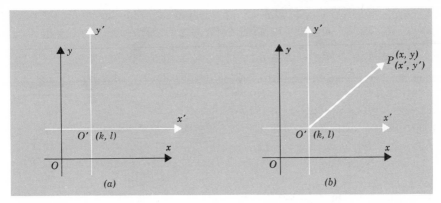

**Figure 3.14**

## Example 3

The solutions to many problems can be simplified by translating the coordinate axes to obtain new axes parallel to the original ones.

In Figure 3.14(*a*) we have translated the *xy*-coordinate axes to obtain an *x'y'*-coordinate system whose origin $O'$ is at the point $(x, y) = (k, l)$. A point $P$ in 2-space now has both $(x, y)$ coordinates and $(x', y')$ coordinates. To see how the two are related, consider the vector $\overrightarrow{O'P}$ (Figure 3.14*b*). In the *xy*-system its initial point is at $(k, l)$ and its terminal point is at $(x, y)$; thus $\overrightarrow{OP} = (x - k, y - l)$. In the *x'y'*-system its initial point is at $(0, 0)$ and its terminal point is at $(x', y')$; thus $\overrightarrow{OP} = (x', y')$. Therefore

$$x' = x - k \qquad y' = y - l$$

These are called the ***translation equations.***

To illustrate, if the new origin is at $(k, l) = (4, 1)$ and the *xy*-coordinates of a point $P$ are $(2, 0)$, then *x'y'*-coordinates of $P$ are $x' = 2 - 4 = -2$ and $y' = 0 - 1 = -1$.

In 3-space the translation equations are

$$x' = x - k \qquad y' = y - l \qquad z' = z - m$$

where $(k, l, m)$ are the *xyz*-coordinates of the new origin.

## EXERCISE SET 3.1

**1.** Draw a right-handed coordinate system, and locate the points whose coordinates are:

(a) $(2, 3, 4)$     (b) $(-2, 3, 4)$     (c) $(2, -3, 4)$     (d) $(2, 3, -4)$
(e) $(-2, -3, 4)$     (f) $(-2, 3, -4)$     (g) $(2, -3, -4)$     (h) $(-2, -3, -4)$
(i) $(0, 2, 0)$     (j) $(0, 0, -2)$     (k) $(2, 0, 2)$     (l) $(-2, 0, 0)$

2. Sketch the following vectors with the initial points located at the origin:
   (a) $\mathbf{v}_1 = (2, 5)$      (b) $\mathbf{v}_2 = (-3, 7)$      (c) $\mathbf{v}_3 = (-5, -4)$
   (d) $\mathbf{v}_4 = (6, -2)$      (e) $\mathbf{v}_5 = (2, 0)$      (f) $\mathbf{v}_6 = (0, -8)$
   (g) $\mathbf{v}_7 = (2, 3, 4)$      (h) $\mathbf{v}_8 = (2, 0, 2)$      (i) $\mathbf{v}_9 = (0, 0, -2)$

3. Find the components of the vector having initial point $P_1$ and terminal point $P_2$.
   (a) $P_1(3, 5), P_2(2, 8)$                    (b) $P_1(7, -2), P_2(0, 0)$
   (c) $P_1(6, 5, 8), P_2(8, -7, -3)$        (d) $P_1(0, 0, 0), P_2(-8, 7, 4)$

4. Find a vector with initial point $P(2, -1, 4)$ that has the same direction as $\mathbf{v} = (7, 6, -3)$.

5. Find a vector, oppositely directed to $\mathbf{v} = (-2, 4, -1)$, having terminal point $Q(2, 0, -7)$.

6. Let $\mathbf{u} = (1, 2, 3)$, $\mathbf{v} = (2, -3, 1)$, and $\mathbf{w} = (3, 2, -1)$. Find the components of
   (a) $\mathbf{u} - \mathbf{w}$      (b) $7\mathbf{v} + 3\mathbf{w}$      (c) $-\mathbf{w} + \mathbf{v}$
   (d) $3(\mathbf{u} - 7\mathbf{v})$      (e) $-3\mathbf{v} - 8\mathbf{w}$      (f) $2\mathbf{v} - (\mathbf{u} + \mathbf{w})$

7. Let $\mathbf{u}$, $\mathbf{v}$, and $\mathbf{w}$ be the vectors in Exercise 6. Find the components of the vector $\mathbf{x}$ that satisfies $2\mathbf{u} - \mathbf{v} + \mathbf{x} = 7\mathbf{x} + \mathbf{w}$.

8. Let $\mathbf{u}$, $\mathbf{v}$, and $\mathbf{w}$ be the vectors in Exercise 6. Find scalars $c_1$, $c_2$, and $c_3$ such that $c_1\mathbf{u} + c_2\mathbf{v} + c_3\mathbf{w} = (6, 14, -2)$.

9. Show that there do not exist scalars $c_1$, $c_2$, and $c_3$ such that
   $c_1(1, 2, -3) + c_2(5, 7, 1) + c_3(6, 9, -2) = (4, 5, 0)$.

10. Find all scalars $c_1$, $c_2$, and $c_3$ such that
    $c_1(2, 7, 8) + c_2(1, -1, 3) + c_3(3, 6, 11) = (0, 0, 0)$.

11. Let $P$ be the point $(2, 3, -2)$ and $Q$ the point $(7, -4, 1)$.
    (a) Find the midpoint of the line segment connecting $P$ and $Q$.
    (b) Find the point on the line segment connecting $P$ and $Q$ that is 3/4 of the way from $P$ to $Q$.

12. Suppose an $xy$-coordinate system is translated to obtain an $x'y'$-coordinate system whose origin $O'$ has $xy$-coordinates $(2, -3)$.
    (a) Find the $x'y'$-coordinates of the point $P$ whose $xy$-coordinates are $(7, 5)$.
    (b) Find the $xy$-coordinates of the point $Q$ whose $x'y'$-coordinates are $(-3, 6)$.
    (c) Draw the $xy$ and $x'y'$-coordinate axes and locate the points $P$ and $Q$.

13. Suppose an $xyz$-coordinate system is translated to obtain an $x'y'z'$-coordinate system. Let $\mathbf{v}$ be a vector whose components are $\mathbf{v} = (v_1, v_2, v_3)$ in the $xyz$-system. Show that $\mathbf{v}$ has the same components in the $x'y'z'$-system.

14. Prove geometrically that if $\mathbf{v} = (v_1, v_2)$, then $k\mathbf{v} = (kv_1, kv_2)$. (Restrict the proof to the case $k > 0$ illustrated in Figure 3.8. The complete proof would involve many cases depending on the quadrant in which the vector falls and the sign of $k$.)

## 3.2 NORM OF A VECTOR; VECTOR ARITHMETIC

In this section we establish the basic rules of vector arithmetic.

**Theorem 1.** *If* **u**, **v** *and* **w** *are vectors in 2 or 3-space and k and l are scalars, then the following relationships hold.*

(*a*) $\mathbf{u} + \mathbf{v} = \mathbf{v} + \mathbf{u}$
(*b*) $(\mathbf{u} + \mathbf{v}) + \mathbf{w} = \mathbf{u} + (\mathbf{v} + \mathbf{w})$
(*c*) $\mathbf{u} + \mathbf{0} = \mathbf{0} + \mathbf{u} = \mathbf{u}$
(*d*) $\mathbf{u} + (-\mathbf{u}) = \mathbf{0}$
(*e*) $k(l\mathbf{u}) = (kl)\mathbf{u}$
(*f*) $k(\mathbf{u} + \mathbf{v}) = k\mathbf{u} + k\mathbf{v}$
(*g*) $(k + l)\mathbf{u} = k\mathbf{u} + l\mathbf{u}$
(*h*) $1\mathbf{u} = \mathbf{u}$

Before discussing the proof, we note that we have developed two approaches to vectors: *geometric,* in which vectors are represented by arrows or directed line segments, and *analytic,* in which vectors are represented by pairs or triples of numbers called components. As a consequence, the results in Theorem 1 can be established either geometrically or analytically. To illustrate, we shall prove part (b) both ways. The remaining proofs are left as exercises.

*Proof of part (b) (analytic).* We shall give the proof for vectors in 3-space. The proof for 2-space is similar. If $\mathbf{u} = (u_1, u_2, u_3)$, $\mathbf{v} = (v_1, v_2, v_3)$, and $\mathbf{w} = (w_1, w_2, w_3)$ are vectors in 3-space, then

$$
\begin{aligned}
(\mathbf{u} + \mathbf{v}) + \mathbf{w} &= ((u_1, u_2, u_3) + (v_1, v_2, v_3)) + (w_1, w_2, w_3) \\
&= (u_1 + v_1, u_2 + v_2, u_3 + v_3) + (w_1, w_2, w_3) \\
&= ((u_1 + v_1) + w_1, (u_2 + v_2) + w_2, (u_3 + v_3) + w_3) \\
&= (u_1 + (v_1 + w_1), u_2 + (v_2 + w_2), u_3 + (v_3 + w_3)) \\
&= (u_1, u_2, u_3) + (v_1 + w_1, v_2 + w_2, v_3 + w_3) \\
&= \mathbf{u} + (\mathbf{v} + \mathbf{w}) \quad \blacksquare
\end{aligned}
$$

*Proof of part (b) (geometric).* Let **u**, **v**, and **w** be represented by $\overrightarrow{PQ}$, $\overrightarrow{QR}$, and $\overrightarrow{RS}$ as shown in Figure 3.15.

Then

$$\mathbf{v} + \mathbf{w} = \overrightarrow{QS} \text{ and } \mathbf{u} + (\mathbf{v} + \mathbf{w}) = \overrightarrow{PS}$$

Also

$$\mathbf{u} + \mathbf{v} = \overrightarrow{PR} \text{ and } (\mathbf{u} + \mathbf{v}) + \mathbf{w} = \overrightarrow{PS}$$

Therefore

$$\mathbf{u} + (\mathbf{v} + \mathbf{w}) = (\mathbf{u} + \mathbf{v}) + \mathbf{w} \quad \blacksquare$$

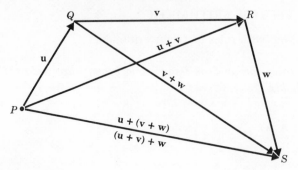

**Figure 3.15**

The length of a vector **v** is often called the ***norm*** of **v** and is denoted by $\|\mathbf{v}\|$. It follows from the theorem of Pythagoras that the norm of a vector $\mathbf{v} = (v_1, v_2)$ in 2-space is

$$\|\mathbf{v}\| = \sqrt{v_1{}^2 + v_2{}^2}$$

(Figure 3.16*a*). From Figure 3.16*b* and two applications of the Theorem of Pythagoras, we have for a vector $\mathbf{v} = (v_1, v_2, v_3)$ in 3-space

$$\|\mathbf{v}\|^2 = (OR)^2 + (RP)^2$$
$$= (OQ)^2 + (OS)^2 + (RP)^2$$
$$= v_1{}^2 + v_2{}^2 + v_3{}^2$$

Thus

$$\|\mathbf{v}\| = \sqrt{v_1{}^2 + v_2{}^2 + v_3{}^2} \tag{3.1}$$

**Figure 3.16**

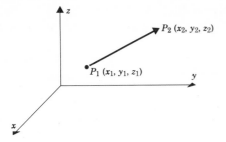

**Figure 3.17**

If $P_1(x_1, y_1, z_1)$ and $P_2(x_2, y_2, z_2)$ are two points in 3-space, then the distance $d$ between them is the norm of the vector $\overrightarrow{P_1P_2}$ (Figure 3.17). Since

$$\overrightarrow{P_1P_2} = (x_2 - x_1, y_2 - y_1, z_2 - z_1)$$

it follows from (3.1) that

$$d = \sqrt{(x_2 - x_1)^2 + (y_2 - y_1)^2 + (z_2 - z_1)^2}$$

Similarly, if $P_1(x_1, y_1)$ and $P_2(x_2, y_2)$ are points in 2-space, then the distance between them is given by

$$d = \sqrt{(x_2 - x_1)^2 + (y_2 - y_1)^2}$$

**Example 4**

The norm of the vector $v = (-3, 2, 1)$ is

$$\|v\| = \sqrt{(-3)^2 + (2)^2 + (1)^2} = \sqrt{14}$$

The distance $d$ between the points $P_1(2, -1, -5)$ and $P_2(4, -3, 1)$ is

$$d = \sqrt{(4 - 2)^2 + (-3 + 1)^2 + (1 + 5)^2} = \sqrt{44} = 2\sqrt{11}$$

**EXERCISE SET 3.2**

1. Compute the norm of $v$ when
   (a) $v = (3, 4)$     (b) $v = (-1, 7)$     (c) $v = (0, -3)$
   (d) $v = (1, 1, 1)$     (e) $v = (-8, 7, 4)$     (f) $v = (9, 0, 0)$

2. Compute the distance between $P_1$ and $P_2$.
   (a) $P_1(2, 3), P_2(4, 6)$          (b) $P_1(-2, 7), P_2(0, -3)$
   (c) $P_1(8, -4, 2), P_2(-6, -1, 0)$     (d) $P_1(1.1, 1), P_2(6, -7, 3)$

3. Let $u = (1, -3, 2)$, $v = (1, 1, 0)$, and $w = (2, 2, -4)$. Find:
   (a) $\|u + v\|$     (b) $\|u\| + \|v\|$     (c) $\|-2u\| + 2\|u\|$

   (d) $\|3u - 5v + w\|$     (e) $\dfrac{1}{\|w\|}w$     (f) $\left\|\dfrac{1}{\|w\|}w\right\|$

**4.** Find all scalars $k$ such that $\|k\mathbf{v}\| = 3$, where $\mathbf{v} = (1, 2, 4)$.

**5.** Verify parts (b), (e), (f), and (g) of Theorem 1 for $\mathbf{u} = (1, -3, 7)$, $\mathbf{v} = (6, 6, 9)$, $\mathbf{w} = (-8, 1, 2)$, $k = -3$, and $l = 6$.

**6.** Show that if $\mathbf{v}$ is nonzero, then $\dfrac{1}{\|\mathbf{v}\|}\mathbf{v}$ has norm 1.

**7.** Use Exercise 6 to find a vector of norm 1 having the same direction as $\mathbf{v} = (1, 1, 1)$.

**8.** Let $\mathbf{p}_0 = (x_0, y_0, z_0)$ and $\mathbf{p} = (x, y, z)$. Describe the set of all points $(x, y, z)$ for which $\|\mathbf{p} - \mathbf{p}_0\| = 1$.

**9.** Prove geometrically that if $u$ and $v$ are vectors in 2 or 3-space, then $\|\mathbf{u} + \mathbf{v}\| \leq \|\mathbf{u}\| + \|\mathbf{v}\|$.

**10.** Prove parts (a), (c), and (e) of Theorem 1 analytically.

**11.** Prove parts (d), (g), and (h) of Theorem 1 analytically.

**12.** Prove part (f) of Theorem 1 geometrically.

**13.** Let $\mathbf{a}$, $\mathbf{b}$, $\mathbf{c}$, and $\mathbf{d}$ be distinct nonzero vectors in 2 or 3-space with the same initial point. Show that if $\mathbf{b} - \mathbf{a} = \mathbf{c} - \mathbf{d}$, then the terminal points are vertices of a parallelogram.

## 3.3 DOT PRODUCT; PROJECTIONS

In this section we introduce a kind of multiplication of vectors in 2-space and 3-space. Arithmetic properties of this multiplication are established and some applications are given.

Let $\mathbf{u}$ and $\mathbf{v}$ be two nonzero vectors in 2-space or 3-space, and assume these vectors have been positioned so their initial points coincide. By the *angle between* $\mathbf{u}$ *and* $\mathbf{v}$, we shall mean the angle $\theta$ determined by $\mathbf{u}$ and $\mathbf{v}$ that satisfies $0 \leq \theta \leq \pi$ (Figure 3.18).

**Definition.** If $\mathbf{u}$ and $\mathbf{v}$ are vectors in 2-space or 3-space and $\theta$ is the angle between $\mathbf{u}$ and $\mathbf{v}$, then the *dot product* or *Euclidean inner product* $\mathbf{u} \cdot \mathbf{v}$ is defined by

$$\mathbf{u} \cdot \mathbf{v} = \begin{cases} \|\mathbf{u}\|\,\|\mathbf{v}\|\,\cos\theta, & \textit{if } \mathbf{u} \neq \mathbf{0} \textit{ and } \mathbf{v} \neq \mathbf{0} \\ 0 & , \textit{if } \mathbf{u} = \mathbf{0} \textit{ or } \mathbf{v} = \mathbf{0} \end{cases}$$

**Figure 3.18**

**Example 5**

As shown in Figure 3.19, the angle between the vectors $\mathbf{u} = (0, 0, 1)$ and $\mathbf{v} = (0, 2, 2)$ is $45°$. Thus

$$\mathbf{u} \cdot \mathbf{v} = \|\mathbf{u}\| \, \|\mathbf{v}\| \cos \theta = (\sqrt{0^2 + 0^2 + 1^2})(\sqrt{0^2 + 2^2 + 2^2})\left(\frac{1}{\sqrt{2}}\right) = 2$$

Let $\mathbf{u} = (u_1, u_2, u_3)$ and $\mathbf{v} = (v_1, v_2, v_3)$ be two nonzero vectors. If, as in Figure 3.20, $\theta$ is the angle between $\mathbf{u}$ and $\mathbf{v}$, then the law of cosines yields

$$\|\overrightarrow{PQ}\|^2 = \|\mathbf{u}\|^2 + \|\mathbf{v}\|^2 - 2\|\mathbf{u}\| \, \|\mathbf{v}\| \cos \theta \tag{3.2}$$

Since $\overrightarrow{PQ} = \mathbf{v} - \mathbf{u}$, we can rewrite (3.2) as

$$\|\mathbf{u}\| \, \|\mathbf{v}\| \cos \theta = \frac{1}{2} \left( \|\mathbf{u}\|^2 + \|\mathbf{v}\|^2 - \|\mathbf{v} - \mathbf{u}\|^2 \right)$$

or

$$\mathbf{u} \cdot \mathbf{v} = \frac{1}{2}(\|\mathbf{u}\|^2 + \|\mathbf{v}\|^2 - 2\|\mathbf{v} - \mathbf{u}\|^2)$$

Substituting

$$\|\mathbf{u}\|^2 = u_1{}^2 + u_2{}^2 + u_3{}^2 \qquad \|\mathbf{v}\|^2 = v_1{}^2 + v_2{}^2 + v_3{}^2$$

and

$$\|\mathbf{v} - \mathbf{u}\|^2 = (v_1 - u_1)^2 + (v_2 - u_2)^2 + (v_3 - u_3)^2$$

we obtain after simplifying

$$\mathbf{u} \cdot \mathbf{v} = u_1 v_1 + u_2 v_2 + u_3 v_3 \tag{3.3}$$

If $\mathbf{u} = (u_1, u_2)$ and $\mathbf{v} = (v_1, v_2)$ are two vectors in 2-space, then the formula corresponding to (3.3) is

$$\mathbf{u} \cdot \mathbf{v} = u_1 v_1 + u_2 v_2$$

**Figure 3.19**

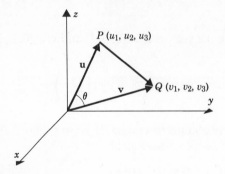

**Figure 3.20**

**Example 6**

Consider the vectors

$$\mathbf{u} = (2, -1, 1) \text{ and } \mathbf{v} = (1, 1, 2)$$

Find $\mathbf{u} \cdot \mathbf{v}$ and determine the angle $\theta$ between $\mathbf{u}$ and $\mathbf{v}$.

*Solution.*

$$\mathbf{u} \cdot \mathbf{v} = u_1 v_1 + u_2 v_2 + u_3 v_3 = (2)(1) + (-1)(1) + (1)(2) = 3$$

Also $\|\mathbf{u}\| = \|\mathbf{v}\| = \sqrt{6}$; so that

$$\cos \theta = \frac{\mathbf{u} \cdot \mathbf{v}}{\|\mathbf{u}\| \, \|\mathbf{v}\|} = \frac{3}{\sqrt{6}\sqrt{6}} = \frac{1}{2}$$

Thus $\theta = 60°$.

**Example 7**

Find the angle between a diagonal of a cube and one of its edges.

*Solution.* Let $k$ be the length of an edge and introduce a coordinate system as shown in Figure 3.21.

If we let $\mathbf{u}_1 = (k, 0, 0)$, $\mathbf{u}_2 = (0, k, 0)$, and $\mathbf{u}_3 = (0, 0, k)$, then the diagonal of the cube is the vector

$$\mathbf{d} = (k, k, k) = \mathbf{u}_1 + \mathbf{u}_2 + \mathbf{u}_3$$

The angle $\theta$ between $\mathbf{d}$ and the edge $\mathbf{u}_1$ satisfies

$$\cos \theta = \frac{\mathbf{u}_1 \cdot \mathbf{d}}{\|\mathbf{u}_1\| \, \|\mathbf{d}\|} = \frac{k^2}{(k)(\sqrt{3k^2})} = \frac{1}{\sqrt{3}}$$

Thus

$$\theta = \cos^{-1}\left(\frac{1}{\sqrt{3}}\right) \approx 54° \, 41'.$$

The next theorem shows how the dot product can be used to obtain information about the angle between two vectors; it also establishes an important relationship between the norm and the dot product.

**Theorem 2.** *Let* **u** *and* **v** *be vectors in 2 or 3-space.*

(a) $\mathbf{v} \cdot \mathbf{v} = \|\mathbf{v}\|^2$; *that is,* $\|\mathbf{v}\| = (\mathbf{v} \cdot \mathbf{v})^{1/2}$
(b) *If* **u** *and* **v** *are nonzero vectors and* $\theta$ *is the angle between them, then*

$$\begin{array}{lll} \theta \text{ is acute} & \text{if and only if} & \mathbf{u} \cdot \mathbf{v} > 0 \\ \theta \text{ is obtuse} & \text{if and only if} & \mathbf{u} \cdot \mathbf{v} < 0 \\ \theta = \dfrac{\pi}{2} & \text{if and only if} & \mathbf{u} \cdot \mathbf{v} = 0 \end{array}$$

*Proof.*

(a) Since the angle $\theta$ between **v** and **v** is 0, $\cos\theta = 1$.
 Therefore $\mathbf{v} \cdot \mathbf{v} = \|\mathbf{v}\| \, \|\mathbf{v}\| \cos\theta = \|\mathbf{v}\|^2$.
(b) Since $\|\mathbf{u}\| > 0$, $\|\mathbf{v}\| > 0$, and $\mathbf{u} \cdot \mathbf{v} = \|\mathbf{u}\| \, \|\mathbf{v}\| \cos\theta$, $\mathbf{u} \cdot \mathbf{v}$ has the same sign as $\cos\theta$. Since $\theta$ always satisfies $0 \le \theta \le \pi$, $\theta$ is acute if and only if $\cos\theta > 0$, $\theta$ is obtuse if and only if $\cos\theta < 0$, and $\theta = \pi/2$ if and only if $\cos\theta = 0$. ∎

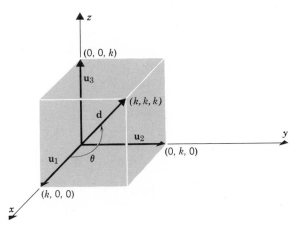

**Figure 3.21**

**Example 8**

If $\mathbf{u} = (1, -2, 3)$, $\mathbf{v} = (-3, 4, 2)$, and $\mathbf{w} = (3, 6, 3)$, then

$$\mathbf{u} \cdot \mathbf{v} = (1)(-3) + (-2)(4) + (3)(2) = -5$$
$$\mathbf{v} \cdot \mathbf{w} = (-3)(3) + (4)(6) + (2)(3) = 21$$
$$\mathbf{u} \cdot \mathbf{w} = (1)(3) + (-2)(6) + (3)(3) = 0$$

Therefore $\mathbf{u}$ and $\mathbf{v}$ make an obtuse angle, $\mathbf{v}$ and $\mathbf{w}$ make an acute angle, and $\mathbf{u}$ and $\mathbf{w}$ are perpendicular.

The main arithmetic properties of the dot product are listed in the next theorem.

**Theorem 3.** *If* $\mathbf{u}$, $\mathbf{v}$ *and* $\mathbf{w}$ *are vectors in 2 or 3-space and* $k$ *is a scalar, then*

(*a*) $\mathbf{u} \cdot \mathbf{v} = \mathbf{v} \cdot \mathbf{u}$
(*b*) $\mathbf{u} \cdot (\mathbf{v} + \mathbf{w}) = \mathbf{u} \cdot \mathbf{v} + \mathbf{u} \cdot \mathbf{w}$
(*c*) $k(\mathbf{u} \cdot \mathbf{v}) = (k\mathbf{u}) \cdot \mathbf{v} = \mathbf{u} \cdot (k\mathbf{v})$
(*d*) $\mathbf{v} \cdot \mathbf{v} > 0$ *if* $\mathbf{v} \neq 0$ *and* $\mathbf{v} \cdot \mathbf{v} = 0$ *if* $\mathbf{v} = 0$

*Proof.* We shall prove (*c*) for vectors in 3-space and leave the remaining proofs as exercises. Let $\mathbf{u} = (u_1, u_2, u_3)$ and $\mathbf{v} = (v_1, v_2, v_3)$; then

$$k(\mathbf{u} \cdot \mathbf{v}) = k(u_1 v_1 + u_2 v_2 + u_3 v_3)$$
$$= (k u_1) v_1 + (k u_2) v_2 + (k u_3) v_3$$
$$= (k\mathbf{u}) \cdot \mathbf{v}$$

Similarly $\qquad k(\mathbf{u} \cdot \mathbf{v}) = \mathbf{u} \cdot (k\mathbf{v})$ ▮

Based on part (*b*) of Theorem 2, we define two vectors $\mathbf{u}$ and $\mathbf{v}$ to be **orthogonal** (written $\mathbf{u} \perp \mathbf{v}$) if $\mathbf{u} \cdot \mathbf{v} = 0$. If we agree that the zero vector makes an angle of $\pi/2$ with every vector, then two vectors are orthogonal if and only if they are geometrically perpendicular.

The dot product is useful in problems where it is of interest to "decompose" a vector into a sum of two perpendicular vectors. If $\mathbf{u}$ and $\mathbf{v}$ are nonzero vectors in 2 or 3-space, then it is always possible to write $\mathbf{u}$ as

$$\mathbf{u} = \mathbf{w}_1 + \mathbf{w}_2$$

where $\mathbf{w}_1$ is a scalar multiple of $\mathbf{v}$, and $\mathbf{w}_2$ is perpendicular to $\mathbf{v}$ (Figure 3.22).

The vectors $\mathbf{w}_1$ and $\mathbf{w}_2$ can be obtained as follows. Since $\mathbf{w}_1$ is a scalar multiple of $\mathbf{v}$, we can write it in the form $\mathbf{w}_1 = k\mathbf{v}$. Thus

$$\mathbf{u} = \mathbf{w}_1 + \mathbf{w}_2 = k\mathbf{v} + \mathbf{w}_2 \tag{3.4}$$

Since $\mathbf{w}_2$ is perpendicular to $\mathbf{v}$, we have $\mathbf{w}_2 \cdot \mathbf{v} = 0$. Thus by taking the dot product of both sides of (3.4) with $\mathbf{v}$, we obtain

$$\mathbf{u} \cdot \mathbf{v} = k(\mathbf{v} \cdot \mathbf{v}) = k\|\mathbf{v}\|^2$$

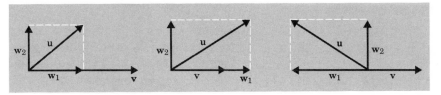

**Figure 3.22**

Therefore $k = \dfrac{\mathbf{u} \cdot \mathbf{v}}{\|\mathbf{v}\|^2}$ and consequently

$$\mathbf{w}_1 = \frac{\mathbf{u} \cdot \mathbf{v}}{\|\mathbf{v}\|^2} \mathbf{v}$$

Solving $\mathbf{u} = \mathbf{w}_1 + \mathbf{w}_2$ for $\mathbf{w}_2$ gives

$$\mathbf{w}_2 = \mathbf{u} - \frac{\mathbf{u} \cdot \mathbf{v}}{\|\mathbf{v}\|^2} \mathbf{v}$$

The vector $\mathbf{w}_1$ is called the ***orthogonal projection of*** $\mathbf{u}$ ***on*** $\mathbf{v}$; the vector $\mathbf{w}_2$ is called the ***component of*** $\mathbf{u}$ ***orthogonal to*** $\mathbf{v}$.

**Example 9**

Consider the vectors

$$\mathbf{u} = (2, -1, 3) \text{ and } \mathbf{v} = (4, -1, 2)$$

Since

$$\mathbf{u} \cdot \mathbf{v} = (2)(4) + (-1)(-1) + (3)(2) = 15$$

and

$$\|\mathbf{v}\|^2 = 4^2 + (-1)^2 + 2^2 = 21$$

the orthogonal projection of $\mathbf{u}$ on $\mathbf{v}$ is

$$\mathbf{w}_1 = \frac{\mathbf{u} \cdot \mathbf{v}}{\|\mathbf{v}\|^2} \mathbf{v} = \frac{15}{21}(4, -1, 2) = \left(\frac{20}{7}, \frac{-5}{7}, \frac{10}{7}\right)$$

The component of $\mathbf{u}$ orthogonal to $\mathbf{v}$ is

$$\mathbf{w}_2 = \mathbf{u} - \mathbf{w}_1 = (2, -1, 3) - \left(\frac{20}{7}, \frac{-5}{7}, \frac{10}{7}\right) = \left(\frac{-6}{7}, \frac{-2}{7}, \frac{11}{7}\right)$$

As a check, the reader may wish to verify that $\mathbf{w}_2$ is perpendicular to $\mathbf{v}$ by showing that $\mathbf{w}_2 \cdot \mathbf{v} = 0$.

## EXERCISE SET 3.3

1. Find $\mathbf{u} \cdot \mathbf{v}$ for:
   (a) $\mathbf{u} = (1, 2), \mathbf{v} = (6, -8)$      (b) $\mathbf{u} = (-7, -3), \mathbf{v} = (0, 1)$
   (c) $\mathbf{u} = (1, -3, 7), \mathbf{v} = (8, -2, -2)$      (d) $\mathbf{u} = (-3, 1, 2), \mathbf{v} = (4, 2, -5)$

2. In each part of Exercise 1, find the cosine of the angle $\theta$ between $\mathbf{u}$ and $\mathbf{v}$.

3. Determine if $\mathbf{u}$ and $\mathbf{v}$ make an acute angle, make an obtuse angle, or are orthogonal.
   (a) $\mathbf{u} = (7, 3, 5), \mathbf{v} = (-8, 4, 2)$      (b) $\mathbf{u} = (6, 1, 3), \mathbf{v} = (4, 0, -6)$
   (c) $\mathbf{u} = (1, 1, 1), \mathbf{v} = (-1, 0, 0)$      (d) $\mathbf{u} = (4, 1, 6), \mathbf{v} = (-3, 0, 2)$

4. Find the orthogonal projection of $\mathbf{u}$ on $\mathbf{v}$ if:
   (a) $\mathbf{u} = (2, 1), \mathbf{v} = (-3, 2)$      (b) $\mathbf{u} = (2, 6), \mathbf{v} = (-9, 3)$
   (c) $\mathbf{u} = (-7, 1, 3), \mathbf{v} = (5, 0, 1)$      (d) $\mathbf{u} = (0, 0, 1), \mathbf{v} = (8, 3, 4)$

5. In each part of Exercise 4, find the component of $\mathbf{u}$ orthogonal to $\mathbf{v}$.

6. Verify Theorem 3 for $\mathbf{u} = (6, -1, 2)$, $\mathbf{v} = (2, 7, 4)$, and $k = -5$.

7. Find two vectors of norm 1 that are orthogonal to $(3, -2)$.

8. Let $\mathbf{u} = (1, 2)$, $\mathbf{v} = (4, -2)$, and $\mathbf{w} = (6, 0)$. Find:
   (a) $\mathbf{u} \cdot (7\mathbf{v} + \mathbf{w})$      (b) $\|(\mathbf{u} \cdot \mathbf{w})\mathbf{w}\|$
   (c) $\|\mathbf{u}\|(\mathbf{v} \cdot \mathbf{w})$      (d) $(\|\mathbf{u}\|\mathbf{v}) \cdot \mathbf{w}$

9. Explain why each of the following expressions makes no sense.
   (a) $\mathbf{u} \cdot (\mathbf{v} \cdot \mathbf{w})$      (b) $(\mathbf{u} \cdot \mathbf{v}) + \mathbf{w}$
   (c) $\|\mathbf{u} \cdot \mathbf{v}\|$      (d) $k \cdot (\mathbf{u} + \mathbf{v})$

10. Use vectors to find the cosines of the interior angles of the triangle with vertices $(-1, 0)$, $(-2, 1)$, and $(1, 4)$.

11. Establish the identity
$$\|\mathbf{u} + \mathbf{v}\|^2 + \|\mathbf{u} - \mathbf{v}\|^2 = 2\|\mathbf{u}\|^2 + 2\|\mathbf{v}\|^2$$

12. Establish the identity
$$\mathbf{u} \cdot \mathbf{v} = \frac{1}{4}\|\mathbf{u} + \mathbf{v}\|^2 - \frac{1}{4}\|\mathbf{u} - \mathbf{v}\|^2$$

13. Find the angle between a diagonal of a cube and one of its faces.

14. The *direction cosines* of a vector $\mathbf{v}$ in 3-space are the numbers $\cos \alpha$, $\cos \beta$, and $\cos \gamma$ where $\alpha$, $\beta$, and $\gamma$ are the angles between $\mathbf{v}$ and the positive $x$, $y$, and $z$ axes. Show that if $\mathbf{v} = (a, b, c)$, then $\cos \alpha = a/\sqrt{a^2 + b^2 + c^2}$. Find $\cos \beta$ and $\cos \gamma$.

15. Show that if $\mathbf{v}$ is orthogonal to $\mathbf{w}_1$ and $\mathbf{w}_2$ then $\mathbf{v}$ is orthogonal to $k_1\mathbf{w}_1 + k_2\mathbf{w}_2$ for all scalars $k_1$ and $k_2$.

16. Let $\mathbf{u}$ and $\mathbf{v}$ be nonzero vectors in 2 or 3-space. If $k = \|\mathbf{u}\|$ and $l = \|\mathbf{v}\|$, show that the vector
$$\mathbf{w} = \frac{1}{k + l}(k\mathbf{v} + l\mathbf{u})$$
bisects the angle between $\mathbf{u}$ and $\mathbf{v}$.

## 3.4 CROSS PRODUCT

In many applications of vectors to problems in geometry, physics, and engineering, it is of interest to construct a vector in 3-space that is perpendicular to two given vectors. In this section we introduce a second kind of vector multiplication that facilitates this construction.

**Definition.** If $\mathbf{u} = (u_1, u_2, u_3)$ and $\mathbf{v} = (v_1, v_2, v_3)$ are vectors in 3-space, then the *cross product* $\mathbf{u} \times \mathbf{v}$ is the vector defined by

$$\mathbf{u} \times \mathbf{v} = (u_2 v_3 - u_3 v_2, u_3 v_1 - u_1 v_3, u_1 v_2 - u_2 v_1)$$

Unlike the dot product, the cross product of two vectors is another vector.

To avoid memorizing the rather unwieldy defining formula for the cross product $\mathbf{u} \times \mathbf{v}$, we suggest using the following mnemonic device. Form the $2 \times 3$ matrix

$$C = \begin{bmatrix} u_1 & u_2 & u_3 \\ v_1 & v_2 & v_3 \end{bmatrix}$$

where the entries in the first row are the components of the first factor $\mathbf{u}$ and those in the second row are the components of the second factor $\mathbf{v}$. The vector $\mathbf{u} \times \mathbf{v}$ is then

$$\mathbf{u} \times \mathbf{v} = \left( \det \begin{bmatrix} u_2 & u_3 \\ v_2 & v_3 \end{bmatrix}, -\det \begin{bmatrix} u_1 & u_3 \\ v_1 & v_3 \end{bmatrix}, \det \begin{bmatrix} u_1 & u_2 \\ v_1 & v_2 \end{bmatrix} \right)$$

where the matrix in the first component of $\mathbf{u} \times \mathbf{v}$ is obtained by deleting the first column of $C$, the matrix in the second component by deleting the second column of $C$, and the matrix in the third component by deleting the third column of $C$.

**Example 10**

Find $\mathbf{u} \times \mathbf{v}$, where $\mathbf{u} = (1, 2, -2)$ and $\mathbf{v} = (3, 0, 1)$.

*Solution.*

$$C = \begin{bmatrix} 1 & 2 & -2 \\ 3 & 0 & 1 \end{bmatrix}$$

$$\mathbf{u} \times \mathbf{v} = \left( \det \begin{bmatrix} 2 & -2 \\ 0 & 1 \end{bmatrix}, -\det \begin{bmatrix} 1 & -2 \\ 3 & 1 \end{bmatrix}, \det \begin{bmatrix} 1 & 2 \\ 3 & 0 \end{bmatrix} \right)$$

$$= (2, -7, -6)$$

The following theorem shows that $\mathbf{u} \times \mathbf{v}$ is orthogonal to both $\mathbf{u}$ and $\mathbf{v}$ and establishes an important identity relating the cross product with the dot product.

**Theorem 4.** *If* **u** *and* **v** *are vectors in 3-space, then:*

(a) $\mathbf{u} \cdot (\mathbf{u} \times \mathbf{v}) = 0$ $\qquad\qquad$ (**u** × **v** *is orthogonal to* **u**)
(b) $\mathbf{v} \cdot (\mathbf{u} \times \mathbf{v}) = 0$ $\qquad\qquad$ (**u** × **v** *is orthogonal to* **v**)
(c) $\|\mathbf{u} \times \mathbf{v}\|^2 = \|\mathbf{u}\|^2 \|\mathbf{v}\|^2 - (\mathbf{u} \cdot \mathbf{v})^2$ $\quad$ *(Lagrange's identity)*

*Proof.* Let $\mathbf{u} = (u_1, u_2, u_3)$ and $\mathbf{v} = (v_1, v_2, v_3)$.

(a) $\mathbf{u} \cdot (\mathbf{u} \times \mathbf{v}) = (u_1, u_2, u_3) \cdot (u_2 v_3 - u_3 v_2, u_3 v_1 - u_1 v_3, u_1 v_2 - u_2 v_1)$
$\qquad\qquad = u_1(u_2 v_3 - u_3 v_2) + u_2(u_3 v_1 - u_1 v_3) + u_3(u_1 v_2 - u_2 v_1)$
$\qquad\qquad = 0$

(b) Similar to (a).

(c) Since

$$\|\mathbf{u} \times \mathbf{v}\|^2 = (u_2 v_3 - u_3 v_2)^2 + (u_3 v_1 - u_1 v_3)^2 + (u_1 v_2 - u_2 v_1)^2 \qquad (3.5)$$

and

$$\|\mathbf{u}\|^2 \|\mathbf{v}\|^2 - (\mathbf{u} \cdot \mathbf{v})^2$$
$$= (u_1{}^2 + u_2{}^2 + u_3{}^2)(v_1{}^2 + v_2{}^2 + v_3{}^2) - (u_1 v_1 + u_2 v_2 + u_3 v_3)^2 \quad (3.6)$$

Lagrange's identity can be established by "multiplying out" the right sides of (3.5) and (3.6) and verifying their equality. ▮

**Example 11**

Consider the vectors

$$\mathbf{u} = (1, 2, -2) \text{ and } \mathbf{v} = (3, 0, 1)$$

In Example 10 we showed that

$$\mathbf{u} \times \mathbf{v} = (2, -7, -6)$$

Since

$$\mathbf{u} \cdot (\mathbf{u} \times \mathbf{v}) = (1)(2) + (2)(-7) + (-2)(-6) = 0$$

and

$$\mathbf{v} \cdot (\mathbf{u} \times \mathbf{v}) = (3)(2) + (0)(-7) + (1)(-6) = 0$$

**u** × **v** is orthogonal to both **u** and **v** as guaranteed by Theorem 4.

The main arithmetic properties of the cross product are listed in the next theorem.

**Theorem 5.** *If* **u**, **v** *and* **w** *are any vectors in 3-space and k is any scalar, then:*

(a) $\mathbf{u} \times \mathbf{v} = -(\mathbf{v} \times \mathbf{u})$
(b) $\mathbf{u} \times (\mathbf{v} + \mathbf{w}) = (\mathbf{u} \times \mathbf{v}) + (\mathbf{u} \times \mathbf{w})$
(c) $(\mathbf{u} + \mathbf{v}) \times \mathbf{w} = (\mathbf{u} \times \mathbf{w}) + (\mathbf{v} \times \mathbf{w})$
(d) $k(\mathbf{u} \times \mathbf{v}) = (k\mathbf{u}) \times \mathbf{v} = \mathbf{u} \times (k\mathbf{v})$
(e) $\mathbf{u} \times \mathbf{0} = \mathbf{0} \times \mathbf{u} = \mathbf{0}$
(f) $\mathbf{u} \times \mathbf{u} = \mathbf{0}$

The proofs follow immediately from the definition of the cross product and are left as exercises.

**Example 12**

It is not true, in general, that $\mathbf{u} \times (\mathbf{v} \times \mathbf{w}) = (\mathbf{u} \times \mathbf{v}) \times \mathbf{w}$. To see this, consider the vectors

$$\mathbf{i} = (1, 0, 0) \qquad \mathbf{j} = (0, 1, 0) \qquad \mathbf{k} = (0, 0, 1)$$

(These vectors each have norm 1 and are called the *standard unit vectors* in 3-space. See. Figure 3.23.)

From the definition of cross product, it follows that

$$\mathbf{i} \times \mathbf{j} = (1, 0, 0) \times (0, 1, 0) = (0, 0, 1) = \mathbf{k}$$

Similarly

$$\mathbf{j} \times \mathbf{k} = \mathbf{i} \qquad \text{and} \qquad \mathbf{k} \times \mathbf{i} = \mathbf{j}$$

By parts (e) and (f) of Theorem 5

$$\mathbf{i} \times (\mathbf{j} \times \mathbf{j}) = \mathbf{i} \times \mathbf{0} = \mathbf{0}$$

On the other hand

$$(\mathbf{i} \times \mathbf{j}) \times \mathbf{j} = \mathbf{k} \times \mathbf{j} = -(\mathbf{j} \times \mathbf{k}) = -\mathbf{i}$$

Therefore

$$\mathbf{i} \times (\mathbf{j} \times \mathbf{j}) \neq (\mathbf{i} \times \mathbf{j}) \times \mathbf{j}$$

We know from Theorem 4 that $\mathbf{u} \times \mathbf{v}$ is orthogonal to both $\mathbf{u}$ and $\mathbf{v}$. If $\mathbf{u}$ and $\mathbf{v}$ are nonzero vectors, it can be shown that the direction of $\mathbf{u} \times \mathbf{v}$ can be determined using the following "right-hand rule*" (Figure 3.24). Let $\theta$ be the angle between

---

*Recall that we agreed to consider only right-handed coordinate systems in this text. Had we used left-handed systems instead, a "left-hand rule" would apply here.

**Figure 3.23**

**Figure 3.24**

**u** and **v**, and suppose **u** is rotated through the angle $\theta$ until it coincides with **v**. If the fingers of the right hand are cupped so they point in the direction of rotation, then the thumb indicates (roughly) the direction of **u** $\times$ **v**.

The reader may find it instructive to practice this rule with the products

$$\mathbf{i} \times \mathbf{j} = \mathbf{k} \quad \mathbf{j} \times \mathbf{k} = \mathbf{i} \quad \mathbf{k} \times \mathbf{i} = \mathbf{j}$$

discussed in Example 12.

If **u** and **v** are nonzero vectors in 3-space, then the norm of **u** $\times$ **v** has a useful geometric interpretation. Lagrange's identity, given in Theorem 4, states that

$$\|\mathbf{u} \times \mathbf{v}\|^2 = \|\mathbf{u}\|^2 \|\mathbf{v}\|^2 - (\mathbf{u} \cdot \mathbf{v})^2 \tag{3.7}$$

If $\theta$ denotes the angle between **u** and **v**, then $\mathbf{u} \cdot \mathbf{v} = \|\mathbf{u}\| \|\mathbf{v}\| \cos \theta$, so that (3.7) can be rewritten as

$$\begin{aligned} \|\mathbf{u} \times \mathbf{v}\|^2 &= \|\mathbf{u}\|^2 \|\mathbf{v}\|^2 - \|\mathbf{u}\|^2 \|\mathbf{v}\|^2 \cos^2 \theta \\ &= \|\mathbf{u}\|^2 \|\mathbf{v}\|^2 (1 - \cos^2 \theta) \\ &= \|\mathbf{u}\|^2 \|\mathbf{v}\|^2 \sin^2 \theta \end{aligned}$$

Thus

$$\|\mathbf{u} \times \mathbf{v}\| = \|\mathbf{u}\| \|\mathbf{v}\| \sin \theta \tag{3.8}$$

But $\|\mathbf{v}\| \sin \theta$ is the altitude of the parallelogram determined by **u** and **v** (Figure 3.25). Thus from (3.8), the area $A$ of this parallelogram is given by

$$A = (\text{base})(\text{altitude}) = \|\mathbf{u}\| \|\mathbf{v}\| \sin \theta = \|\mathbf{u} \times \mathbf{v}\|$$

In other words, *the norm of* **u** $\times$ **v** *is equal to the area of the parallelogram determined by* **u** *and* **v**.

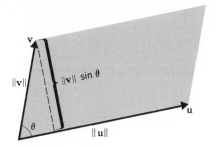

**Figure 3.25**

**Example 13**

Find the area of the triangle determined by the points $P_1(2, 2, 0)$, $P_2(-1, 0, 2)$, and $P_3(0, 4, 3)$.

*Solution.* The area $A$ of the triangle is $1/2$ the area of the parallelogram determined by the vectors $\overrightarrow{P_1P_2}$ and $\overrightarrow{P_1P_3}$ (Figure 3.26).

As illustrated in Example 2, $\overrightarrow{P_1P_2} = (-3, -2, 2)$ and $\overrightarrow{P_1P_3} = (-2, 2, 3)$. It follows that

$$\overrightarrow{P_1P_2} \times \overrightarrow{P_1P_3} = (-10, 5, -10)$$

and consequently

$$A = \frac{1}{2} \|\overrightarrow{P_1P_2} \times \overrightarrow{P_1P_3}\| = \frac{1}{2}(15) = \frac{15}{2}$$

A vector was defined to be a directed line segment or arrow in 2-space or 3-space. Coordinate systems and components were introduced later in order to simplify computations with vectors. It is important to keep in mind that a vector has a

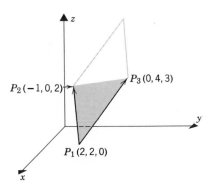

**Figure 3.26**

"mathematical existence" regardless of whether a coordinate system has been introduced. Further, the components of a vector are not determined by the vector alone;
they depend as well on the coordinate system chosen. For example, in Figure 3.27
we have indicated a fixed plane vector **v** and two different coordinate systems. In
the $xy$-coordinate system the components of **v** are $(1, 1)$, and in the $x'y'$-system they
are $(\sqrt{2}, 0)$.

This raises an important question about our definition of cross product. Since we
defined the cross product $\mathbf{u} \times \mathbf{v}$ in terms of the components of **u** and **v**, and since
these components depend on the coordinate system chosen, it seems possible that
two *fixed* vectors **u** and **v** might have different cross products in different coordinate
systems. Fortunately, this is not the case. To see this, we need only recall that:

(i) $\mathbf{u} \times \mathbf{v}$ is perpendicular to both **u** and **v**.
(ii) The direction of $\mathbf{u} \times \mathbf{v}$ is determined by the right-hand rule.
(iii) $\|\mathbf{u} \times \mathbf{v}\| = \|\mathbf{u}\| \, \|\mathbf{v}\| \sin \theta$.

A moment's reflection will make it evident that these three properties completely
determine the direction and length of $\mathbf{u} \times \mathbf{v}$. Since these properties depend only
on the lengths and relative orientation of **u** and **v** and not on the particular right-
handed coordinate system being used, the vector $\mathbf{u} \times \mathbf{v}$ will remain unchanged if
a different right-hand coordinate system is introduced. This fact is described by
stating that the definition of $\mathbf{u} \times \mathbf{v}$ is ***coordinate free.*** This result is of importance
to physicists and engineers who often work with many coordinate systems in the
same problem.

**Example 14**

Let an $x'y'z'$-coordinate system be obtained by rotating an $xyz$-coordinate system
about the $z$-axis through an angle of $-45°$ as shown in Figure 3.28.

**Figure 3.27**

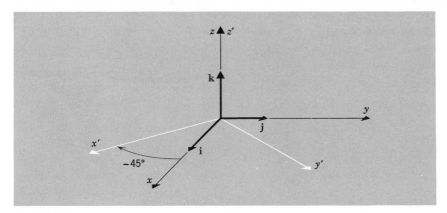

**Figure 3.28**

If **i**, **j**, and **k** are the standard unit vectors in the *xyz*-system (Example 12), then the components of these vectors in the *x'y'z'*-system are

$$\mathbf{i} = \left( \frac{1}{\sqrt{2}}, \frac{1}{\sqrt{2}}, 0 \right) \quad \mathbf{j} = \left( -\frac{1}{\sqrt{2}}, \frac{1}{\sqrt{2}}, 0 \right) \quad \mathbf{k} = (1, 0, 0)$$

Calculating **i** × **j** with components in the *x'y'z'*-system yields

$$\mathbf{i} \times \mathbf{j} = \left( \frac{1}{\sqrt{2}}, \frac{1}{\sqrt{2}}, 0 \right) \times \left( -\frac{1}{\sqrt{2}}, \frac{1}{\sqrt{2}}, 0 \right)$$

$$= \left( \det \begin{bmatrix} \frac{1}{\sqrt{2}} & 0 \\ \frac{1}{\sqrt{2}} & 0 \end{bmatrix}, -\det \begin{bmatrix} \frac{1}{\sqrt{2}} & 0 \\ -\frac{1}{\sqrt{2}} & 0 \end{bmatrix}, \det \begin{bmatrix} \frac{1}{\sqrt{2}} & \frac{1}{\sqrt{2}} \\ -\frac{1}{\sqrt{2}} & \frac{1}{\sqrt{2}} \end{bmatrix} \right)$$

$$= (0, 0, 1) = \mathbf{k}.$$

This result agrees with that obtained in Example 12 using the components in the *xyz*-system.

## EXERCISE SET 3.4

**1.** Let $\mathbf{u} = (2, -1, 3)$, $\mathbf{v} = (0, 1, 7)$, and $\mathbf{w} = (1, 4, 5)$. Compute:
   (a) $\mathbf{v} \times \mathbf{w}$   (b) $\mathbf{u} \times (\mathbf{v} \times \mathbf{w})$   (c) $(\mathbf{u} \times \mathbf{v}) \times \mathbf{w}$
   (d) $(\mathbf{u} \times \mathbf{v}) \times (\mathbf{v} \times \mathbf{w})$   (e) $\mathbf{u} \times (\mathbf{v} - 2\mathbf{w})$   (f) $(\mathbf{u} \times \mathbf{v}) - 2\mathbf{w}$

**2.** In each part find a vector orthogonal to both **u** and **v**.
   (a) $\mathbf{u} = (-7, 3, 1)$     $\mathbf{v} = (2, 0, 4)$
   (b) $\mathbf{u} = (-1, -1, -1)$     $\mathbf{v} = (2, 0, 2)$

**3.** In each part find the area of the triangle having vertices $P$, $Q$, and $R$.
  (a) $P(1, 5, -2)$    $Q(0, 0, 0)$    $R(3, 5, 1)$
  (b) $P(2, 0, -3)$    $Q(1, 4, 5)$    $R(7, 2, 9)$

**4.** Verify Theorem 4 for the vectors $\mathbf{u} = (1, -5, 6)$ and $\mathbf{v} = (2, 1, 2)$.

**5.** Verify Theorem 5 for $\mathbf{u} = (2, 0, -1)$, $\mathbf{v} = (6, 7, 4)$, $\mathbf{w} = (1, 1, 1)$ and $k = -3$.

**6.** What is wrong with the expression $\mathbf{u} \times \mathbf{v} \times \mathbf{w}$?

**7.** Let $\mathbf{u} = (-1, 3, 2)$ and $\mathbf{w} = (1, 1, -1)$. Find all $\mathbf{x}$ vectors that satisfy $\mathbf{u} \times \mathbf{x} = \mathbf{w}$.

**8.** Let $\mathbf{u} = (u_1, u_2, u_3)$, $\mathbf{v} = (v_1, v_2, v_3)$, and $\mathbf{w} = (w_1, w_2, w_3)$ Show that

$$\mathbf{u} \cdot (\mathbf{v} \times \mathbf{w}) = \begin{vmatrix} u_1 & u_2 & u_3 \\ v_1 & v_2 & v_3 \\ w_1 & w_2 & w_3 \end{vmatrix}$$

**9.** Use the result of Exercise 8 to compute $\mathbf{u} \cdot (\mathbf{v} \times \mathbf{w})$ when $\mathbf{u} = (-1, 4, 7)$, $\mathbf{v} = (6, -7, 3)$, and $\mathbf{w} = (4, 0, 1)$.

**10.** Let $\mathbf{u}$ and $\mathbf{v}$ be vectors whose components in the $xyz$-system of Figure 3.28 are $\mathbf{u} = (1, 1, 1)$ and $\mathbf{v} = (1, 1, 0)$.
  (a) Find the components of $\mathbf{u}$ and $\mathbf{v}$ in the $x'y'z'$-system of Figure 3.28.
  (b) Compute $\mathbf{u} \times \mathbf{v}$ using the components in the $xyz$-system.
  (c) Compute $\mathbf{u} \times \mathbf{v}$ using the components in the $x'y'z'$-system.
  (d) Show that the vectors obtained in (b) and (c) are the same.

**11.** Prove the following identities.
  (a) $(\mathbf{u} + k\mathbf{v}) \times \mathbf{v} = \mathbf{u} \times \mathbf{v}$
  (b) $(\mathbf{u} \times \mathbf{v}) \cdot \mathbf{z} = \mathbf{u} \cdot (\mathbf{v} \times \mathbf{z})$

**12.** Let $\mathbf{u}$, $\mathbf{v}$, and $\mathbf{w}$ be nonzero vectors in 3-space, no two of which are collinear. Show:
  (a) $\mathbf{u} \times (\mathbf{v} \times \mathbf{w})$ lies in the plane determined by $\mathbf{v}$ and $\mathbf{w}$ (assuming that the vectors are positioned so they have the same initial points)
  (b) $(\mathbf{u} \times \mathbf{v}) \times \mathbf{w}$ lies in the plane determined by $\mathbf{u}$ and $\mathbf{v}$.

**13.** Prove that $\mathbf{x} \times (\mathbf{y} \times \mathbf{z}) = (\mathbf{x} \cdot \mathbf{z})\mathbf{y} - (\mathbf{x} \cdot \mathbf{y})\mathbf{z}$. (**Hint.** First prove the result in the case where $\mathbf{z} = \mathbf{i} = (1, 0, 0)$, then when $\mathbf{z} = \mathbf{j} = (0, 1, 0)$, and then when $\mathbf{z} = \mathbf{k} = (0, 0, 1)$. Finally prove it for an arbitrary vector $\mathbf{z} = (z_1, z_2, z_3)$ by writing $\mathbf{z} = z_1\mathbf{i} + z_2\mathbf{j} + z_3\mathbf{k}$.)

**14.** Prove parts (a) and (b) of Theorem 5.

**15.** Prove parts (c) and (d) of Theorem 5.

**16.** Prove parts (e) and (f) of Theorem 5.

## 3.5 LINES AND PLANES IN 3-SPACE

In this section we shall use vectors to derive equations of lines and planes in 3-space. We shall also use these equations to solve some basic geometric problems.

In plane analytic geometry a line can be specified by giving its slope and one of its points. Similarly, a plane in 3-space can be determined by giving its inclination and specifying one of its points. A convenient method for describing the inclination is to specify a vector (called a *normal*) that is perpendicular to the plane.

Suppose we want the equation of the plane passing through the point $P_0(x_0, y_0, z_0)$ and having the nonzero vector $\mathbf{n} = (a, b, c)$ as a normal. It is evident from Figure 3.29 that the plane consists precisely of those points $P(x, y, z)$ for which the vector $\overrightarrow{P_0P}$ is orthogonal to $\mathbf{n}$; that is, for which

$$\mathbf{n} \cdot \overrightarrow{P_0P} = 0 \tag{3.9}$$

Since $\overrightarrow{P_0P} = (x - x_0, y - y_0, z - z_0)$, (3.9) can be rewritten as

$$a(x - x_0) + b(y - y_0) + c(z - z_0) = 0 \tag{3.10}$$

We shall call this a *point-normal* form of the equation of a plane.

### Example 15

Find the equation of the plane passing through the point $(3, -1, 7)$ and perpendicular to the vector $\mathbf{n} = (4, 2, -5)$.

*Solution.* From (3.10) a point-normal form is

$$4(x - 3) + 2(y + 1) - 5(z - 7) = 0$$

By multiplying out and collecting terms, (3.10) can be rewritten in the form

$$ax + by + cz + d = 0 \tag{3.11}$$

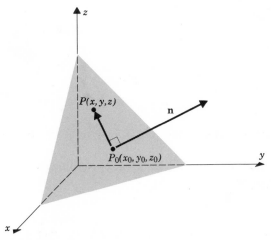

**Figure 3.29**

where $a$, $b$, $c$, and $d$ are constants, and $a$, $b$, and $c$ are not all zero. To illustrate, the equation in Example 15 can be rewritten as

$$4x + 2y - 5z + 25 = 0$$

As our next theorem shows, every equation having the form of (3.11) represents a plane in 3-space.

**Theorem 6.** *If $a$, $b$, $c$, and $d$ are constants and $a$, $b$, and $c$ are not all zero, then the graph of the equation*

$$ax + by + cz + d = 0$$

*is a plane having the vector $\mathbf{n} = (a, b, c)$ as a normal.*

*Proof.* By hypothesis $a$, $b$, and $c$ are not all zero. Assume, for the moment, that $a \neq 0$. The equation $ax + by + cz + d = 0$ can be rewritten as $a(x + (d/a)) + by + cz = 0$. But this is a point-normal form of the plane passing through the point $(-d/a, 0, 0)$ and having $\mathbf{n} = (a, b, c)$ as a normal.

If $a = 0$, then either $b \neq 0$ or $c \neq 0$. A straightforward modification of the above argument will handle these other cases. ▪

Equation (3.11) is a linear equation in $x$, $y$ and $z$; it is called the **general form** of the equation of a plane.

Just as the solutions to a system of linear equations

$$ax + by = k_1$$
$$cx + dy = k_2$$

correspond to points of intersection of the lines $ax + by = k_1$ and $cx + dy = k_2$ in the $xy$-plane, so the solutions of a system

$$ax + by + cz = k_1$$
$$dx + ey + fz = k_2 \tag{3.12}$$
$$gx + hy + iz = k_3$$

correspond to points of intersection of the planes $ax + by + cz = k_1$, $dx + ey + fz = k_2$ and $gx + hy + iz = k_3$.

In Figure 3.30 we have illustrated some of the geometric possibilities when (3.12) has zero, one, or infinitely many solutions.

### Example 16

Find the equation of the plane through the points $P_1(1, 2, -1)$, $P_2(2, 3, 1)$ and $P_3(3, -1, 2)$.

**Figure 3.30** (*a*) No solutions (3 parallel planes). (*b*) No solutions (2 planes parallel). (*c*) Infinitely many solutions (3 coincident planes). (*d*) Infinitely many solutions (3 planes intersecting in a line). (*e*) One solution (3 planes intersecting at a point).

*Solution.* Since the three points lie in the plane, their coordinates must satisfy the general equation $ax + by + cz + d = 0$ of the plane. Thus

$$a + 2b - c + d = 0$$
$$2a + 3b + c + d = 0$$
$$3a - b + 2c + d = 0$$

Solving this system gives

$$a = -\frac{9}{16}t \qquad b = -\frac{1}{16}t \qquad c = \frac{5}{16}t \qquad d = t$$

Letting $t = -16$, for example, yields the desired equation

$$9x + y - 5z - 16 = 0$$

We note that any other choice of $t$ gives a multiple of this equation so that any value of $t \neq 0$ would do equally well.

*Alternate solution.* Since $P_1(1, 2, -1)$, $P_2(2, 3, 1)$, and $P_3(3, -1, 2)$ lie in the plane, the vectors $\overrightarrow{P_1P_2} = (1, 1, 2)$ and $\overrightarrow{P_1P_3} = (2, -3, 3)$ are parallel to the plane. Therefore $\overrightarrow{P_1P_2} \times \overrightarrow{P_1P_3} = (9, 1, -5)$ is normal to the plane, since it is perpendicular to both $\overrightarrow{P_1P_2}$ and $\overrightarrow{P_1P_3}$. From this and the fact that $P_1$ lies in the plane, a point normal form for the equation of the plane is

$$9(x - 1) + (y - 2) - 5(z + 1) = 0$$

or

$$9x + y - 5z - 16 = 0$$

We shall now show how to obtain equations for lines in 3-space. Suppose $l$ is the line in 3-space through the point $P_0(x_0, y_0, z_0)$ and is parallel to the nonzero vector $\mathbf{v} = (a, b, c)$. It is clear (Figure 3.31) that $l$ consists precisely of those points $P(x, y, z)$ for which the vector $\overrightarrow{P_0P}$ is parallel to $\mathbf{v}$, that is, for which there is a scalar $t$ such that

$$\overrightarrow{P_0P} = t\mathbf{v} \tag{3.13}$$

In terms of components (3.13) can be written

$$(x - x_0, y - y_0, z - z_0) = (ta, tb, tc)$$

or equivalently

$$
\begin{aligned}
x &= x_0 + ta \\
y &= y_0 + tb \qquad \text{where } -\infty < t < +\infty \\
z &= z_0 + tc
\end{aligned}
$$

These equations are called **parametric equations** for $l$ since the line $l$ is traced out by $P(x, y, z)$ as the parameter $t$ varies from $-\infty$ to $+\infty$.

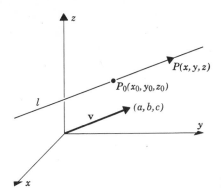

**Figure 3.31**

## Example 17

The line through the point $(1, 2, -3)$ and parallel to the vector $\mathbf{v} = (4, 5, -7)$ has parametric equations

$$
\begin{aligned}
x &= 1 + 4t \\
y &= 2 + 5t \qquad \text{where } -\infty < t < +\infty \\
z &= -3 - 7t
\end{aligned}
$$

## Example 18

Find parametric equations for the line $l$ passing through the points $P_1(2, 4, -1)$ and $P_2(5, 0, 7)$.

*Solution.* Since the vector $\overrightarrow{P_1 P_2} = (3, -4, 8)$ is parallel to $l$ and $P_1(2, 4, -1)$ lies on $l$, the line $l$ is given by

$$
\begin{aligned}
x &= 2 + 3t \\
y &= 4 - 4t \qquad \text{where } -\infty < t < +\infty \\
z &= -1 + 8t
\end{aligned}
$$

## Example 19

Find parametric equations for the line of intersection of the planes

$$3x + 2y - 4z - 6 = 0 \quad \text{and} \quad x - 3y - 2z - 4 = 0$$

*Solution.* The line of intersection consists of all points $(x, y, z)$ that satisfy the two equations in the system

$$
\begin{aligned}
3x + 2y - 4z &= 6 \\
x - 3y - 2z &= 4
\end{aligned}
$$

Solving this system gives

$$x = \frac{26}{11} + \frac{16}{11} t \qquad y = -\frac{6}{11} - \frac{2}{11} t \qquad z = t$$

The parametric equations for $l$ are therefore

$$x = \frac{26}{11} + \frac{16}{11} t$$

$$y = -\frac{6}{11} - \frac{2}{11} t \qquad \text{where } -\infty < t < +\infty$$

$$z \qquad = \quad t$$

In some problems, a line

$$\begin{aligned} x &= x_0 + at \\ y &= y_0 + bt \qquad \text{where } -\infty < t < +\infty \\ z &= z_0 + ct \end{aligned} \qquad (3.14)$$

is given, and it is of interest to find two planes whose intersection is the given line. Since there are infinitely many planes through the line, there are always infinitely many such pairs of planes. To find two such planes when $a$, $b$, and $c$ are all different from zero, we can rewrite each equation in (3.14) in the form

$$\frac{x - x_0}{a} = t \qquad \frac{y - y_0}{b} = t \qquad \frac{z - z_0}{c} = t$$

Eliminating the parameter $t$ shows that the line consists of all points $(x, y, z)$ that satisfy the equations

$$\frac{x - x_0}{a} = \frac{y - y_0}{b} = \frac{z - z_0}{c}$$

called **symmetric equations** for the line. The line is therefore the intersection of the planes

$$\frac{x - x_0}{a} = \frac{y - y_0}{b} \qquad \text{and} \qquad \frac{y - y_0}{b} = \frac{z - z_0}{c}$$

**Example 20**

Find two planes whose intersection is the line

$$\begin{aligned} x &= 3 + 2t \\ y &= -4 + 7t \qquad \text{where } -\infty < t < +\infty \\ z &= 1 + 3t \end{aligned}$$

*Solution.* Since the symmetric equations for this line are

$$\frac{x-3}{2} = \frac{y+4}{7} = \frac{z-1}{3}$$

it is the intersection of the planes

$$\frac{x-3}{2} = \frac{y+4}{7} \quad \text{and} \quad \frac{y+4}{7} = \frac{z-1}{3}$$

or equivalently

$$7x - 2y - 29 = 0 \quad \text{and} \quad 3y - 7z + 19 = 0$$

## EXERCISE SET 3.5

1. In each part find a point-normal form of the equation of the plane passing through $P$ and having $\mathbf{n}$ as a normal.
   (a) $P(2, 6, 1)$; $\mathbf{n} = (1, 4, 2)$    (b) $P(-1, -1, 2)$; $\mathbf{n} = (-1, 7, 6)$
   (c) $P(1, 0, 0)$; $\mathbf{n} = (0, 0, 1)$    (d) $P(0, 0, 0)$; $\mathbf{n} = (2, 3, 4)$

2. Write the equations of the planes in Exercise 1 in general form.

3. Find a point-normal form of:
   (a) $2x - 3y + 7z - 10 = 0$    (b) $x + 3z = 0$

4. In each part find an equation for the plane passing through the given points.
   (a) $(-2, 1, 1)$    $(0, 2, 3)$    $(1, 0, -1)$
   (b) $(3, 2, 1)$    $(2, 1, -1)$    $(-1, 3, 2)$

5. In each part find parametric equations for the line passing through $P$ and parallel to $\mathbf{n}$.
   (a) $P(2, 4, 6)$; $\mathbf{n} = (1, 2, 5)$    (b) $P(-3, 2, -4)$; $\mathbf{n} = (5, -7, -3)$
   (c) $P(1, 1, 5)$; $\mathbf{n} = (0, 0, 1)$    (d) $P(0, 0, 0)$; $\mathbf{n} = (1, 1, 1)$

6. Find symmetric equations for the lines in parts (a) and (b) of Exercise 5.

7. In each part find parametric equations for the line passing through the given points.
   (a) $(6, -1, 5), (7, 2, -4)$    (b) $(0, 0, 0), (-1, -1, -1)$

8. In each part find parametric equations for the line of intersection of the given planes.
   (a) $-2x + 3y + 7z + 2 = 0$    and    $x + 2y - 3z + 5 = 0$
   (b) $3x - 5y + 2z = 0$    and    $z = 0$

9. In each part find equations for two planes whose intersection is the given line.

   (a) $\begin{array}{l} x = \phantom{-}3 + 4t \\ y = -7 + 2t \\ z = \phantom{-}6 - t \end{array}$    $-\infty < t < +\infty$    (b) $\begin{array}{l} x = 5t \\ y = 3t \\ z = 6t \end{array}$    $-\infty < t < +\infty$

10. Find equations for the $xy$-plane, the $xz$-plane and the $yz$-plane.

**11.** Show that the line

$$x = 0$$
$$y = t \qquad -\infty < t < +\infty$$
$$z = t$$

(a) lies in the plane $6x + 4y - 4z = 0$
(b) is parallel to and below the plane $5x - 3y + 3z = 1$
(c) is parallel to and above the plane $6x + 2y - 2z = 3$.

**12.** Find the point of intersection of the line

$$x - 4 = 5t$$
$$y + 2 = \ \ t$$
$$z - 4 = -t$$

and the plane $3x - y + 7z + 8 = 0$.

**13.** Find a plane passing through the point $(2, -7, 6)$ and parallel to the plane $5x - 2y + z - 9 = 0$.

**14.** Show that the line

$$x - 4 = \ \ 2t$$
$$y \ \ \ = -t \qquad -\infty < t < +\infty$$
$$z + 1 = -4t$$

is parallel to the plane $3x + 2y + z - 7 = 0$.

**15.** Show that the lines

$$x + 1 = 4t \qquad\qquad x + 13 \doteq 12t$$
$$y - 3 = \ \ t \quad \text{and} \quad y - 1 = \ \ 6t$$
$$z - 1 = 0 \qquad\qquad\quad z - 2 = \ \ 3t$$

intersect. Find the point of intersection.

**16.** Find an equation for the plane determined by the lines in Exercise 15.

# Chapter 4 | *Vector Spaces*

## 4.1 EUCLIDEAN n-SPACE

The idea of using pairs of numbers to locate points in the plane and triples of numbers to locate points in 3-space was first clearly spelled out in the mid-seventeenth century. By the latter part of the nineteenth century mathematicians and physicists began to realize that there was no need to stop with triples. It was recognized that quadruples of numbers $(a_1, a_2, a_3, a_4)$ could be regarded as points in "4-dimensional" space, quintuples $(a_1, a_2, \ldots, a_5)$ as points in "5-dimensional" space, etc. Although our geometric visualization does not extend beyond 3-space, it is nevertheless possible to extend many familiar ideas beyond 3-space by working with analytic or numerical properties of points and vectors rather than the geometric properties. In this section we shall make these ideas more precise.

**Definition.** If $n$ is a positive integer, then an ***ordered n-tuple*** is a sequence of $n$ real numbers $(a_1, a_2, \ldots, a_n)$. The set of all ordered $n$-tuples is called ***n-space*** and is denoted by $R^n$.

(When $n = 2$ or 3, it is usual to use the terms "ordered pair" and "ordered triple" rather than ordered 2-tuple and 3-tuple.)

It might have occurred to the reader in his study of 3-space that both vectors and points are represented analytically by ordered triples of real numbers. For vectors the numbers are the components, and for points they are the coordinates. Therefore, although points and vectors (arrows) in 3-space are geometrically different kinds of objects, they are the same kind of object analytically, namely, ordered triples of real numbers. Since ordered $n$-tuples are generalizations of ordered pairs and triples of real numbers, and since ordered pairs and triples of real numbers can be regarded as either points or vectors, $n$-tuples can equally well be viewed as generalized points or generalized vectors—the distinction is mathematically unimportant. Although we shall generally describe $n$-tuples as vectors, we shall not hesitate to call them points if it seems more appropriate. Thus the 5-tuple $(-2, 4, 0, 1, 6)$ might be described either as a point in $R^5$ or a vector in $R^5$.

**Definition.** Two vectors $\mathbf{u} = (u_1, u_2, \ldots, u_n)$ and $\mathbf{v} = (v_1, v_2, \ldots, v_n)$ in $R^n$ are called *equal* if

$$u_1 = v_1, u_2 = v_2, \ldots, u_n = v_n$$

The *sum* $\mathbf{u} + \mathbf{v}$ is defined by

$$\mathbf{u} + \mathbf{v} = (u_1 + v_1, u_2 + v_2, \ldots, u_n + v_n)$$

and if $k$ is any scalar, the *scalar multiple* $k\mathbf{u}$ is defined by

$$k\mathbf{u} = (ku_1, ku_2, \ldots, ku_n)$$

The operations of addition and scalar multiplication in this definition are called the *standard operations* on $R^n$.

We define the *zero vector* in $R^n$ to be the vector

$$\mathbf{0} = (0, 0, \ldots, 0)$$

If $\mathbf{u} = (u_1, u_2, \ldots, u_n)$ is any vector in $R^n$, then the *negative* (or *additive inverse*) of $\mathbf{u}$ is denoted by $-\mathbf{u}$ and is defined by

$$-\mathbf{u} = (-u_1, -u_2, \ldots, -u_n)$$

We define subtraction of vectors in $R^n$ by $\mathbf{v} - \mathbf{u} = \mathbf{v} + (-\mathbf{u})$, or in terms of components

$$\mathbf{v} - \mathbf{u} = \mathbf{v} + (-\mathbf{u}) = (v_1, v_2, \ldots, v_n) + (-u_1, -u_2, \ldots, -u_n)$$
$$= (v_1 - u_1, v_2 - u_2, \ldots, v_n - u_n)$$

The most important arithmetic properties of addition and scalar multiplication of vectors in $R^n$ are listed in the next theorem. The proofs are all easy and are left as exercises.

**Theorem 1.** *If* $\mathbf{u} = (u_1, u_2, \ldots, u_n)$, $\mathbf{v} = (v_1, v_2, \ldots, v_n)$ *and* $\mathbf{w} = (w_1, w_2, \ldots, w_n)$ *are vectors in* $R^n$ *and* $k$ *and* $l$ *are scalars, then:*

(a) $\mathbf{u} + \mathbf{v} = \mathbf{v} + \mathbf{u}$
(b) $\mathbf{u} + (\mathbf{v} + \mathbf{w}) = (\mathbf{u} + \mathbf{v}) + \mathbf{w}$
(c) $\mathbf{u} + \mathbf{0} = \mathbf{0} + \mathbf{u} = \mathbf{u}$
(d) $\mathbf{u} + (-\mathbf{u}) = \mathbf{0}$, *that is,* $\mathbf{u} - \mathbf{u} = \mathbf{0}$
(e) $k(l\mathbf{u}) = (kl)\mathbf{u}$
(f) $k(\mathbf{u} + \mathbf{v}) = k\mathbf{u} + k\mathbf{v}$
(g) $(k + l)\mathbf{u} = k\mathbf{u} + l\mathbf{u}$
(h) $1\mathbf{u} = \mathbf{u}$

This theorem enables us to manipulate vectors in $R^n$ without expressing the vectors in terms of components in much the same way that we manipulate real numbers. For example, to solve the vector equation $\mathbf{x} + \mathbf{u} = \mathbf{v}$ for $\mathbf{x}$, we can add $-\mathbf{u}$ to both sides and proceed as follows.

$$(\mathbf{x} + \mathbf{u}) + (-\mathbf{u}) = \mathbf{v} + (-\mathbf{u})$$
$$\mathbf{x} + (\mathbf{u} - \mathbf{u}) = \mathbf{v} - \mathbf{u}$$
$$\mathbf{x} + \mathbf{0} = \mathbf{v} - \mathbf{u}$$
$$\mathbf{x} = \mathbf{v} - \mathbf{u}$$

The reader will find it useful to name the parts of Theorem 1 that justify each of the steps in this computation.

To extend the notions of distance, norm, and angle to $R^n$, we begin with the following generalization of the dot product on $R^2$ and $R^3$ (Section 3.3).

**Definition.** If $\mathbf{u} = (u_1, u_2, \ldots, u_n)$ and $\mathbf{v} = (v_1, v_2, \ldots, v_n)$ are any vectors in $R^n$, then the *Euclidean inner product* $\mathbf{u} \cdot \mathbf{v}$ is defined by

$$\mathbf{u} \cdot \mathbf{v} = u_1 v_1 + u_2 v_2 + \cdots + u_n v_n$$

Observe that when $n = 2$ or 3, the Euclidean inner product is the ordinary dot product.

## Example 1

The Euclidean inner product of the vectors

$$\mathbf{u} = (-1, 3, 5, 7) \quad \text{and} \quad \mathbf{v} = (5, -4, 7, 0)$$

in $R^4$ is

$$\mathbf{u} \cdot \mathbf{v} = (-1)(5) + (3)(-4) + (5)(7) + (7)(0) = 18$$

The main arithmetic properties of the Euclidean inner product are listed in the next theorem.

**Theorem 2.** *If* $\mathbf{u}$, $\mathbf{v}$, *and* $\mathbf{w}$ *are vectors in* $R^n$ *and* $k$ *is any scalar, then:*

(a) $\mathbf{u} \cdot \mathbf{v} = \mathbf{v} \cdot \mathbf{u}$
(b) $(\mathbf{u} + \mathbf{v}) \cdot \mathbf{w} = \mathbf{u} \cdot \mathbf{w} + \mathbf{v} \cdot \mathbf{w}$
(c) $k(\mathbf{u} \cdot \mathbf{v}) = (k\mathbf{u}) \cdot \mathbf{v} = \mathbf{u} \cdot (k\mathbf{v})$
(d) $\mathbf{v} \cdot \mathbf{v} \geq 0$. *Further,* $\mathbf{v} \cdot \mathbf{v} = 0$ *if and only if* $\mathbf{v} = \mathbf{0}$

We shall prove parts (b) and (d) and leave proof of the rest as exercises.

*Proof (b).* Let $\mathbf{u} = (u_1, u_2, \ldots, u_n)$, $\mathbf{v} = (v_1, v_2, \ldots, v_n)$ and $\mathbf{w} = (w_1, w_2, \ldots, w_n)$.

Then

$$(\mathbf{u} + \mathbf{v}) \cdot \mathbf{w} = (u_1 + v_1, u_2 + v_2, \ldots, u_n + v_n) \cdot (w_1, w_2, \ldots, w_n)$$
$$= (u_1 + v_1)w_1 + (u_2 + v_2)w_2 + \cdots + (u_n + v_n)w_n$$
$$= (u_1 w_1 + u_2 w_2 + \cdots + u_n w_n) + (v_1 w_1 + v_2 w_2 + \cdots + v_n w_n)$$
$$= \mathbf{u} \cdot \mathbf{w} + \mathbf{v} \cdot \mathbf{w}$$

(*d*). $\mathbf{v} \cdot \mathbf{v} = v_1^2 + v_2^2 + \cdots + v_n^2 \geq 0$. Further, equality holds if and only if $v_1 = v_2 = \cdots = v_n = 0$; that is, if and only if $\mathbf{v} = \mathbf{0}$. ∎

## Example 2

Theorem 2 allows us to perform computations with Euclidean inner products in much the same way that we perform them with ordinary arithmetic products. For example,

$$(3\mathbf{u} + 2\mathbf{v}) \cdot (4\mathbf{u} + \mathbf{v}) = (3\mathbf{u} + 2\mathbf{v}) \cdot 4\mathbf{u} + (3\mathbf{u} + 2\mathbf{v}) \cdot \mathbf{v}$$
$$= (3\mathbf{u}) \cdot (4\mathbf{u}) + (2\mathbf{v}) \cdot (4\mathbf{u}) + (3\mathbf{u}) \cdot \mathbf{v} + (2\mathbf{v}) \cdot \mathbf{v}$$
$$= 12(\mathbf{u} \cdot \mathbf{u}) + 8(\mathbf{v} \cdot \mathbf{u}) + 3(\mathbf{u} \cdot \mathbf{v}) + 2(\mathbf{v} \cdot \mathbf{v})$$
$$= 12(\mathbf{u} \cdot \mathbf{u}) + 11(\mathbf{u} \cdot \mathbf{v}) + 2(\mathbf{v} \cdot \mathbf{v})$$

The reader should determine which parts of Theorem 2 were used in each step.

By analogy with the familiar formulas in $R^2$ and $R^3$, we define the *Euclidean norm* (or *Euclidean length*) of a vector $\mathbf{u} = (u_1, u_2, \ldots, u_n)$ in $R^n$ by

$$\|\mathbf{u}\| = (\mathbf{u} \cdot \mathbf{u})^{1/2} = \sqrt{u_1^2 + u_2^2 + \cdots + u_n^2}$$

Similarly, the *Euclidean distance* between the points $\mathbf{u} = (u_1, u_2, \ldots, u_n)$ and $\mathbf{v} = (v_1, v_2, \ldots, v_n)$ in $R^n$ is defined by

$$d(\mathbf{u}, \mathbf{v}) = \|\mathbf{u} - \mathbf{v}\| = \sqrt{(u_1 - v_1)^2 + (u_2 - v_2)^2 + \cdots + (u_n - v_n)^2}$$

## Example 3

If $\mathbf{u} = (1, 3, -2, 7)$ and $\mathbf{v} = (0, 7, 2, 2)$ then

$$\|\mathbf{u}\| = \sqrt{(1)^2 + (3)^2 + (-2)^2 + (7)^2} = \sqrt{63} = 3\sqrt{7}$$

and

$$d(\mathbf{u}, \mathbf{v}) = \sqrt{(1 - 0)^2 + (3 - 7)^2 + (-2 - 2)^2 + (7 - 2)^2} = \sqrt{58}$$

Since so many of the familiar ideas from 2-space and 3-space carry over, it is common to refer to $R^n$ with the operations of addition, scalar multiplication, and inner product that we have defined here as *Euclidean n-space.*

We conclude this section by noting that many writers prefer to use the matrix notation

$$\begin{bmatrix} u_1 \\ u_2 \\ \vdots \\ u_n \end{bmatrix}$$

rather than $(u_1, u_2, \ldots, u_n)$ to denote vectors in $R^n$. Since both notations are useful, we shall use whichever seems preferable in the problem being considered. With this in mind, we shall, from here on, denote $n \times 1$ matrices by lower case boldface letters. Thus a system of linear equations will be written

$$A\mathbf{x} = \mathbf{b}$$

rather than $AX = B$ as before.

## EXERCISE SET 4.1

1. Let $\mathbf{u} = (2, 0, -1, 3)$, $\mathbf{v} = (5, 4, 7, -1)$, and $\mathbf{w} = (6, 2, 0, 9)$. Find:
   (a) $\mathbf{u} - \mathbf{v}$       (b) $7\mathbf{v} + 3\mathbf{w}$       (c) $-\mathbf{w} + \mathbf{v}$
   (d) $3(\mathbf{u} - 7\mathbf{v})$       (e) $-3\mathbf{v} - 8\mathbf{w}$       (f) $2\mathbf{v} - (\mathbf{u} + \mathbf{w})$

2. Let $\mathbf{u}$, $\mathbf{v}$, and $\mathbf{w}$ be the vectors in Exercise 1. Find the vector $\mathbf{x}$ that satisfies $2\mathbf{u} - \mathbf{v} + \mathbf{x} = 7\mathbf{x} + \mathbf{w}$.

3. Let $\mathbf{u}_1 = (-1, 3, 2, 0)$, $\mathbf{u}_2 = (2, 0, 4, -1)$, $\mathbf{u}_3 = (7, 1, 1, 4)$, and $\mathbf{u}_4 = (6, 3, 1, 2)$. Find scalars $c_1$, $c_2$, $c_3$, and $c_4$ such that $c_1\mathbf{u}_1 + c_2\mathbf{u}_2 + c_3\mathbf{u}_3 + c_4\mathbf{u}_4 = (0, 5, 6, -3)$.

4. Show that there do not exist scalars $c_1$, $c_2$, and $c_3$ such that $c_1(1, 0, -2, 1) + c_2(2, 0, 1, 2) + c_3(1, -2, 2, 3) = (1, 0, 1, 0)$.

5. Compute the Euclidean norm of $\mathbf{v}$ when
   (a) $\mathbf{v} = (4, -3)$         (b) $\mathbf{v} = (1, -1, 3)$
   (c) $\mathbf{v} = (2, 0, 3, -1)$       (d) $\mathbf{v} = (-1, 1, 1, 3, 6)$

6. Let $\mathbf{u} = (3, 0, 1, 2)$, $\mathbf{v} = (-1, 2, 7, -3)$, and $\mathbf{w} = (2, 0, 1, 1)$. Find
   (a) $\|\mathbf{u} + \mathbf{v}\|$         (b) $\|\mathbf{u}\| + \|\mathbf{v}\|$       (c) $\|-2\mathbf{u}\| + 2\|\mathbf{u}\|$
   (d) $\|3\mathbf{u} - 5\mathbf{v} + \mathbf{w}\|$       (e) $\dfrac{1}{\|\mathbf{w}\|}\mathbf{w}$       (f) $\left\|\dfrac{1}{\|\mathbf{w}\|}\mathbf{w}\right\|$

7. Show that if $\mathbf{v}$ is a nonzero vector in $R^n$, then $(1/\|\mathbf{v}\|)\mathbf{v}$ has norm 1.

8. Find all scalars $k$ such that $\|k\mathbf{v}\| = 3$, where $\mathbf{v} = (-1, 2, 0, 3)$.

9. Find the Euclidean inner product $\mathbf{u} \cdot \mathbf{v}$ when:
   (a) $\mathbf{u} = (-1, 3)$, $\mathbf{v} = (7, 2)$       (b) $\mathbf{u} = (3, 7, 1)$, $\mathbf{v} = (-1, 0, 2)$
   (c) $\mathbf{u} = (1, -1, 2, 3)$, $\mathbf{v} = (3, 3, -6, 4)$
   (d) $\mathbf{u} = (1, 3, 2, 6, -1)$, $\mathbf{v} = (0, 0, 2, 4, 1)$

10. (a) Find two vectors in $R^2$ with Euclidean norm 1 whose Euclidean inner products with $(-2, 4)$ are zero.

(b) Show that there are infinitely many vectors in $R^3$ with Euclidean norm 1 whose Euclidean inner product with $(-1, 7, 2)$ is zero.

11. Find the Euclidean distance between **u** and **v** when:
   (a) $\mathbf{u} = (2, -1)$, $\mathbf{v} = (3, 2)$      (b) $\mathbf{u} = (1, 1, -1)$, $\mathbf{v} = (2, 6, 0)$
   (c) $\mathbf{u} = (2, 0, 1, 3)$, $\mathbf{v} = (-1, 4, 6, 6)$
   (d) $\mathbf{u} = (6, 0, 1, 3, 0)$, $\mathbf{v} = (-1, 4, 2, 8, 3)$

12. Establish the identity

$$\|\mathbf{u} + \mathbf{v}\|^2 + \|\mathbf{u} - \mathbf{v}\|^2 = 2\|\mathbf{u}\|^2 + 2\|\mathbf{v}\|^2$$

for vectors in $R^n$. Interpret this result geometrically in $R^2$.

13. Establish the identity

$$\mathbf{u} \cdot \mathbf{v} = \frac{1}{4}\|\mathbf{u} + \mathbf{v}\|^2 - \frac{1}{4}\|\mathbf{u} - \mathbf{v}\|^2$$

for vectors in $R^n$.

14. Verify parts (b), (e), (f), and (g) of Theorem 1 when $\mathbf{u} = (1, 0, -1, 2)$, $\mathbf{v} = (3, -1, 2, 4)$, $\mathbf{w} = (2, 7, 3, 0)$, $k = 6$, and $l = -2$.

15. Verify parts (b) and (c) of Theorem 2 for the values of **u**, **v**, **w**, and $k$ in Exercise 14.

16. Prove (a) through (d) of Theorem 1.

17. Prove (e) through (h) of Theorem 1.

18. Prove (a) and (c) of Theorem 2.

## 4.2 GENERAL VECTOR SPACES

In this section the concept of a vector will undergo a radical generalization. We shall state a set of axioms which, if satisfied by a class of objects, will entitle those objects to be called "vectors." The axioms will be chosen by abstracting the most important properties of vectors in $R^n$; as consequence, vectors in $R^n$ will automatically satisfy these axioms. Thus our new concept of a vector will include our old vectors and many new kinds of vectors as well.

**Definition.** A real *vector space* is a set $V$ of elements called *vectors* together with two operations, addition and multiplication by scalars (real numbers). By addition we mean that given two vectors **u** and **v** in $V$, there is a rule for determining an object $\mathbf{u} + \mathbf{v}$ called the *sum* of **u** and **v**. By scalar multiplication we mean that given a vector **u** and a real number $k$, there is a rule for determining an object $k\mathbf{u}$ called the *scalar multiple* of **u** by $k$. Further, the following axioms must be satisfied by all vectors **u**, **v**, **w** in $V$ and all scalars $k$ and $l$.

(1) If **u** and **v** are vectors in $V$, then **u** + **v** is in $V$

(2) **u** + **v** = **v** + **u**

(3) **u** + (**v** + **w**) = (**u** + **v**) + **w**

(4) There is a vector **0** in $V$ called the *zero vector* such that **0** + **u** = **u** + **0** = **u** for all **u** in $V$.

(5) For each **u** in $V$, there is a vector −**u** in $V$ called the *negative* of **u** such that **u** + (−**u**) = (−**u**) + **u** = **0**

(6) If $k$ is any real number and **u** is any vector in $V$, then $k$**u** is in $V$

(7) $k$(**u** + **v**) = $k$**u** + $k$**v**

(8) $(k + l)$**u** = $k$**u** + $l$**u**

(9) $k(l$**u**$) = (kl)($**u**$)$

(10) $1$**u** = **u**

For some applications it is necessary to consider vector spaces where the scalars are complex numbers rather than real numbers. Such vector spaces are called *complex vector spaces.* In this text, however, *all our scalars will be real.*

The reader should keep in mind that, in the definition of a vector space, neither the nature of the vectors nor the operations is specified. Any kinds of objects whatsoever can serve as vectors; all that is required is that the vector-space axioms be satisfied. The following examples will give some idea of the diversity of possible vector spaces.

## Example 4

The set $V = R^n$ with the standard operations of addition and scalar multiplication defined in the previous section is a vector space. Axioms (1) and (6) follow from the definitions of the standard operations on $R^n$; the remaining axioms follow from Theorem 1.

## Example 5

Let $V$ be any plane through the origin in $R^3$. We shall show that the points in $V$ form a vector space under the standard addition and scalar multiplication operations for vectors in $R^3$.

From Example 4, we know that $R^3$ itself is a vector space under these operations. Axioms (2), (3), (7), (8), (9), and (10) hold for all points in $R^3$ and consequently for all points in the plane $V$. We therefore need only show that Axioms (1), (4), (5), and (6) are satisfied.

Since the plane $V$ passes through the origin, it has an equation of the form

$$ax + by + cz = 0 \tag{4.1}$$

(Theorem 6 in Chapter 3.) Thus, if **v** = $(v_1, v_2, v_3)$ and **w** = $(w_1, w_2, w_3)$ are points in $V$, then $av_1 + bv_2 + cv_3 = 0$ and $aw_1 + bw_2 + cw_3 = 0$. Adding these equations gives

$$a(v_1 + w_1) + b(v_2 + w_2) + c(v_3 + w_3) = 0$$

This equality tells us that the coordinates of the point $\mathbf{v} + \mathbf{w} = (v_1 + w_1, v_2 + w_2, v_3 + w_3)$ satisfy (4.1); thus $\mathbf{v} + \mathbf{w}$ lies in the plane $V$. This proves that Axiom (1) is satisfied. Multiplying $av_1 + bv_2 + cv_3 = 0$ through by $-1$ gives

$$a(-v_1) + b(-v_2) + c(-v_3) = 0$$

Thus $-\mathbf{v} = (-v_1, -v_2, -v_3)$ lies in $V$. This establishes Axiom (5). The verifications of Axioms (4) and (6) are left as exercises.

## Example 6

The points on a line $V$ passing through the origin in $R^3$ form a vector space under the standard addition and scalar multiplication operations for vectors in $R^3$.

The argument is similar to that used in Example 5 and is based on the fact that the points of $V$ satisfy parametric equations of the form

$$\begin{aligned} x &= at \\ y &= bt \qquad -\infty < t < +\infty \\ z &= ct \end{aligned}$$

(Section 3.5). The details are left as an exercise.

## Example 7

The set $V$ of all $m \times n$ matrices with real entries, together with the operations of matrix addition and scalar multiplication, is a vector space. The $m \times n$ zero matrix is the zero vector $\mathbf{0}$, and if $\mathbf{u}$ is the $m \times n$ matrix $A$, then the matrix $-A$ will serve as $-\mathbf{u}$ in Axiom (5). Most of the remaining axioms are satisfied by virtue of Theorem 2 in Section 1.5. We shall denote this vector space by the symbol $M_{mn}$.

## Example 8

Let $V$ be the set of real-valued functions defined on the entire real line. If $\mathbf{f} = f(x)$ and $\mathbf{g} = g(x)$ are two such functions and $k$ is any real number, define the sum function $\mathbf{f} + \mathbf{g}$ and the scalar multiple $k\mathbf{f}$ by

$$\begin{aligned} (\mathbf{f} + \mathbf{g})(x) &= f(x) + g(x) \\ (k\mathbf{f})(x) &= kf(x) \end{aligned}$$

In other words, the value of the function $\mathbf{f} + \mathbf{g}$ at $x$ is obtained by adding together the values of $\mathbf{f}$ and $\mathbf{g}$ at $x$ (Figure 4.1*a*). Similarly, the value of $k\mathbf{f}$ at $x$ is $k$ times the value of $\mathbf{f}$ at $x$ (Figure 4.1*b*). The set $V$ is a vector space under these operations.

The zero vector in this space is the zero constant function, that is, the function whose graph is a horizontal line through the origin. The verification of the remaining axioms is an exercise.

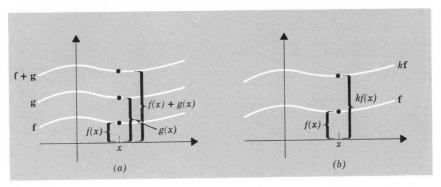

**Figure 4.1**

## Example 9

Let $V$ be the set of all points $(x, y)$ in $R^2$ that lie in the first quadrant; that is, such that $x \geq 0$ and $y \geq 0$. The set $V$ *fails* to be a vector space under the standard operations on $R^2$, since Axioms (5) and (6) are not satisfied. To see this, observe that $\mathbf{v} = (1, 1)$ lies in $V$, but $(-1)\mathbf{v} = -\mathbf{v} = (-1, -1)$ does not (Figure 4.2).

As we progress, we shall add more examples of vector spaces to our list. We conclude this section with a theorem that gives a useful list of vector properties.

**Theorem 3.** *Let $V$ be a vector space, $\mathbf{u}$ a vector in $V$, and $k$ a scalar; then:*
  *(a)* $0\mathbf{u} = \mathbf{0}$
  *(b)* $k\mathbf{0} = \mathbf{0}$
  *(c)* $(-1)\mathbf{u} = -\mathbf{u}$
  *(d)* *If* $k\mathbf{u} = \mathbf{0}$, *then* $k = 0$ *or* $\mathbf{u} = \mathbf{0}$

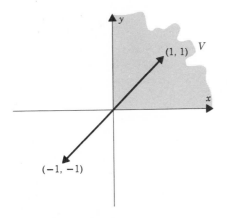

**Figure 4.2**

*Proof.* We shall prove parts (a) and (b) and leave proofs of the remaining parts as exercises.

(*a*) Let $\mathbf{w} = 0\mathbf{u}$. We must show that $\mathbf{w} = \mathbf{0}$. Adding $\mathbf{w}$ to itself and using Axiom (8) gives

$$\mathbf{w} + \mathbf{w} = 0\mathbf{u} + 0\mathbf{u} = (0 + 0)\mathbf{u} = 0\mathbf{u} = \mathbf{w}$$

Adding $-\mathbf{w}$ to both sides gives

$$(\mathbf{w} + \mathbf{w}) + (-\mathbf{w}) = \mathbf{w} + (-\mathbf{w}) = \mathbf{0}$$

On the other hand, Axiom (3) gives

$$(\mathbf{w} + \mathbf{w}) + (-\mathbf{w}) = \mathbf{w} + (\mathbf{w} + (-\mathbf{w})) = \mathbf{w} + \mathbf{0} = \mathbf{w}$$

Therefore $\mathbf{w} = \mathbf{0}$.

(*b*) To show $(-1)\mathbf{u} = -\mathbf{u}$, we must demonstrate that $\mathbf{u} + (-1)\mathbf{u} = \mathbf{0}$. To see this, observe that

$$\mathbf{u} + (-1)\mathbf{u} = 1\mathbf{u} + (-1)\mathbf{u} = (1 + (-1))\mathbf{u} = 0\mathbf{u} = \mathbf{0} \quad \blacksquare$$

## EXERCISE SET 4.2

In Exercises 1–14 a set of objects is given together with operations of addition and scalar multiplication. Determine which sets are vector spaces under the given operations. For those which are not, list all axioms that fail to hold.

**1.** The set of all triples of real numbers $(x, y, z)$ with the operations $(x, y, z) + (x', y', z') = (x + x', y + y', z + z')$ and $k(x, y, z) = (kx, y, z)$.

**2.** The set of all triples of real numbers $(x, y, z)$ with the operations $(x, y, z) + (x', y', z') = (x + x', y + y', z + z')$ and $k(x, y, z) = (0, 0, 0)$.

**3.** The set of all pairs of real numbers $(x, y)$ with the operations $(x, y) + (x', y') = (x + x', y + y')$ and $k(x, y) = (2kx, 2ky)$.

**4.** The set of all real numbers $x$ with the standard operations of addition and multiplication.

**5.** The set of all pairs of real numbers of the form $(x, 0)$ with the standard operations on $R^2$.

**6.** The set of all pairs of real numbers of the form $(x, y)$, where $x \geq 0$, with the standard operations on $R^2$.

**7.** The set of all $n$-tuples of real numbers of the form $(x, x, \ldots, x)$ with the standard operations on $R^n$.

**8.** The set of all pairs of real numbers $(x, y)$ with the operations $(x, y) + (x', y') = (x + x' + 1, y + y' + 1)$ and $k(x, y) = (kx, ky)$.

9. The set of all positive real numbers $x$ with the operations $x + x' = xx'$ and $kx = x^k$.

10. The set of all $2 \times 2$ matrices of the form

$$\begin{bmatrix} a & 1 \\ 1 & b \end{bmatrix}$$

with the ordinary matrix operations.

11. The set of all $2 \times 2$ matrices of the form

$$\begin{bmatrix} a & 0 \\ 0 & b \end{bmatrix}$$

with the ordinary matrix operations.

12. The set of all real-valued functions $f$ defined everywhere on the real line and such that $f(1) = 0$, with the operations defined in Example 8.

13. The set of all $2 \times 2$ matrices of the form

$$\begin{bmatrix} a & a + b \\ a + b & b \end{bmatrix}$$

with the ordinary matrix operations.

14. The set whose only element is the moon. The operations are moon + moon = moon and $k(\text{moon}) = \text{moon}$, where $k$ is a real number.

15. Prove that a line passing through the origin in $R^3$ is a vector space under the standard operations on $R^3$.

16. Complete the unfinished details in Example 5.

17. Complete the unfinished details in Example 8.

18. Prove part (c) of Theorem 3.

19. Prove part (d) of Theorem 3.

20. Prove that a vector space cannot have more than one zero vector.

21. Prove that a vector has exactly one negative.

## 4.3 SUBSPACES

If $V$ is a vector space, then certain subsets of $V$ themselves form vector spaces under the vector addition and scalar multiplication defined on $V$. In this section we shall study such subsets in detail.

**Definition.** A subset $W$ of a vector space $V$ is called a ***subspace*** of $V$ if $W$ is itself a vector space under the addition and scalar multiplication defined on $V$.

For example, lines and planes passing through the origin are subspaces of $R^3$ (Examples 5 and 6).

Our first theorem provides a convenient method for determining whether a given set of vectors in a vector space is actually a subspace.

**Theorem 4.** *If W is a set of one or more vectors from a vector space V, then W is a subspace of V if and only if the following conditions hold.*

(a) *If* **u** *and* **v** *are vectors in W, then* **u** + **v** *is in W.*
(b) *If k is any scalar and* **u** *is any vector in W, then k***u** *is in W.*

[Conditions (a) and (b) are often described by saying that *W* is **closed under addition** and **closed under scalar multiplication.**]

*Proof.* If *W* is a subspace of *V*, then all the vector-space axioms are satisfied; in particular Axioms (1) and (6) hold. But these are precisely conditions (a) and (b).

Conversely, assume conditions (a) and (b) hold. Since these conditions are vector-space Axioms (1) and (6), we need only show that *W* satisfies the remaining eight axioms. Axioms (2), (3), (7), (8), (9), and (10) are automatically satisfied by the vectors in *W* since they are satisfied by all vectors in *V*. Therefore, to complete the proof, we need only verify that Axioms (4) and (5) are satisfied by *W*.

Let **u** be any vector in *W*. By condition (b), $k$**u** is in *W* for every scalar $k$. Setting $k = 0$, it follows that $0$**u** = **0** is in *W*, and setting $k = -1$, it follows that $(-1)$**u** = $-$**u** is in *W*. ▨

Every vector space *V* has at least two subspaces. *V* itself is a subspace and the set $\{$**0**$\}$ consisting of just the zero vector in *V* is a subspace called the **zero subspace.** The following examples provide less trivial illustrations of subspaces.

**Example 10**

Show that the set *W* of all $2 \times 2$ matrices having zeros on the main diagonal is a subspace of the vector space $M_{22}$ of all $2 \times 2$ matrices.

*Solution.* Let

$$A = \begin{bmatrix} 0 & a_{12} \\ a_{21} & 0 \end{bmatrix} \qquad B = \begin{bmatrix} 0 & b_{12} \\ b_{21} & 0 \end{bmatrix}$$

be any two matrices in *W* and $k$ any scalar. Then

$$kA = \begin{bmatrix} 0 & ka_{12} \\ ka_{21} & 0 \end{bmatrix} \quad \text{and} \quad A + B = \begin{bmatrix} 0 & a_{12} + b_{12} \\ a_{21} + b_{21} & 0 \end{bmatrix}$$

Since $kA$ and $A + B$ have zeros on the main diagonal, they lie in *W*. Thus *W* is a subspace of $M_{22}$.

**Example 11**

Let $n$ be a positive integer, and let $W$ be the set of all real polynomial functions having degree $\leq n$ together with the zero function; that is, all functions expressible in the form

$$p(x) = a_0 + a_1 x + \cdots + a_n x^n \qquad (4.2)$$

where $a_0, a_1, \ldots, a_n$ are real numbers. The set $W$ is a subspace of the vector space of all real-valued functions discussed in Example 8. To see this, let **p** and **q** be the polynomials

$$p(x) = a_0 + a_1 x + \cdots + a_n x^n$$

and

$$q(x) = b_0 + b_1 x + \cdots + b_n x^n$$

Then

$$(\mathbf{p} + \mathbf{q})(x) = p(x) + q(x) = (a_0 + b_0) + (a_1 + b_1)x + \cdots + (a_n + b_n)x^n$$

and

$$(k\mathbf{p})(x) = kp(x) = (ka_0) + (ka_1)x + \cdots + (ka_n)x^n$$

have the form given in (4.2). Therefore, $\mathbf{p} + \mathbf{q}$ and $k\mathbf{p}$ lie in $W$. We shall denote the vector space $W$ in this example by the symbol $P_n$.

**Example 12**

**(For readers who have studied calculus.)**

Recall from calculus that if **f** and **g** are continuous functions and $k$ is a constant, then $\mathbf{f} + \mathbf{g}$ and $k\mathbf{f}$ are also continuous functions. It follows that the set of all continuous functions is a subspace of the vector space of all real-valued functions. This space is denoted by $C(-\infty, +\infty)$. A closely related example is the vector space of all functions that are continuous on a closed interval $a \leq x \leq b$. This space is denoted by $C[a, b]$.

**Example 13**

Consider a system of $m$ linear equations in $n$ unknowns

$$
\begin{aligned}
a_{11}x_1 + a_{12}x_2 + \cdots + a_{1n}x_n &= b_1 \\
a_{21}x_1 + a_{22}x_2 + \cdots + a_{2n}x_n &= b_2 \\
&\vdots \\
a_{m1}x_1 + a_{m2}x_2 + \cdots + a_{mn}x_n &= b_m
\end{aligned}
$$

or, in matrix notation, $A\mathbf{x} = \mathbf{b}$. A vector*

$$\mathbf{s} = \begin{bmatrix} s_1 \\ s_2 \\ \vdots \\ s_n \end{bmatrix}$$

in $R^n$ is called a **solution vector** of the system if $x_1 = s_1$, $x_2 = s_2$, $\ldots$, $x_n = s_n$ is a solution of the system. In this example we shall show that the set of solution vectors of a *homogeneous* system is a subspace of $R^n$.

Let $A\mathbf{x} = \mathbf{0}$ be a homogeneous system of $m$ linear equations in $n$ unknowns, let $W$ be the set of solution vectors, and let $\mathbf{s}$ and $\mathbf{s}'$ be vectors in $W$. To show that $W$ is closed under addition and scalar multiplication, we must demonstrate that $\mathbf{s} + \mathbf{s}'$ and $k\mathbf{s}$ are also solution vectors, where $k$ is any scalar. Since $\mathbf{s}$ and $\mathbf{s}'$ are solution vectors we have

$$A\mathbf{s} = \mathbf{0} \qquad \text{and} \qquad A\mathbf{s}' = \mathbf{0}$$

Therefore

$$A(\mathbf{s} + \mathbf{s}') = A\mathbf{s} + A\mathbf{s}' = \mathbf{0} + \mathbf{0} = \mathbf{0}$$

and

$$A(k\mathbf{s}) = k(A\mathbf{s}) = k\mathbf{0} = \mathbf{0}$$

These equations show that $\mathbf{s} + \mathbf{s}'$ and $k\mathbf{s}$ satisfy the equation $A\mathbf{x} = \mathbf{0}$. Thus $\mathbf{s} + \mathbf{s}'$ and $k\mathbf{s}$ are solution vectors.

The subspace $W$ in this example is called the **solution space** of the system $A\mathbf{x} = \mathbf{0}$.

In many problems a vector space $V$ is given, and it is of importance to find the "smallest" subspace of $V$ that contains a specified set of vectors $\{\mathbf{v}_1, \mathbf{v}_2, \ldots, \mathbf{v}_r\}$. The following definition provides the key to the construction of such subspaces.

**Definition.** A vector $\mathbf{w}$ is called a **linear combination** of the vectors $\mathbf{v}_1, \mathbf{v}_2, \ldots, \mathbf{v}_r$ if it can be expressed in the form

$$\mathbf{w} = k_1\mathbf{v}_1 + k_2\mathbf{v}_2 + \cdots + k_r\mathbf{v}_r$$

where $k_1, k_2, \ldots, k_r$ are scalars.

**Example 14** ✓

Consider the vectors $\mathbf{u} = (1, 2, -1)$ and $\mathbf{v} = (6, 4, 2)$ in $R^3$. Show that $\mathbf{w} = (9, 2, 7)$ is a linear combination of $\mathbf{u}$ and $\mathbf{v}$ and that $\mathbf{w}' = (4, -1, 8)$ is *not* a linear combination of $\mathbf{u}$ and $\mathbf{v}$.

*Solution.* In order for $\mathbf{w}$ to be a linear combination of $\mathbf{u}$ and $\mathbf{v}$, there must be scalars $k_1$ and $k_2$ such that $\mathbf{w} = k_1\mathbf{u} + k_2\mathbf{v}$; that is

$$(9, 2, 7) = k_1(1, 2, -1) + k_2(6, 4, 2)$$

*In this example we are using the matrix notation for vectors in $R^n$.

or

$$(9, 2, 7) = (k_1 + 6k_2, 2k_1 + 4k_2, -k_1 + 2k_2)$$

Equating corresponding components gives

$$k_1 + 6k_2 = 9$$
$$2k_1 + 4k_2 = 2$$
$$-k_1 + 2k_2 = 7$$

Solving this system yields $k_1 = -3$, $k_2 = 2$ so that

$$\mathbf{w} = -3\mathbf{u} + 2\mathbf{v}$$

Similarly, for $\mathbf{w}'$ to be a linear combination of $\mathbf{u}$ and $\mathbf{v}$, there must be scalars $k_1$ and $k_2$ such that $\mathbf{w}' = k_1\mathbf{u} + k_2\mathbf{v}$; that is

$$(4, -1, 8) = k_1(1, 2, -1) + k_2(6, 4, 2)$$

or

$$(4, -1, 8) = (k_1 + 6k_2, 2k_1 + 4k_2, -k_1 + 2k_2)$$

Equating corresponding components gives

$$k_1 + 6k_2 = \phantom{-}4$$
$$2k_1 + 4k_2 = -1$$
$$-k_1 + 2k_2 = \phantom{-}8$$

This system of equations is inconsistent, so that no such scalars exist. Consequently, $\mathbf{w}$ is not a linear combination of $\mathbf{u}$ and $\mathbf{v}$.

The next theorem shows that if $S = \{\mathbf{v}_1, \mathbf{v}_2, \ldots, \mathbf{v}_r\}$ is a finite set of vectors in a vector space $V$, then the set $W$ of all possible linear combinations of these vectors forms the smallest subspace of $V$ that contains all the vectors in $S$. It is called the *linear space spanned* by $\{\mathbf{v}_1, \mathbf{v}_2, \ldots, \mathbf{v}_r\}$, or more simply the *space spanned* by $\{\mathbf{v}_1, \mathbf{v}_2, \ldots, \mathbf{v}_r\}$.

**Theorem 5.** *If $\mathbf{v}_1, \mathbf{v}_2, \ldots, \mathbf{v}_r$ are vectors in a vector space $V$, then:*

(a) *The set $W$ of all linear combinations of $\mathbf{v}_1, \mathbf{v}_2, \ldots, \mathbf{v}_r$ is a subspace of $V$.*

(b) *$W$ is the smallest subspace of $V$ that contains $\mathbf{v}_1, \mathbf{v}_2, \ldots, \mathbf{v}_r$ in the sense that every other subspace of $V$ that contains $\mathbf{v}_1, \mathbf{v}_2, \ldots, \mathbf{v}_r$ must contain $W$.*

*Proof.*

(a) To show that $W$ is a subspace of $V$, we must prove it is closed under addition and scalar multiplication. If $\mathbf{u}$ and $\mathbf{v}$ are vectors in $W$, then

$$\mathbf{u} = c_1\mathbf{v}_1 + c_2\mathbf{v}_2 + \cdots + c_r\mathbf{v}_r$$

and

$$\mathbf{v} = k_1\mathbf{v}_1 + k_2\mathbf{v}_2 + \cdots + k_r\mathbf{v}_r$$

where $c_1, c_2, \ldots, c_r, k_1, k_2, \ldots, k_r$ are scalars. Therefore

$$\mathbf{u} + \mathbf{v} = (c_1 + k_1)\mathbf{v}_1 + (c_2 + k_2)\mathbf{v}_2 + \cdots + (c_r + k_r)\mathbf{v}_r$$

and, for any scalar $k$

$$k\mathbf{u} = (kc_1)\mathbf{v}_1 + (kc_2)\mathbf{v}_2 + \cdots + (kc_r)\mathbf{v}_r$$

Thus $\mathbf{u} + \mathbf{v}$ and $k\mathbf{u}$ are linear combinations of $\mathbf{v}_1, \mathbf{v}_2, \ldots, \mathbf{v}_r$, and consequently lie in $W$. Therefore $W$ is closed under addition and scalar multiplication.

(b) Each vector $\mathbf{v}_i$ is a linear combination of $\mathbf{v}_1, \mathbf{v}_2, \ldots, \mathbf{v}_r$ since we can write

$$\mathbf{v}_i = 0\mathbf{v}_1 + 0\mathbf{v}_2 + \cdots + 1\mathbf{v}_i + \cdots + 0\mathbf{v}_r$$

The subspace $W$ therefore contains each of the vectors $\mathbf{v}_1, \mathbf{v}_2, \ldots, \mathbf{v}_r$. Let $W'$ be any other subspace that contains $\mathbf{v}_1, \mathbf{v}_2, \ldots, \mathbf{v}_r$. Since $W'$ is closed under addition and scalar multiplication, it must contain all linear combinations $c_1\mathbf{v}_1 + c_2\mathbf{v}_2 + \cdots + c_r\mathbf{v}_r$ of $\mathbf{v}_1, \mathbf{v}_2, \ldots, \mathbf{v}_r$; thus $W'$ contains each vector of $W$. ∎

The linear space $W$ spanned by a set of vectors $S = \{\mathbf{v}_1, \mathbf{v}_2, \ldots, \mathbf{v}_r\}$ shall be denoted by $\text{lin}(S)$ or $\text{lin}\{\mathbf{v}_1, \mathbf{v}_2, \ldots, \mathbf{v}_r\}$, and we shall say that $\mathbf{v}_1, \mathbf{v}_2, \ldots, \mathbf{v}_r$ *span* $W$.

**Example 15**

If $\mathbf{v}_1$ and $\mathbf{v}_2$ are noncollinear vectors in $R^3$ with initial points at the origin, then $\text{lin}\{\mathbf{v}_1, \mathbf{v}_2\}$, which consists of all linear combinations $k_1\mathbf{v}_1 + k_2\mathbf{v}_2$, is the plane determined by $\mathbf{v}_1$ and $\mathbf{v}_2$ (Figure 4.3a).

Similarly, if $\mathbf{v}$ is a nonzero vector in $R^3$, then $\text{lin}\{\mathbf{v}\}$, which is the set of all linear combinations $k\mathbf{v}$, is the line determined by $\mathbf{v}$ (Figure 4.3b).

**Figure 4.3**

**Example 16** ✓

Determine if $v_1 = (1, 1, 2)$, $v_2 = (1, 0, 1)$, and $v_3 = (2, 1, 3)$ span $R^3$.

*Solution.* We must determine whether an arbitrary vector $\mathbf{b} = (b_1, b_2, b_3)$ in $R^3$ can be expressed as a linear combination

$$\mathbf{b} = k_1 \mathbf{v}_1 + k_2 \mathbf{v}_2 + k_3 \mathbf{v}_3$$

of the vectors $v_1$, $v_2$, and $v_3$. Expressing this equation in terms of components gives

$$(b_1, b_2, b_3) = (k_1 + k_2 + 2k_3, k_1 + k_3, 2k_1 + k_2 + 3k_3)$$

or

$$\begin{aligned} k_1 + k_2 + 2k_3 &= b_1 \\ k_1 \qquad + k_3 &= b_2 \\ 2k_1 + k_2 + 3k_3 &= b_3 \end{aligned}$$

The problem thus reduces to determining whether or not this system is consistent for all values of $b_1$, $b_2$, and $b_3$. By parts (a) and (d) of the corollary to Theorem 12 in Section 1.8, this system will be consistent for all $b_1$, $b_2$, and $b_3$ if and only if the matrix of coefficients

$$A = \begin{bmatrix} 1 & 1 & 2 \\ 1 & 0 & 1 \\ 2 & 1 & 3 \end{bmatrix} = 0$$

*what is the Trivial Solution how*

is invertible. But $\det(A) = 0$, so that $A$ is not invertible, and consequently $v_1$, $v_2$, and $v_3$ do not span $R^3$.

## EXERCISE SET 4.3

1. Use Theorem 4 to determine which of the following are subspaces of $R^3$.
   (a) all vectors of the form $(a, 0, 0)$
   (b) all vectors of the form $(a, 1, 1)$
   (c) all vectors of the form $(a, b, c)$, where $b = a + c$
   (d) all vectors of the form $(a, b, c)$, where $b = a + c + 1$

2. Use Theorem 4 to determine which of the following are subspaces of $M_{22}$.
   (a) all matrices of the form

   $$\begin{bmatrix} a & b \\ c & d \end{bmatrix}$$

   where $a$, $b$, $c$, and $d$ are integers
   (b) all matrices of the form

   $$\begin{bmatrix} a & b \\ c & d \end{bmatrix}$$

   where $a + d = 0$
   (c) all $2 \times 2$ matrices $A$ such that $A = A^t$
   (d) all $2 \times 2$ matrices $A$ such that $\det(A) = 0$

3. Use Theorem 4 to determine which of the following are subspaces of $P_3$.
   (a) all polynomials $a_0 + a_1x + a_2x^2 + a_3x^3$ for which $a_0 = 0$
   (b) all polynomials $a_0 + a_1x + a_2x^2 + a_3x^3$ for which $a_0 + a_1 + a_2 + a_3 = 0$
   (c) all polynomials $a_0 + a_1x + a_2x^2 + a_3x^3$ for which $a_0, a_1, a_2,$ and $a_2$ are integers
   (d) all polynomials of the form $a_0 + a_1x$, where $a_0$ and $a_1$ are real numbers.

4. Use Theorem 4 to determine which of the following are subspaces of the space of all real-valued functions $f$ defined on the entire real line.
   (a) all $f$ such that $f(x) \leq 0$ for all $x$
   (b) all $f$ such that $f(0) = 0$
   (c) all $f$ such that $f(0) = 2$
   (d) all constant functions
   (e) all $f$ of the form $k_1 + k_2 \sin x$, where $k_1$ and $k_2$ are real numbers

5. Which of the following are linear combinations of $u = (1, -1, 3)$ and $v = (2, 4, 0)$?
   (a) $(3, 3, 3)$     (b) $(4, 2, 6)$
   (c) $(1, 5, 6)$     (d) $(0, 0, 0)$

6. Express the following as linear combinations of $u = (2, 1, 4)$, $v = (1, -1, 3)$, and $w = (3, 2, 5)$.
   (a) $(5, 9, 5)$     (b) $(2, 0, 6)$
   (c) $(0, 0, 0)$     (d) $(2, 2, 3)$

7. Express the following as linear combinations of $p_1 = 2 + x + 4x^2$, $p_2 = 1 - x + 3x^2$, and $p_3 = 3 + 2x + 5x^2$?
   (a) $5 + 9x + 5x^2$     (b) $12 + 6x^2$
   (c) $0$                 (d) $2 + 2x + 3x^2$

8. Which of the following are linear combinations of
$$A = \begin{bmatrix} 1 & 2 \\ -1 & 3 \end{bmatrix}, B = \begin{bmatrix} 0 & 1 \\ 2 & 4 \end{bmatrix}, \text{ and } C = \begin{bmatrix} 4 & -2 \\ 0 & -2 \end{bmatrix}?$$
   (a) $\begin{bmatrix} 6 & 3 \\ 0 & 8 \end{bmatrix}$     (b) $\begin{bmatrix} -1 & 7 \\ 5 & 1 \end{bmatrix}$     (c) $\begin{bmatrix} 0 & 0 \\ 0 & 0 \end{bmatrix}$     (d) $\begin{bmatrix} 6 & -1 \\ -8 & -8 \end{bmatrix}$

9. In each part determine whether the given vectors span $R^3$.
   (a) $v_1 = (1, 1, 1)$, $v_2 = (2, 2, 0)$, $v_3 = (3, 0, 0)$
   (b) $v_1 = (2, -1, 3)$, $v_2 = (4, 1, 2)$, $v_3 = (8, -1, 8)$
   (c) $v_1 = (3, 1, 4)$, $v_2 = (2, -3, 5)$, $v_3 = (5, -2, 9)$, $v_4 = (1, 4, -1)$
   (d) $v_1 = (1, 3, 3)$, $v_2 = (1, 3, 4)$, $v_3 = (1, 4, 3)$, $v_4 = (6, 2, 1)$

10. Determine which of the following lie in the space spanned by
$$f = \cos^2 x \quad \text{and} \quad g = \sin^2 x$$
   (a) $\cos 2x$     (b) $3 + x^2$
   (c) $1$           (d) $\sin x$

11. Determine if the following polynomials span $P_3$
$$p_1 = 1 + 2x - x^2 \quad p_2 = 3 + x^2$$
$$p_3 = 5 + 4x - x^2 \quad p_4 = -2 + 2x - 2x^2$$

12. Let $v_1 = (2, 1, 0, 3)$, $v_2 = (3, -1, 5, 2)$, and $v_3 = (-1, 0, 2, 1)$. Which of the following vectors are in $\lin\{v_1, v_2, v_3\}$?
    (a) $(2, 3, -7, 3)$    (b) $(0, 0, 0, 0)$
    (c) $(1, 1, 1, 1)$    (d) $(-4, 6, -13, 4)$

13. Find an equation for the plane spanned by the vectors $u = (1, 1, -1)$ and $v = (2, 3, 5)$.

14. Find parametric equations for the line spanned by the vector $u = (2, 7, -1)$.

15. Show that the solution vectors of a consistent nonhomogeneous system of $m$ linear equations in $n$ unknowns do not form a subspace of $R^n$.

16. **(For readers who have studied calculus.)** Show that the following sets of functions are subspaces of the vector space in Example 8.
    (a) all everywhere continuous functions
    (b) all everywhere differentiable functions
    (c) all everywhere differentiable functions that satisfy $f' + 2f = 0$

## 4.4  LINEAR INDEPENDENCE

From Section 4.3, a vector space $V$ is spanned by a set of vectors $S = \{v_1, v_2, \ldots, v_r\}$ if each vector in $V$ is a linear combination of $v_1, v_2, \ldots, v_r$. Spanning sets are useful in a variety of problems since it is often possible to study a vector space $V$ by first studying the vectors in a spanning set $S$, and then extending the results to the rest of $V$. Therefore, it is desirable to keep the spanning set $S$ as small as possible. The problem of finding smallest spanning sets for a vector space depends on the notion of linear independence, which we shall study in this section.

If $S = \{v_1, v_2, \ldots, v_r\}$ is a set of vectors, then the vector equation

$$k_1 v_1 + k_2 v_2 + \cdots + k_r v_r = 0$$

has at least one solution, namely

$$k_1 = 0, k_2 = 0, \ldots, k_r = 0$$

If this is the only solution, then $S$ is called a **linearly independent** set. If there are other solutions, then $S$ is called a **linearly dependent** set.

### Example 17

The set of vectors $S = \{v_1, v_2, v_3\}$, where $v_1 = (2, -1, 0, 3)$, $v_2 = (1, 2, 5, -1)$, $v_3 = (7, -1, 5, 8)$ is linearly dependent, since $3v_1 + v_2 - v_3 = 0$.

### Example 18

The polynomials $p_1 = 1 - x$, $p_2 = 5 + 3x - 2x^2$, and $p_3 = 1 + 3x - x^2$ form a linearly dependent set in $P_2$ since $3p_1 - p_2 + 2p_3 = 0$.

### Example 19

Consider the vectors $e_1 = (1, 0, 0)$, $e_2 = (0, 1, 0)$, and $e_3 = (0, 0, 1)$ in $R^3$. In terms of components the vector equation

$$k_1 e_1 + k_2 e_2 + k_3 e_3 = 0$$

becomes

$$k_1(1, 0, 0) + k_2(0, 1, 0) + k_3(0, 0, 1) = (0, 0, 0)$$

or equivalently

$$(k_1, k_2, k_3) = (0, 0, 0)$$

Thus $k_1 = 0$, $k_2 = 0$, $k_3 = 0$; the set $S = \{e_1, e_2, e_3\}$ is therefore linearly independent. A similar argument can be used to show that the vectors $e_1 = (1, 0, 0, \ldots, 0)$, $e_2 = (0, 1, 0, \ldots, 0), \ldots, e_n = (0, 0, 0, \ldots, 1)$ form a linearly independent set in $R^n$.

### Example 20

Determine whether the vectors

$$v_1 = (1, -2, 3) \qquad v_2 = (5, 6, -1) \qquad v_3 = (3, 2, 1)$$

form a linearly dependent or linearly independent set.

*Solution.* In terms of components the vector equation

$$k_1 v_1 + k_2 v_2 + k_3 v_3 = 0$$

becomes

$$k_1(1, -2, 3) + k_2(5, 6, -1) + k_3(3, 2, 1) = (0, 0, 0)$$

or equivalently

$$(k_1 + 5k_2 + 3k_3, -2k_1 + 6k_2 + 2k_3, 3k_1 - k_2 + k_3) = (0, 0, 0)$$

Equating corresponding components gives

$$\begin{aligned} k_1 + 5k_2 + 3k_3 &= 0 \\ -2k_1 + 6k_2 + 2k_3 &= 0 \\ 3k_1 - k_2 + k_3 &= 0 \end{aligned}$$

Thus $v_1$, $v_2$, and $v_3$ form a linearly dependent set if this system has a nontrivial solution, or a linearly independent set if it has only the trivial solution. Solving this system yields

$$x_1 = -\frac{1}{2}t \qquad x_2 = -\frac{1}{2}t \qquad x_3 = t$$

Thus the system has nontrivial solutions; consequently, $v_1$, $v_2$, and $v_3$ form a linearly dependent set.

The term "linearly dependent" suggests that the vectors "depend" on each other in some way. To see how, assume $S = \{v_1, v_2, \ldots, v_r\}$ is a linearly dependent set. Therefore the vector equation

$$k_1v_1 + k_2v_2 + \cdots + k_rv_r = 0$$

has a solution other than $k_1 = k_2 = \cdots = k_r = 0$. To be specific, assume $k_1 \neq 0$. Multiplying both sides by $1/k_1$ and solving for $v_1$ yields

$$v_1 = \left(-\frac{k_2}{k_1}\right)v_2 + \cdots + \left(-\frac{k_r}{k_1}\right)v_r$$

Thus $v_1$ has been expressed as a linear combination of the remaining vectors $v_2, \ldots, v_r$. We leave it as an exercise to show that *a set with two* or *more vectors is linearly dependent if and only if at least one of the vectors is a linear combination of the remaining vectors.*

The next two examples give a geometric interpretation of linear dependence in $R^2$ and $R^3$.

## Example 21

Two vectors, $v_1$ and $v_2$, form a linearly dependent set if and only if one of the vectors is a scalar multiple of the other. To see why, assume $S = \{v_1, v_2\}$ is linearly dependent. Since the vector equation $k_1v_1 + k_2v_2 = 0$ has a solution other than $k_1 = k_2 = 0$, this equation can be rewritten as

$$v_1 = \left(-\frac{k_2}{k_1}\right)v_2 \qquad \text{or} \qquad v_2 = \left(-\frac{k_1}{k_2}\right)v_1$$

But this tells us that $v_1$ is a scalar multiple of $v_2$ or $v_2$ is a scalar multiple of $v_1$. The converse is left as an exercise.

It follows that two vectors in $R^2$ or $R^3$ are linearly dependent if and only if they lie on the same line through the origin (Figure 4.4).

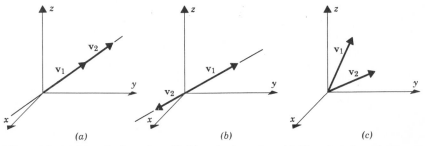

**Figure 4.4** (*a*) Linearly dependent. (*b*) Linearly dependent. (*c*) Linearly independent.

**Example 22**

If $v_1$, $v_2$ and $v_3$ are three vectors in $R^3$, then the set $S = \{v_1, v_2, v_3\}$ is linearly dependent if and only if the three vectors lie in the same plane through the origin (Figure 4.5). To see this in the case where no two of the vectors are collinear, recall that $v_1$, $v_2$, and $v_3$ are linearly dependent if and only if at least one of the vectors is a linear combination of the others. Specifically, assume $v_1$ is a linear combination of $v_2$ and $v_3$. Thus $v_1$ lies in the space spanned by $v_2$ and $v_3$, which is the plane they determine (Example 15). Therefore, $v_1$, $v_2$, and $v_3$ lie in the same plane. The complete details of the proof are left as an exercise.

We conclude this section with a theorem that shows that a linearly independent set in $R^n$ can contain at most $n$ vectors.

**Theorem 6.** *Let $S = \{v_1, v_2, \ldots, v_r\}$ be a set of vectors in $R^n$. If $r > n$, then $S$ is linearly dependent.*

*Proof.* Suppose

$$v_1 = (v_{11}, v_{12}, \ldots, v_{1n})$$
$$v_2 = (v_{21}, v_{22}, \ldots, v_{2n})$$
$$\vdots$$
$$v_r = (v_{r1}, v_{r2}, \ldots, v_{rn})$$

Consider the equation

$$k_1 v_1 + k_2 v_2 + \cdots + k_r v_r = 0$$

If, as illustrated in Example 20, we express both sides of this equation in terms of components and then equate corresponding components, we obtain the system

$$v_{11}k_1 + v_{21}k_2 + \cdots + v_{r1}k_r = 0$$
$$v_{12}k_1 + v_{22}k_2 + \cdots + v_{r2}k_r = 0$$
$$\vdots \qquad \vdots \qquad \qquad \vdots \qquad \vdots$$
$$v_{1n}k_1 + v_{2n}k_2 + \cdots + v_{rn}k_r = 0$$

This is a homogeneous system of $n$ equations in the $r$ unknowns $k_1, \ldots, k_r$. Since $r > n$, it follows from Theorem 1 in Section 1.3 that the system has nontrivial solutions. Therefore, $S = \{v_1, v_2, \ldots, v_r\}$ is a linearly dependent set. ∎

In particular, this theorem tells us that a set in $R^2$ with more than two vectors is linearly dependent, and a set in $R^3$ with more than three vectors is linearly dependent.

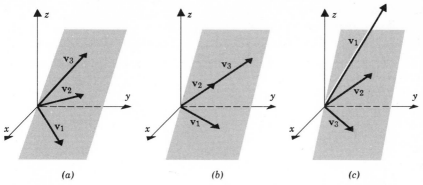

**Figure 4.5** (*a*) Linearly dependent. (*b*) Linearly dependent. (*c*) Linearly independent.

## EXERCISE SET 4.4

1. Explain why the following are linearly dependent sets of vectors. (Solve this problem by inspection.)
   (a) $\mathbf{u}_1 = (1, 2)$ and $\mathbf{u}_2 = (-3, -6)$ in $R^2$
   (b) $\mathbf{u}_1 = (2, 3)$, $\mathbf{u}_2 = (-5, 8)$, $\mathbf{u}_3 = (6, 1)$ in $R^2$
   (c) $\mathbf{p}_1 = 2 + 3x - x^2$ and $\mathbf{p}_2 = 6 + 9x - 3x^2$ in $P_2$
   (d) $A = \begin{bmatrix} 1 & 3 \\ 2 & 0 \end{bmatrix}$ and $B = \begin{bmatrix} -1 & -3 \\ -2 & 0 \end{bmatrix}$ in $M_{22}$

2. Which of the following sets of vectors in $R^3$ are linearly dependent?
   (a) $(2, -1, 4), (3, 6, 2), (2, 10, -4)$
   (b) $(3, 1, 1), (2, -1, 5), (4, 0, -3)$
   (c) $(6, 0, -1), (1, 1, 4)$
   (d) $(1, 3, 3), (0, 1, 4), (5, 6, 3), (7, 2, -1)$

3. Which of the following sets of vectors in $R^4$ are linearly dependent?
   (a) $(1, 2, 1, -2), (0, -2, -2, 0), (0, 2, 3, 1), (3, 0, -3, 6)$
   (b) $(4, -4, 8, 0), (2, 2, 4, 0), (6, 0, 0, 2), (6, 3, -3, 0)$
   (c) $(4, 4, 0, 0), (0, 0, 6, 6), (-5, 0, 5, 5)$
   (d) $(3, 0, 4, 1), (6, 2, -1, 2), (-1, 3, 5, 1), (-3, 7, 8, 3)$

4. Which of the following sets of vectors in $P_2$ are linearly dependent?
   (a) $2 - x + 4x^2, 3 + 6x + 2x^2, 2 + 10x - 4x^2$
   (b) $3 + x + x^2, 2 - x + 5x^2, 4 - 3x^2$
   (c) $6 - x^2, 1 + x + 4x^2$
   (d) $1 + 3x + 3x^2, x + 4x^2, 5 + 6x + 3x^2, 7 + 2x - x^2$

5. Let $V$ be the vector space of all real-valued functions defined on the entire real line. Which of the following sets of vectors in $V$ are linearly dependent?
   (a) $2, 4 \sin^2 x, \cos^2 x$      (b) $x, \cos x$
   (c) $1, \sin x, \sin 2x$      (d) $\cos 2x, \sin^2 x, \cos^2 x$
   (e) $(1 + x)^2, x^2 + 2x, 3$      (f) $0, x, x^2$

6. Assume that $v_1$, $v_2$, and $v_3$ are vectors in $R^3$ that have their initial points at the origin. In each part determine whether the three vectors lie in a plane.
   (a) $v_1 = (1, 0, -2)$, $v_2 = (3, 1, 2)$, $v_3 = (1, -1, 0)$
   (b) $v_1 = (2, -1, 4)$, $v_2 = (4, 2, 3)$, $v_3 = (2, 7, -6)$

7. Assume that $v_1$, $v_2$, and $v_3$ are vectors in $R^3$ that have their initial points at the origin. In each part determine whether the three vectors lie on the same line.
   (a) $v_1 = (3, -6, 9)$, $v_2 = (2, -4, 6)$, $v_3 = (1, 1, 1)$
   (b) $v_1 = (2, -1, 4)$, $v_2 = (4, 2, 3)$, $v_3 = (2, 7, -6)$
   (c) $v_1 = (4, 6, 8)$, $v_2 = (2, 3, 4)$, $v_3 = (-2, -3, -4)$

8. For which real values of $\lambda$ do the following vectors form a linearly dependent set in $R^3$?

$$v_1 = \left(\lambda, -\frac{1}{2}, -\frac{1}{2}\right), v_2 = \left(-\frac{1}{2}, \lambda, -\frac{1}{2}\right), v_3 = \left(-\frac{1}{2}, -\frac{1}{2}, \lambda\right)$$

9. Let $S = \{v_1, v_2, \ldots, v_n\}$ be a set of vectors in a vector space $V$. Show that if one of the vectors is zero, then $S$ is linearly dependent.

10. If $\{v_1, v_2, v_3\}$ is a linearly independent set of vectors, show that $\{v_1, v_2\}$, $\{v_1, v_3\}$, $\{v_2, v_3\}$, $\{v_1\}$, $\{v_2\}$, and $\{v_3\}$ are also linearly independent sets.

11. If $S = \{v_1, v_2, \ldots, v_n\}$ is a linearly independent set of vectors, show that every subset of $S$ with one or more vectors is also linearly independent.

12. If $\{v_1, v_2, v_3\}$ is a linearly dependent set of vectors in a vector space $V$, show that $\{v_1, v_2, v_3, v_4\}$ is also linearly dependent, where $v_4$ is any other vector in $V$.

13. If $\{v_1, v_2, \ldots, v_r\}$ is a linearly dependent set of vectors in a vector space $V$, show that $\{v_1, v_2, \ldots, v_{r+1}, \ldots, v_n\}$ is also linearly dependent, where $v_{r+1}, \ldots, v_n$ are any other vectors in $V$.

14. Show that every set with more than three vectors from $P_2$ is linearly dependent.

15. Show that if $\{v_1, v_2\}$ is linearly independent and $v_3$ does not lie in lin$\{v_1, v_2\}$, then $\{v_1, v_2, v_3\}$ is linearly independent.

16. Prove that a set with two or more vectors is linearly dependent if and only if at least one of the vectors is expressible as a linear combination of the remaining vectors.

17. Prove that three vectors in $R^3$ are linearly dependent if and only if they lie in a plane.

18. **(For readers who have studied calculus.)** Let $V$ be the vector space of real-valued functions defined on the entire real line. If $f$, $g$, and $h$ are vectors in $V$ that are twice differentiable, then the function $w = w(x)$ defined by

$$w(x) = \begin{vmatrix} f(x) & g(x) & h(x) \\ f'(x) & g'(x) & h'(x) \\ f''(x) & g''(x) & h''(x) \end{vmatrix}$$

is called the **Wronskian** of $f$, $g$, and $h$. Prove that $f$, $g$, and $h$ form a linearly independent set if the Wronskian is not the zero vector in $V$ [i.e., $w(x)$ is not identically zero].

19. **(For readers who have studied calculus.)** Use the Wronskian (Exercise 18) to show that the following sets of vectors are linearly independent.
    (a) $1, x, e^x$      (b) $\sin x, \cos x, x \sin x$
    (c) $e^x, xe^x, x^2 e^x$     (d) $1, x, x^2$

## 4.5 BASIS AND DIMENSION

We usually think of a line as being one dimensional, a plane as two dimensional, and space around us as three dimensional. It is the primary purpose of this section to make this intuitive notion of dimension precise.

**Definition.** If $V$ is any vector space and $S = \{\mathbf{v}_1, \mathbf{v}_2, \ldots, \mathbf{v}_r\}$ is a finite set of vectors in $V$, then $S$ is called a **basis** for $V$ if

(i) $S$ is linearly independent;
(ii) $S$ spans $V$

### Example 23

Let $\mathbf{e}_1 = (1, 0, 0, \ldots, 0)$, $\mathbf{e}_2 = (0, 1, 0, \ldots, 0), \ldots, \mathbf{e}_n = (0, 0, 0, \ldots, 1)$. In Example 19 we pointed out that $S = \{\mathbf{e}_1, \mathbf{e}_2, \ldots, \mathbf{e}_n\}$ is a linearly independent set in $R^n$. Since any vector $\mathbf{v} = (v_1, v_2, \ldots, v_n)$ in $R^n$ can be written as $\mathbf{v} = v_1 \mathbf{e}_1 + v_2 \mathbf{e}_2 + \cdots + v_n \mathbf{e}_n$, $S$ spans $R^n$ and is therefore a basis. It is called the **standard basis for $R^n$**.

### Example 24

Let $\mathbf{v}_1 = (1, 2, 1)$, $\mathbf{v}_2 = (2, 9, 0)$, and $\mathbf{v}_3 = (3, 3, 4)$. Show that the set $S = \{\mathbf{v}_1, \mathbf{v}_2, \mathbf{v}_3\}$ is a basis for $R^3$.

*Solution.* To show that $S$ spans $R^3$, we must show that an arbitrary vector $\mathbf{b} = (b_1, b_2, b_3)$ can be expressed as a linear combination

$$\mathbf{b} = k_1 \mathbf{v}_1 + k_2 \mathbf{v}_2 + k_3 \mathbf{v}_3 \tag{4.3}$$

of the vectors in $S$. Expressing (4.3) in terms of components gives

$$(b_1, b_2, b_3) = (k_1 + 2k_2 + 3k_3, 2k_1 + 9k_2 + 3k_3, k_1 + 4k_3)$$

or

$$\begin{aligned} k_1 + 2k_2 + 3k_3 &= b_1 \\ 2k_1 + 9k_2 + 3k_3 &= b_2 \\ k_1 \qquad\;\; + 4k_3 &= b_3 \end{aligned} \tag{4.4}$$

Thus, to show $S$ spans $V$, we must demonstrate that system (4.4) has a solution for all choices of $\mathbf{b} = (b_1, b_2, b_3)$. To prove that $S$ is linearly independent, we must show that the only solution of

$$k_1\mathbf{v}_1 + k_2\mathbf{v}_2 + k_3\mathbf{v}_3 = \mathbf{0} \qquad (4.5)$$

is $k_1 = k_2 = k_3 = 0$.

As above, if (4.5) is expressed in terms of components, the verification of independence reduces to showing the homogeneous system

$$\begin{aligned} k_1 + 2k_2 + 3k_3 &= 0 \\ 2k_1 + 9k_2 + 3k_3 &= 0 \\ k_1 \qquad\quad + 4k_3 &= 0 \end{aligned} \qquad (4.6)$$

has only the trivial solution. By parts (a), (b), and (d) of the corollary to Theorem 12 in Section 1.8, we can simultaneously prove $S$ is linearly independent and spans $R^3$ by demonstrating that the matrix of coefficients

$$A = \begin{bmatrix} 1 & 2 & 3 \\ 2 & 9 & 3 \\ 1 & 0 & 4 \end{bmatrix}$$

in systems (4.4) and (4.6) is invertible. Since

$$\det(A) = \det \begin{bmatrix} 1 & 2 & 3 \\ 2 & 9 & 3 \\ 1 & 0 & 4 \end{bmatrix} = -1$$

it follows from Theorem 6 in Section 2.3 that $A$ is invertible. Thus $S$ is linearly independent.

## Example 25

The set $S = \{1, x, x^2, \ldots, x^n\}$ is a basis for the vector space $P_n$ introduced in Example 11. It is evident that $S$ spans $P_n$, since a typical vector (polynomial) in $P_n$ has the form

$$\mathbf{p} = a_0 + a_1 x + \cdots + a_n x^n$$

which is a linear combination of the vectors in $S$. To see that $S$ is linearly independent, assume that some linear combination of vectors in $S$ is the zero vector, that is

$$c_0 + c_1 x + \cdots + c_n x^n \equiv 0 \qquad (4.7)$$

We must show that $c_0 = c_1 = \cdots = c_n = 0$. From algebra a nonzero polynomial of degree $n$ has at most $n$ distinct roots. Since (4.7) is an identity, every value of $x$ is a root of the left-hand side. This implies that $c_1 = c_2 = \cdots = c_n = 0$; for otherwise, $c_0 + c_1 x + \cdots + c_n x^n$ could have at most $n$ roots. The set $S$ is therefore linearly independent.

The basis $S$ in this example is called the **standard basis for $P_n$.**

**Example 26**

Let

$$M_1 = \begin{bmatrix} 1 & 0 \\ 0 & 0 \end{bmatrix} \quad M_2 = \begin{bmatrix} 0 & 1 \\ 0 & 0 \end{bmatrix} \quad M_3 = \begin{bmatrix} 0 & 0 \\ 1 & 0 \end{bmatrix} \quad M_4 = \begin{bmatrix} 0 & 0 \\ 0 & 1 \end{bmatrix}$$

The set $S = \{M_1, M_2, M_3, M_4\}$ is a basis for the vector space $M_{22}$ of $2 \times 2$ matrices. To see that $S$ spans $M_{22}$, note that a typical vector (matrix)

$$\begin{bmatrix} a & b \\ c & d \end{bmatrix}$$

can be written as

$$\begin{bmatrix} a & b \\ c & d \end{bmatrix} = a \begin{bmatrix} 1 & 0 \\ 0 & 0 \end{bmatrix} + b \begin{bmatrix} 0 & 1 \\ 0 & 0 \end{bmatrix} + c \begin{bmatrix} 0 & 0 \\ 1 & 0 \end{bmatrix} + d \begin{bmatrix} 0 & 0 \\ 0 & 1 \end{bmatrix}$$

$$= aM_1 + bM_2 + cM_3 + dM_4$$

To see that $S$ is linearly independent, assume that

$$aM_1 + bM_2 + cM_3 + dM_4 = 0$$

that is

$$a \begin{bmatrix} 1 & 0 \\ 0 & 0 \end{bmatrix} + b \begin{bmatrix} 0 & 1 \\ 0 & 0 \end{bmatrix} + c \begin{bmatrix} 0 & 0 \\ 1 & 0 \end{bmatrix} + d \begin{bmatrix} 0 & 0 \\ 0 & 1 \end{bmatrix} = \begin{bmatrix} 0 & 0 \\ 0 & 0 \end{bmatrix}$$

Then

$$\begin{bmatrix} a & b \\ c & d \end{bmatrix} = \begin{bmatrix} 0 & 0 \\ 0 & 0 \end{bmatrix}$$

Thus $a = b = c = d = 0$ so that $S$ is linearly independent.

**Example 27**

If $S = \{v_1, v_2, \ldots, v_r\}$ is a *linearly independent* set in a vector space $V$, then $S$ is a basis for the subspace $\text{lin}(S)$ since $S$ is independent and by definition of $\text{lin}(S)$, $S$ spans $\text{lin}(S)$.

A vector space $V$ is called **finite dimensional** if it contains a finite set of vectors $\{v_1, v_2, \ldots, v_n\}$ which forms a basis. If no such vectors exist, $V$ is called **infinite dimensional.** (In addition, we shall regard the zero vector space as finite dimensional even though it has no linearly independent sets, and consequently no basis.)

**Example 28**

By examples 23, 25, and 26, $R^n$, $P_n$ and $M_{22}$ are finite dimensional vector spaces.

The next theorem provides the key to the concept of dimension of a vector space. As a consequence, we shall obtain one of the most important results in linear algebra.

**Theorem 7.** *If $S = \{v_1, v_2, \ldots, v_n\}$ is a basis for a finite dimensional vector space $V$, then every set with more than $n$ vectors is linearly dependent.*

*Proof.* Let $S' = \{w_1, w_2, \ldots, w_m\}$ be any set of $m$ vectors in $V$, where $m > n$. We wish to show that $S'$ is linearly dependent. Since $S = \{v_1, v_2, \ldots, v_n\}$ is a basis, each $w_i$ can be expressed as a linear combination of the vectors in $S$, say,

$$\begin{aligned}
w_1 &= a_{11}v_1 + a_{21}v_2 + \cdots + a_{n1}v_n \\
w_2 &= a_{12}v_1 + a_{22}v_2 + \cdots + a_{n2}v_n \\
&\ \vdots \qquad\ \vdots \qquad\ \vdots \qquad\ \ \vdots \\
w_m &= a_{1m}v_1 + a_{2m}v_2 + \cdots + a_{nm}v_n
\end{aligned} \qquad (4.8)$$

To show that $S'$ is linearly dependent, we must find scalars $k_1, k_2, \ldots, k_m$, not all zero, such that

$$k_1 w_1 + k_2 w_2 + \cdots + k_m w_m = 0 \qquad (4.9)$$

Using the equations in (4.8), we can rewrite (4.9) as

$$\begin{aligned}
(k_1 a_{11} + k_2 a_{12} + \cdots + k_m a_{1m})v_1 & \\
+ (k_1 a_{21} + k_2 a_{22} + \cdots + k_m a_{2m})v_2 & \\
\ddots \qquad\qquad\qquad & \\
+ (k_1 a_{n1} + k_2 a_{n2} + \cdots + k_m a_{nm})v_n &= 0
\end{aligned}$$

Since the left-hand side of this equation is a linear combination of the vectors $v_1$, $v_2, \ldots, v_n$, it follows from the independence of $S$ that each of the coefficients is zero; that is,

$$\begin{aligned}
a_{11}k_1 + a_{12}k_2 + \cdots + a_{1m}k_m &= 0 \\
a_{21}k_1 + a_{22}k_2 + \cdots + a_{2m}k_m &= 0 \\
\vdots \qquad\ \vdots \qquad\quad\ \vdots \qquad\ \ \vdots \\
a_{n1}k_1 + a_{n2}k_2 + \cdots + a_{nm}k_m &= 0
\end{aligned} \qquad (4.10)$$

The problem of proving that $S'$ is a linearly dependent set thus reduces to showing there are $k_1, k_2, \ldots, k_m$, not all zero, that satisfy system (4.10). Since (4.10) has more unknowns than equations, the proof is complete, since Theorem 1 in Section 1.3 guarantees the existence of nontrivial solutions. ▊

As a consequence, we obtain the following result.

**Theorem 8.** *Any two bases for a finite dimensional vector space have the same number of vectors.*

*Proof.* Let $S = \{v_1, v_2, \ldots, v_n\}$ and $S' = \{w_1, w_2, \ldots, w_m\}$ be two bases for a finite dimensional vector space $V$. Since $S$ is a basis and $S'$ is a linearly independent set, Theorem 7 implies that $m \leq n$. Similarly, since $S'$ is a basis and $S$ is linearly independent, we also have $n \leq m$. Therefore $m = n$. ▨

### Example 29

The standard basis for $R^n$ contains $n$ vectors (Example 23). Therefore every basis for $R^n$ contains $n$ vectors.

### Example 30

The standard basis for $P_n$ (Example 25) contains $n + 1$ vectors; thus every basis for $P_n$ contains $n + 1$ vectors.

The number of vectors in a basis for a finite dimensional vector space is a particularly important quantity. By Example 29, every basis for $R^2$ has two vectors and every basis for $R^3$ has three vectors. Since $R^2$ (the plane) is intuitively two dimensional and $R^3$ is intuitively three dimensional, the dimension of these spaces is the same as the number of vectors that occur in their bases. This suggests the following definition.

**Definition.** The ***dimension*** of a finite dimensional vector space $V$ is defined to be the number of vectors in a basis for $V$. (For convenience, we define the zero vector space to have dimension zero.)

From Examples 29 and 30, $R^n$ is an $n$-dimensional vector space and $P_n$ is an $n + 1$-dimensional vector space.

### Example 31

Determine a basis and the dimension for the solution space of the homogeneous system

$$
\begin{aligned}
2x_1 + 2x_2 - x_3 \quad\quad + x_5 &= 0 \\
-x_1 - x_2 + 2x_3 - 3x_4 + x_5 &= 0 \\
x_1 + x_2 - 2x_3 \quad\quad - x_5 &= 0 \\
x_3 + x_4 + x_5 &= 0
\end{aligned}
$$

*Solution.* In Example 8 of Section 1.3 it was shown that the solutions are given by

$$x_1 = -s - t \quad\quad x_2 = s \quad\quad x_3 = -t \quad\quad x_4 = 0 \quad\quad x_5 = t$$

The solution vectors $(x_1, x_2, x_3, x_4, x_5)$ can therefore be written as

$$
\begin{aligned}
(x_1, x_2, x_3, x_4, x_5) &= (-s - t, s, -t, 0, t) \\
&= (-s, s, 0, 0, 0) + (-t, 0, -t, 0, t) \\
&= s(-1, 1, 0, 0, 0) + t(-1, 0, -1, 0, 1)
\end{aligned}
$$

Therefore the vectors

$$
\mathbf{v}_1 = (-1, 1, 0, 0, 0) \text{ and } \mathbf{v}_2 = (-1, 0, -1, 0, 1)
$$

span the solution space. Since they are also linearly independent, $\{\mathbf{v}_1, \mathbf{v}_2\}$ is a basis, and the solution space is two dimensional.

We conclude this section by discussing some additional useful results on basis and dimension. The proofs are left as exercises.

**Theorem 9**

(a) *If $S = \{\mathbf{v}_1, \mathbf{v}_2, \ldots, \mathbf{v}_n\}$ is a set of $n$ linearly independent vectors in an $n$-dimensional space $V$, then $S$ is a basis for $V$.*

(b) *If $S = \{\mathbf{v}_1, \mathbf{v}_2, \ldots, \mathbf{v}_n\}$ is a set of $n$ vectors that spans an $n$-dimensional space $V$, then $S$ is a basis for $V$.*

(c) *If $S = \{\mathbf{v}_1, \mathbf{v}_2, \ldots, \mathbf{v}_r\}$ is a linearly independent set in an $n$-dimensional space $V$ and $r < n$, then $S$ can be enlarged to a basis for $V$; that is, there are vectors $\mathbf{v}_{r+1}, \ldots, \mathbf{v}_n$ such that $\{\mathbf{v}_1, \mathbf{v}_2, \ldots, \mathbf{v}_r, \mathbf{v}_{r+1}, \ldots, \mathbf{v}_n\}$ is a basis for $V$.*

If the dimension $n$ of a finite dimensional vector space $V$ is known, then parts (a) and (b) simplify the problem of determining whether or not a given set $S$ of $n$ vectors in $V$ is a basis. To show $S$ is a basis, it is sufficient to show either $S$ spans $V$, *or* $S$ is linearly independent. Of course, if the dimension of $V$ is not known, then it is necessary to verify both that $S$ spans $V$ *and* $S$ is linearly independent.

The usefulness of part (c) will become evident later.

**Example 32**

Show that $\mathbf{v}_1 = (-3, 7)$ and $\mathbf{v}_2 = (5, 5)$ form a basis for $R^2$.

*Solution.* Since neither vector is a scalar multiple of the other, $S = \{\mathbf{v}_1, \mathbf{v}_2\}$ is linearly independent. Since $R^2$ is 2-dimensional, $S$ is a basis for $R^2$ by part (a) of Theorem 9.

**EXERCISE SET 4.5**

1. Explain why the following sets of vectors are *not* bases for the indicated vector spaces. (Solve this problem by inspection.)

(a) $\mathbf{u}_1 = (1, 2)$, $\mathbf{u}_2 = (0, 3)$, $\mathbf{u}_3 = (2, 7)$ for $R^2$

(b) $\mathbf{u}_1 = (-1, 3, 2)$, $\mathbf{u}_2 = (6, 1, 1)$ for $R^3$

(c) $\mathbf{p}_1 = 1 + x + x^2$, $\mathbf{p}_2 = x - 1$ for $P_2$

(d) $A = \begin{bmatrix} 1 & 1 \\ 2 & 3 \end{bmatrix}$ $B = \begin{bmatrix} 6 & 0 \\ -1 & 4 \end{bmatrix}$ $C = \begin{bmatrix} 3 & 0 \\ 1 & 7 \end{bmatrix}$

$D = \begin{bmatrix} 5 & 1 \\ 4 & 2 \end{bmatrix}$ $E = \begin{bmatrix} 7 & 1 \\ 2 & 9 \end{bmatrix}$ for $M_{22}$

2. Which of the following sets of vectors are bases for $R^2$?

(a) $(2, 1)$, $(3, 0)$          (b) $(4, 1)$, $(-7, -8)$

(c) $(0, 0)$, $(1, 3)$          (d) $(3, 9)$, $(-4, -12)$

3. Which of the following sets of vectors are bases for $R^3$?

(a) $(1, 0, 0)$, $(2, 2, 0)$, $(3, 3, 3)$      (b) $(3, 1, -4)$, $(2, 5, 6)$, $(1, 4, 8)$

(c) $(2, -3, 1)$, $(4, 1, 1)$, $(0, -7, 1)$      (d) $(1, 6, 4)$, $(2, 4, -1)$, $(-1, 2, 5)$

4. Which of the following sets of vectors are bases for $P_2$?

(a) $1 - 3x + 2x^2$, $1 + x + 4x^2$, $1 - 7x$

(b) $4 + 6x + x^2$, $-1 + 4x + 2x^2$, $5 + 2x - x^2$

(c) $1 + x + x^2$, $x + x^2$, $x^2$

(d) $-4 + x + .3x^2$, $6 + 5x + 2x^2$, $8 + 4x + x^2$

5. Show that the following set of vectors is a basis for $M_{22}$.

$$\begin{bmatrix} 3 & 6 \\ 3 & -6 \end{bmatrix}, \begin{bmatrix} 0 & -1 \\ -1 & 0 \end{bmatrix}, \begin{bmatrix} 0 & -8 \\ -12 & -4 \end{bmatrix}, \begin{bmatrix} 1 & 0 \\ -1 & 2 \end{bmatrix}$$

6. Let $V$ be the space spanned by $\mathbf{v}_1 = \cos^2 x$, $\mathbf{v}_2 = \sin^2 x$, $\mathbf{v}_3 = \cos 2x$.

(a) Show that $S = \{\mathbf{v}_1, \mathbf{v}_2, \mathbf{v}_3\}$ is not a basis for $V$.

(b) Find a basis for $V$.

In Exercises 7–12 determine the dimension of and a basis for the solution space of the system.

7. $\begin{aligned} 2x_1 + x_2 + 3x_3 &= 0 \\ x_1 + 2x_2 \quad\quad &= 0 \\ x_2 + x_3 &= 0 \end{aligned}$      8. $\begin{aligned} 3x_1 + x_2 + x_3 + x_4 &= 0 \\ 5x_1 - x_2 + x_3 - x_4 &= 0 \end{aligned}$

9. $\begin{aligned} 3x_1 + x_2 + 2x_3 &= 0 \\ 4x_1 \quad\quad + 5x_3 &= 0 \\ x_1 - 3x_2 + 4x_3 &= 0 \end{aligned}$      10. $\begin{aligned} x_1 - 3x_2 + x_3 &= 0 \\ 2x_1 - 6x_2 + 2x_3 &= 0 \\ 3x_1 - 9x_2 + 3x_3 &= 0 \end{aligned}$

11. $\begin{aligned} 2x_1 - 4x_2 + x_3 + x_4 &= 0 \\ x_1 - 5x_2 + 2x_3 \quad\quad &= 0 \\ -2x_2 - 2x_3 - x_4 &= 0 \\ x_1 + 3x_2 \quad\quad + x_4 &= 0 \\ x_1 - 2x_2 - x_3 + x_4 &= 0 \end{aligned}$      12. $\begin{aligned} x + y + z &= 0 \\ 3x + 2y - z &= 0 \\ 2x - 4y + z &= 0 \\ 4x + 8y - 3z &= 0 \\ 2x + y - 2z &= 0 \end{aligned}$

13. Determine bases for the following subspaces of $R^3$.

(a) The plane $3x - 2y + 5z = 0$

(b) The plane $x - y = 0$

(c) The line 
$$\begin{array}{l} x = 2t \\ y = -t \\ z = 4t \end{array} \qquad -\infty < t < +\infty$$

(d) All vectors of the form $(a, b, c)$, where $b = a + c$

**14.** Determine the dimensions of the following subspaces of $R^4$.
(a) All vectors of the form $(a, b, c, 0)$
(b) All vectors of the form $(a, b, c, d)$, where $d = a + b$ and $c = a - b$.
(c) All vectors of the form $(a, b, c, d)$, where $a = b = c = d$

**15.** Determine the dimension of the subspace of $P_3$ consisting of all polynomials $a_0 + a_1 x + a_2 x^2 + a_3 x^3$ for which $a_0 = 0$.

**16.** Let $\{v_1, v_2, v_3\}$ be a basis for a vector space $V$. Show that $\{u_1, u_2, u_3\}$ is also a basis, where $u_1 = v_1$, $u_2 = v_1 + v_2$, and $u_3 = v_1 + v_2 + v_3$.

**17.** Show that the vector space of all real-valued functions defined on the entire real line is infinite dimensional. (**Hint:** Assume it is finite dimensional with dimension $n$, and obtain a contradiction by producing $n + 1$ linearly independent vectors.)

**18.** Show that a subspace of a finite dimensional vector space is finite dimensional.

**19.** Let $V$ be a subspace of a finite dimensional vector space $W$. Show that $\dim(V) \leq \dim(W)$. (**Hint:** Use part (c) of Theorem 9.)

**20.** Show that the only subspaces of $R^3$ are lines through the origin, planes through the origin, the zero subspace, and $R^3$ itself. (**Hint:** By Exercise 19 the subspaces of $R^3$ must be 0-dimensional, 1-dimensional, 2-dimensional, or 3-dimensional.)

**21.** Prove part (a) of Theorem 9.

**22.** Prove part (b) of Theorem 9.

**23.** Prove part (c) of Theorem 9.

## 4.6 ROW SPACE OF A MATRIX; COORDINATES; APPLICATIONS TO FINDING BASES

In this section we shall study certain vector spaces associated with matrices. Our results will provide a simple procedure for finding bases by reducing an appropriate matrix to row-echelon form.

**Definition.** Consider the $m \times n$ matrix

$$A = \begin{bmatrix} a_{11} & a_{12} & \cdots & a_{1n} \\ a_{21} & a_{22} & \cdots & a_{2n} \\ \vdots & \vdots & & \vdots \\ a_{m1} & a_{m2} & \cdots & a_{mn} \end{bmatrix}$$

The vectors

$$
\begin{aligned}
\mathbf{r}_1 &= (a_{11}, a_{12}, \ldots, a_{1n}) \\
\mathbf{r}_2 &= (a_{21}, a_{22}, \ldots, a_{2n}) \\
&\;\;\vdots \qquad\qquad \vdots \\
\mathbf{r}_m &= (a_{m1}, a_{m2}, \ldots, a_{mn})
\end{aligned}
$$

formed from the rows of $A$, are called the **row vectors** of $A$. The subspace of $R^n$ spanned by the row vectors is called the **row space** of $A$.

**Example 33**

The row vectors of

$$
A = \begin{bmatrix} 2 & 1 & 0 \\ 3 & -1 & 4 \end{bmatrix}
$$

are $\mathbf{r}_1 = (2, 1, 0)$ and $\mathbf{r}_2 = (3, -1, 4)$.

The next theorem is useful for finding bases for vector spaces. We shall defer its proof to the end of the section.

**Theorem 10.** *Elementary row operations do not change the row space of a matrix.*

Suppose we are interested in finding a basis for the subspace of $R^n$ spanned by a given set of vectors $\{\mathbf{v}_1, \mathbf{v}_2, \ldots, \mathbf{v}_m\}$. If we form the matrix $B$ whose row vectors are $\mathbf{v}_1, \mathbf{v}_2, \ldots, \mathbf{v}_m$, then the problem reduces to finding a basis for the row space of $B$. For example, a basis for the space spanned by the vectors

$$
\mathbf{v}_1 = (1, -2, 0, 0, 3) \qquad \mathbf{v}_2 = (2, -5, -3, -2, 6) \qquad \mathbf{v}_3 = (0, 5, 15, 10, 0)
$$
$$
\mathbf{v}_4 = (2, 6, 18, 8, 6)
$$

can be obtained by finding a basis for the row space of the matrix

$$
B = \begin{bmatrix} 1 & -2 & 0 & 0 & 3 \\ 2 & -5 & -3 & -2 & 6 \\ 0 & 5 & 15 & 10 & 0 \\ 2 & 6 & 18 & 8 & 6 \end{bmatrix}
$$

The next example illustrates how such a basis can be found.

**Example 34**

It follows from Theorem 10 that the row space of a matrix is not changed by reducing the matrix to row-echelon form. The reader should have no trouble showing that

$$
R = \begin{bmatrix} 1 & -2 & 0 & 0 & 3 \\ 0 & 1 & 3 & 2 & 0 \\ 0 & 0 & 1 & 1 & 0 \\ 0 & 0 & 0 & 0 & 0 \end{bmatrix}
$$

is a row-echelon form of $B$. Since $B$ and $R$ have the same row space, a basis for the row space of $R$ will also be a basis for the row space of $B$. It is easy to check that the nonzero row vectors of $R$,

$$\mathbf{r}_1 = (1, -2, 0, 0, 3)$$
$$\mathbf{r}_2 = (0, 1, 3, 2, 0)$$
$$\mathbf{r}_3 = (0, 0, 1, 1, 0)$$

are linearly independent; thus, by Example 27, they form a basis for the row space of $R$, and consequently for the row space of $B$.

In this example the nonzero row vectors in the row-echelon form of $B$ form a basis for the row space of $B$. The next theorem shows that this is always the case. We omit the proof.

**Theorem 11.** *The nonzero row vectors in a row-echelon form of a matrix $A$ form a basis for the row space of $A$.*

We turn now to the problem of finding a basis for the space spanned by a set of vectors $S = \{\mathbf{v}_1, \mathbf{v}_2, \ldots, \mathbf{v}_r\}$ in an arbitrary finite dimensional vector space. Although the technique for finding bases illustrated in Example 34 applies only to vectors in $R^n$, we shall show that this method can be adapted to other spaces as well.

Suppose $S = \{\mathbf{v}_1, \mathbf{v}_2, \ldots, \mathbf{v}_r\}$ is a basis for a finite dimensional vector space $V$. Since $S$ spans $V$, every vector in $V$ is expressible as a linear combination of vectors in $S$. Moreover, the linear independence of $S$ assures that there is only *one* way to express a vector as a linear combination of vectors in $S$. To see why, suppose a vector $\mathbf{v}$ can be written as

$$\mathbf{v} = c_1\mathbf{v}_1 + c_2\mathbf{v}_2 + \cdots + c_n\mathbf{v}_n$$

and also

$$\mathbf{v} = k_1\mathbf{v}_1 + k_2\mathbf{v}_2 + \cdots + k_n\mathbf{v}_n$$

Subtracting the second equation from the first gives

$$\mathbf{0} = (c_1 - k_1)\mathbf{v}_1 + (c_2 - k_2)\mathbf{v}_2 + \cdots + (c_n - k_n)\mathbf{v}_n$$

Since the right side of this equation is a linear combination of vectors in $S$, the linear independence of $S$ implies that

$$c_1 - k_1 = 0, \; c_2 - k_2 = 0, \ldots, c_n - k_n = 0$$

that is,

$$c_1 = k_1, \; c_2 = k_2, \ldots, c_n = k_n$$

In summary, we have the following result.

**Theorem 12.** *If $S = \{v_1, v_2, \ldots, v_n\}$ is a basis for a vector space $V$, then every vector $v$ in $V$ can be expressed in the form $v = c_1 v_1 + c_2 v_2 + \cdots + c_n v_n$ in exactly one way.*

If $S = \{v_1, v_2, \ldots, v_n\}$ is a basis for a finite dimensional vector space $V$, and

$$v = c_1 v_1 + c_2 v_2 + \cdots + c_n v_n$$

is the expression for $v$ in terms of the basis $S$, then the scalars $c_1, c_2, \ldots, c_n$ are called the ***coordinates*** of $v$ relative to the basis $S$. The ***coordinate vector*** of $v$ relative to $S$ is denoted by $(v)_S$ and is the vector in $R^n$ defined by

$$(v)_S = (c_1, c_2, \ldots, c_n)$$

The ***coordinate matrix*** of $v$ relative to $S$ is denoted $[v]_S$ and is the $n \times 1$ matrix defined by

$$\begin{bmatrix} c_1 \\ c_2 \\ \vdots \\ c_n \end{bmatrix}$$

**Example 35**

In Example 24 we showed $S = \{v_1, v_2, v_3\}$ is a basis for $R^3$, where $v_1 = (1, 2, 1)$, $v_2 = (2, 9, 0)$, and $v_3 = (3, 3, 4)$.

    (a) Find the coordinate vector and coordinate matrix of $v = (5, -1, 9)$ with respect to $S$.

    (b) Find the vector $v$ in $R^3$ whose coordinate vector with respect to $S$ is $(v)_S = (-1, 3, 2)$.

*Solution (a).* We must find scalars $c_1, c_2, c_3$ such that

$$v = c_1 v_1 + c_2 v_2 + c_3 v_3$$

that is, such that

$$(5, -1, 9) = c_1(1, 2, 1) + c_2(2, 9, 0) + c_3(3, 3, 4)$$

Equating corresponding components gives

$$\begin{aligned} c_1 + 2c_2 + 3c_3 &= 5 \\ 2c_1 + 9c_2 + 3c_3 &= -1 \\ c_1 \qquad\quad + 4c_3 &= 9 \end{aligned}$$

Solving this system, we obtain $c_1 = 1$, $c_2 = -1$, $c_3 = 2$. Therefore

$$(v)_S = (1, -1, 2) \quad \text{and} \quad [v]_S = \begin{bmatrix} 1 \\ -1 \\ 2 \end{bmatrix}$$

*Solution* (*b*). Using the definition of the coordinate vector $(\mathbf{v})_S$, we obtain

$$\mathbf{v} = (-1)\mathbf{v}_1 + 3\mathbf{v}_2 + 2\mathbf{v}_3 = (11, 31, 7)$$

**Example 36**

Consider the basis $S = \{1, x, x^2\}$ for $P_2$. By inspection the coordinate vector and coordinate matrix with respect to $S$ for a polynomial $\mathbf{p} = a_0 + a_1 x + a_2 x^2$ are

$$(\mathbf{p})_S = (a_0, a_1, a_2) \qquad \text{and} \qquad [\mathbf{p}]_S = \begin{bmatrix} a_0 \\ a_1 \\ a_2 \end{bmatrix}$$

**Example 37**

Consider the basis $S = \{M_1, M_2, M_3, M_4\}$ given in Example 26 for the vector space of $2 \times 2$ matrices. If

$$M = \begin{bmatrix} a & b \\ c & d \end{bmatrix}$$

is any $2 \times 2$ matrix, then $M = aM_1 + bM_2 + cM_3 + dM_4$. Therefore its coordinate vector and matrix with respect to $S$ are

$$(M)_S = (a, b, c, d) \qquad \text{and} \qquad [M]_S = \begin{bmatrix} a \\ b \\ c \\ d \end{bmatrix}$$

**Example 38**

Consider the standard basis $S = \{\mathbf{e}_1, \mathbf{e}_2, \mathbf{e}_3\}$ for $R^3$; that is, $\mathbf{e}_1 = (1, 0, 0), \mathbf{e}_2 = (0, 1, 0),$ $\mathbf{e}_3 = (0, 0, 1)$. If $\mathbf{v} = (v_1, v_2, v_3)$ is any vector in $R^3$, then

$$\mathbf{v} = v_1\mathbf{e}_1 + v_2\mathbf{e}_2 + v_3\mathbf{e}_3$$

so that

$$(\mathbf{v})_S = (v_1, v_2, v_3) \qquad \text{and} \qquad [\mathbf{v}]_S = \begin{bmatrix} v_1 \\ v_2 \\ v_3 \end{bmatrix}$$

Observe that $\mathbf{v} = (\mathbf{v})_S$. In general, if $S$ is the *standard* basis for $R^n$, then a vector $\mathbf{v}$ is the same as its coordinate vector relative to $S$.

Let $\mathbf{u}$ and $\mathbf{v}$ be vectors in a finite dimensional vector space $V$, and let $k$ be a scalar. We shall now derive some relationships between the coordinate vectors of $\mathbf{u}, \mathbf{v}, \mathbf{u} + \mathbf{v}$, and $k\mathbf{u}$. If $S = \{\mathbf{w}_1, \mathbf{w}_2, \ldots, \mathbf{w}_n\}$ is a basis for $V$, then, by Theorem 12, $\mathbf{u}$ and $\mathbf{v}$ can be written uniquely as

$$\mathbf{u} = c_1\mathbf{w}_1 + c_2\mathbf{w}_2 + \cdots + c_n\mathbf{w}_n \text{ and } \mathbf{v} = d_1\mathbf{w}_1 + d_2\mathbf{w}_2 + \cdots + d_n\mathbf{w}_n \quad (4.11)$$

Therefore

$$\mathbf{u} + \mathbf{v} = (c_1 + d_1)\mathbf{w}_1 + (c_2 + d_2)\mathbf{w}_2 + \cdots + (c_n + d_n)\mathbf{w}_n$$

and

$$k\mathbf{u} = (kc_1)\mathbf{w}_1 + (kc_2)\mathbf{w}_2 + \cdots + (kc_n)\mathbf{w}_n$$

From these equations the coordinate vectors for $\mathbf{u} + \mathbf{v}$ and $k\mathbf{u}$ relative to the basis $S$ are

$$(\mathbf{u} + \mathbf{v})_S = (c_1 + d_1, c_2 + d_2, \ldots, c_n + d_n)$$
$$(k\mathbf{u})_S = (kc_1, kc_2, \ldots, kc_n)$$

From (4.11)

$$(\mathbf{u})_S = (c_1, c_2, \ldots, c_n)$$

and

$$(\mathbf{v})_S = (d_1, d_2, \ldots, d_n)$$

Therefore

$$(\mathbf{u} + \mathbf{v})_S = (\mathbf{u})_S + (\mathbf{v})_S \quad \text{and} \quad (k\mathbf{u})_S = k(\mathbf{u})_S \qquad (4.12)$$

Similarly, for coordinate matrices we have

$$[\mathbf{u} + \mathbf{v}]_S = [\mathbf{u}]_S + [\mathbf{v}]_S \quad \text{and} \quad [k\mathbf{u}]_S = k[\mathbf{u}]_S$$

As illustrated in the previous examples, it is a straightforward matter to compute the coordinate vector for a vector and to recover a vector if its coordinate vector is known. Thus knowing the coordinate vector $(\mathbf{v})_S$ for a vector $\mathbf{v}$ is as good as knowing $\mathbf{v}$ itself. Further, the formulas in (4.12) enable us to carry out computations with vectors from an arbitrary $n$-dimensional vector space by working in $R^n$ with the coordinate vectors. For example, as illustrated in Figure 4.6, the sum of two vectors $\mathbf{u}$ and $\mathbf{v}$ can be obtained *indirectly* by computing the coordinate vectors $(\mathbf{u})_S$ and $(\mathbf{v})_S$, adding these vectors together (a computation in $R^n$), and then recovering the vector $\mathbf{u} + \mathbf{v}$ from its coordinate vector $(\mathbf{u} + \mathbf{v})_S = (\mathbf{u})_S + (\mathbf{v})_S$.

This procedure is admittedly circuitous; it has the advantage, however, of transferring the computations to $R^n$, where it is often possible to take advantage of matrix methods. The following example will further illustrate this point.

**Figure 4.6**

### Example 39

Find a basis for the subspace of $P_4$ spanned by the set of vectors $\{\mathbf{p}_1, \mathbf{p}_2, \mathbf{p}_3, \mathbf{p}_4\}$, where

$$\mathbf{p}_1 = 1 - 2x + 3x^4 \qquad \mathbf{p}_2 = 2 - 5x - 3x^2 - 2x^3 + 6x^4$$
$$\mathbf{p}_3 = 5x + 15x^2 + 10x^3 \qquad \mathbf{p}_4 = 2 + 6x + 18x^2 + 8x^3 + 6x^4$$

*Solution.* Let $S = \{1, x, x^2, x^3, x^4\}$ be the standard basis for $P_4$. By inspection

$$(\mathbf{p}_1)_S = (1, -2, 0, 0, 3) \qquad (\mathbf{p}_2)_S = (2, -5, -3, -2, 6)$$
$$(\mathbf{p}_3)_S = (0, 5, 15, 10, 0) \qquad (\mathbf{p}_4)_S = (2, 6, 18, 8, 6)$$

A basis for the subspace of $R^4$ spanned by these coordinate vectors can be obtained by determining a basis for the rowspace of the matrix

$$\begin{bmatrix} 1 & -2 & 0 & 0 & 3 \\ 2 & -5 & -3 & -2 & 6 \\ 0 & 5 & 15 & 10 & 0 \\ 2 & 6 & 18 & 8 & 6 \end{bmatrix}$$

In Example 34 it was found that the vectors

$$\mathbf{r}_1 = (1, -2, 0, 0, 3) \qquad \mathbf{r}_2 = (0, 1, 3, 2, 0) \qquad \mathbf{r}_3 = (0, 0, 1, 1, 0)$$

form a basis for the rowspace. These vectors are the coordinate vectors with respect to $S$ for the basis vectors we are seeking. The desired basis vectors are therefore

$$\mathbf{q}_1 = 1 - 2x + 3x^4 \qquad \mathbf{q}_2 = x + 3x^2 + 2x^3 \qquad \text{and} \qquad \mathbf{q}_3 = x^2 + x^3$$

In this example, any basis for $P_4$ could have been used in place of the standard basis; the computation of the coordinate vectors would have been more involved, however.

### Optional

We conclude this section with a proof of Theorem 10.

*Proof.* Suppose that the row vectors of a matrix $A$ are $\mathbf{r}_1, \mathbf{r}_2, \ldots, \mathbf{r}_m$ and let $B$ be obtained from $A$ by performing an elementary row operation. We shall show that every vector in the row space of $B$ is also in the row space of $A$, and conversely that every vector in the row space of $A$ is in the row space of $B$. We can then conclude that $A$ and $B$ have the same row space.

Consider the possibilities. If the row operation is a row interchange, then $B$ and $A$ have the same row vectors, and consequently the same row space. If the row operation is multiplication of a row by a scalar or addition of a multiple of one row to another, then the row vectors $\mathbf{r}'_1, \mathbf{r}'_2, \ldots, \mathbf{r}'_m$ of $B$ are linear combinations of $\mathbf{r}_1, \mathbf{r}_2, \ldots, \mathbf{r}_m$; thus they lie in the row space of $A$. Since a vector space is closed under addition and scalar multiplication, all linear combinations of $\mathbf{r}'_1, \mathbf{r}'_2, \ldots, \mathbf{r}'_m$ will also lie in the row space of $A$. Therefore each vector in the row space of $B$ is in the row space of $A$.

Since $B$ is obtained from $A$ by performing a row operation, $A$ can be obtained from $B$ by performing the inverse operation (Section 1.7). The proof can now be completed by using an argument like the one above to show that the row space of $A$ is contained in the row space of $B$. ▮

## EXERCISE SET 4.6

**1.** List the row vectors of the matrix

$$\begin{bmatrix} 2 & -1 & 0 & 1 \\ 3 & 5 & 7 & -1 \\ 1 & 4 & 2 & 7 \end{bmatrix}$$

In Exercises 2–5 find a basis for the row space of the given matrix.

**2.** $\begin{bmatrix} 1 & -3 \\ 2 & -6 \end{bmatrix}$
**3.** $\begin{bmatrix} 1 & 2 & -1 \\ 2 & 4 & 6 \\ 0 & 0 & -8 \end{bmatrix}$
**4.** $\begin{bmatrix} -3 & 5 & 1 & 2 \\ 7 & 2 & 0 & -4 \\ -8 & 3 & 1 & 6 \end{bmatrix}$

**5.** $\begin{bmatrix} 1 & -3 & 2 & 2 & 1 \\ 0 & 3 & 6 & 0 & -2 \\ 2 & -3 & -2 & 4 & 4 \\ 3 & -3 & 6 & 6 & 3 \\ 5 & -3 & 10 & 10 & 5 \end{bmatrix}$

**6.** Find a basis for the subspace of $R^4$ spanned by the given vectors.
   (a) $(1, 1, -4, -3), (2, 0, 2, -2), (2, -1, 3, 2)$
   (b) $(-1, 1, -2, 0), (3, 3, 6, 0), (9, 0, 0, 3)$
   (c) $(1, 1, 0, 0), (0, 0, 1, 1), (-2, 0, 2, 2), (0, -3, 0, 3)$

**7.** Find a basis for the subspace of $P_2$ spanned by the given vectors.
   (a) $-1 + x - 2x^2, 3 + 3x + 6x^2, 9$
   (b) $1 + x, x^2, -2 + 2x^2, -3x$
   (c) $1 + x - 3x^2, 2 + 2x - 6x^2, 3 + 3x - 9x^2$

**8.** Find a basis for the subspace of $M_{22}$ spanned by the given vectors.

   (a) $\begin{bmatrix} -1 & -1 \\ 4 & 3 \end{bmatrix}, \begin{bmatrix} 2 & 0 \\ 2 & -2 \end{bmatrix}, \begin{bmatrix} 2 & -1 \\ 3 & 2 \end{bmatrix}$

   (b) $\begin{bmatrix} -1 & 1 \\ -2 & 0 \end{bmatrix}, \begin{bmatrix} 3 & -3 \\ 6 & 0 \end{bmatrix}, \begin{bmatrix} 9 & 0 \\ 0 & 3 \end{bmatrix}$

**9.** Find the coordinate vector and coordinate matrix for $\mathbf{v}$ relative to $S = \{\mathbf{v}_1, \mathbf{v}_2, \mathbf{v}_3\}$
   (a) $\mathbf{v} = (2, -1, 3), \mathbf{v}_1 = (1, 0, 0), \mathbf{v}_2 = (2, 2, 0), \mathbf{v}_3 = (3, 3, 3)$
   (b) $\mathbf{v} = (5, -12, 3), \mathbf{v}_1 = (1, 2, 3), \mathbf{v}_2 = (-4, 5, 6), \mathbf{v}_3 = (7, -8, 9)$

10. Find the coordinate vector and coordinate matrix for $\mathbf{p}$ relative to $S = \{\mathbf{p}_1, \mathbf{p}_2, \mathbf{p}_3\}$.
$\mathbf{p} = 2 - x + x^2$, $\mathbf{p}_1 = 1 + x$, $\mathbf{p}_2 = 1 + x^2$, $\mathbf{p}_3 = x + x^2$

11. Find the coordinate vector and coordinate matrix for $A$ relative to $S = \{A_1, A_2, A_3, A_4\}$.

$$A = \begin{bmatrix} 2 & 0 \\ -1 & 3 \end{bmatrix} \quad A_1 = \begin{bmatrix} -1 & 1 \\ 0 & 0 \end{bmatrix} \quad A_2 = \begin{bmatrix} 1 & 1 \\ 0 & 0 \end{bmatrix} \quad A_3 = \begin{bmatrix} 0 & 0 \\ 1 & 0 \end{bmatrix}$$

$$A_4 = \begin{bmatrix} 0 & 0 \\ 0 & 1 \end{bmatrix}$$

12. Show that the nonzero row vectors in any row-echelon form of a matrix $A$ form a basis for the row space of $A$.

13. Show that row vectors in an $n \times n$ invertible matrix $A$ form a basis for $R^n$.

14. Let $S$ be a basis for an $n$-dimensional vector space $V$. Show that if $\mathbf{v}_1, \mathbf{v}_2, \ldots, \mathbf{v}_r$ form a linearly independent set of vectors in $V$, then the coordinate vectors $(\mathbf{v}_1)_S, (\mathbf{v}_2)_S, \ldots, (\mathbf{v}_r)_S$ form a linearly independent set in $R^n$.

15. Using the notation from Exercise 14, show that if $\mathbf{v}_1, \mathbf{v}_2, \ldots, \mathbf{v}_r$ span $V$, then the coordinate vectors $(\mathbf{v}_1)_S, (\mathbf{v}_2)_S, \ldots, (\mathbf{v}_r)_S$ span $R^n$.

## 4.7 INNER PRODUCT SPACES

In Section 4.1 we studied the Euclidean inner product on the vector space $R^n$. In this section we introduce the notion of an inner product on an arbitrary vector space. As a consequence of our work, we will be able to meaningfully define notions of angle, length, and distance in more general vector spaces.

In Theorem 2 of Section 4.1, we collected the most important properties of the Euclidean inner product. In a general vector space, an inner product is defined axiomatically using these properties as a motivation.

**Definition.** An *inner product* on a vector space $V$ is a function that associates a real number $\langle \mathbf{u}, \mathbf{v} \rangle$ with each pair of vectors $\mathbf{u}$ and $\mathbf{v}$ in $V$ in such a way that the following axioms are satisfied for all vectors $\mathbf{u}$, $\mathbf{v}$, and $\mathbf{w}$ in $V$ and all scalars $k$.

(1) $\langle \mathbf{u}, \mathbf{v} \rangle = \langle \mathbf{v}, \mathbf{u} \rangle$      (symmetry axiom)
(2) $\langle \mathbf{u} + \mathbf{v}, \mathbf{w} \rangle = \langle \mathbf{u}, \mathbf{w} \rangle + \langle \mathbf{v}, \mathbf{w} \rangle$      (additivity axiom)
(3) $\langle k\mathbf{u}, \mathbf{v} \rangle = k\langle \mathbf{u}, \mathbf{v} \rangle$      (homogeneity axiom)
(4) $\langle \mathbf{v}, \mathbf{v} \rangle \geq 0$ and $\langle \mathbf{v}, \mathbf{v} \rangle = 0$      (positivity axiom)
     if and only if $\mathbf{v} = \mathbf{0}$

A vector space with an inner product is called an *inner product space.*

The following additional properties follow immediately from the four inner product axioms:

(i) $\langle \mathbf{0}, \mathbf{v} \rangle = \langle \mathbf{v}, \mathbf{0} \rangle = 0$

(ii) $\langle \mathbf{u}, \mathbf{v} + \mathbf{w} \rangle = \langle \mathbf{u}, \mathbf{v} \rangle + \langle \mathbf{u}, \mathbf{w} \rangle$

(iii) $\langle \mathbf{u}, k\mathbf{v} \rangle = k\langle \mathbf{u}, \mathbf{v} \rangle$

We prove (ii) and leave (i) and (iii) as exercises.

$$\begin{aligned}
\langle \mathbf{u}, \mathbf{v} + \mathbf{w} \rangle &= \langle \mathbf{v} + \mathbf{w}, \mathbf{u} \rangle &&\text{(by symmetry)} \\
&= \langle \mathbf{v}, \mathbf{u} \rangle + \langle \mathbf{w}, \mathbf{u} \rangle &&\text{(by additivity)} \\
&= \langle \mathbf{u}, \mathbf{v} \rangle + \langle \mathbf{u}, \mathbf{w} \rangle &&\text{(by symmetry)}
\end{aligned}$$

## Example 40

Let $\mathbf{u} = (u_1, u_2, \ldots, u_n)$ and $\mathbf{v} = (v_1, v_2, \ldots, v_n)$; the Euclidean inner product $\langle \mathbf{u}, \mathbf{v} \rangle = \mathbf{u} \cdot \mathbf{v} = u_1v_1 + u_2v_2 + \cdots + u_nv_n$ satisfies all the inner product axioms by Theorem 2.

## Example 41

If $\mathbf{u} = (u_1, u_2)$ and $\mathbf{v} = (v_1, v_2)$ are vectors in $R^2$, then

$$\langle \mathbf{u}, \mathbf{v} \rangle = 3u_1v_1 + 2u_2v_2$$

defines an inner product. To see this, note first that if $\mathbf{u}$ and $\mathbf{v}$ are interchanged in this equation, then the right side remains the same. Therefore

$$\langle \mathbf{u}, \mathbf{v} \rangle = \langle \mathbf{v}, \mathbf{u} \rangle$$

If $\mathbf{w} = (w_1, w_2)$, then

$$\begin{aligned}
\langle \mathbf{u} + \mathbf{v}, \mathbf{w} \rangle &= 3(u_1 + v_1)w_1 + 2(u_2 + v_2)w_2 \\
&= (3u_1w_1 + 2u_2w_2) + (3v_1w_1 + 2v_2w_2) \\
&= \langle \mathbf{u}, \mathbf{w} \rangle + \langle \mathbf{v}, \mathbf{w} \rangle
\end{aligned}$$

which establishes the second axiom.

Next

$$\begin{aligned}
\langle k\mathbf{u}, \mathbf{v} \rangle &= 3(ku_1)v_1 + 2(ku_2)v_2 \\
&= k(3u_1v_1 + 2u_2v_2) \\
&= k\langle \mathbf{u}, \mathbf{v} \rangle
\end{aligned}$$

which establishes the third axiom.

Finally

$$\langle \mathbf{v}, \mathbf{v} \rangle = 3v_1v_1 + 2v_2v_2 = 3v_1{}^2 + 2v_2{}^2$$

Obviously $\langle \mathbf{v}, \mathbf{v} \rangle = 3v_1{}^2 + 2v_2{}^2 \geq 0$. Further $\langle \mathbf{v}, \mathbf{v} \rangle = 3v_1{}^2 + 3v_2{}^2 = 0$ if and only if $v_1 = v_2 = 0$; that is, if and only if $\mathbf{v} = (v_1, v_2) = \mathbf{0}$. Thus the fourth axiom is satisfied.

The inner product in this example is different from the Euclidean inner product on $R^2$; this shows that a vector space can have more than one inner product.

Let **u** and **v** be vectors in a finite dimensional vector space $V$, and let

$$(\mathbf{u})_S = (u_1, u_2, \ldots, u_n)$$

and

$$(\mathbf{v})_S = (v_1, v_2, \ldots, v_n)$$

be the coordinate vectors for **u** and **v** with respect to a fixed basis $S$. We leave it as an exercise to show that

$$\langle \mathbf{u}, \mathbf{v} \rangle = (u)_S \cdot (v)_S = u_1 v_1 + u_2 v_2 + \cdots + u_n v_n$$

defines an inner product on $V$; it is called the **inner product on $V$ determined by the basis $S$.**

### Example 42

Let $S = \{M_1, M_2, M_3, M_4\}$ be the standard basis for $M_{22}$. If

$$U = \begin{bmatrix} u_1 & u_2 \\ u_3 & u_4 \end{bmatrix} \quad \text{and} \quad V = \begin{bmatrix} v_1 & v_2 \\ v_3 & v_4 \end{bmatrix}$$

are any two $2 \times 2$ matrices, then

$$(U)_S = (u_1, u_2, u_3, u_4) \quad \text{and} \quad (V)_S = (v_1, v_2, v_3, v_4)$$

(Example 37). The inner product determined by $S$ is therefore given by the formula

$$\langle U, V \rangle = (U)_S \cdot (V)_S = u_1 v_1 + u_2 v_2 + u_3 v_3 + u_4 v_4$$

For example, if

$$U = \begin{bmatrix} 1 & 2 \\ 3 & 4 \end{bmatrix} \quad \text{and} \quad V = \begin{bmatrix} -1 & 0 \\ 3 & 2 \end{bmatrix}$$

then

$$\langle U, V \rangle = 1(-1) + 2(0) + 3(3) + 4(2) = 16$$

### Example 43

Let $S$ be the standard basis for $P_2$. If

$$\mathbf{p} = a_0 + a_1 x + a_2 x^2 \quad \text{and} \quad \mathbf{q} = b_0 + b_1 x + b_2 x^2$$

are any two vectors in $P_2$, then

$$(\mathbf{p})_S = (a_0, a_1, a_2) \quad \text{and} \quad (\mathbf{q})_S = (b_0, b_1, b_2)$$

The inner product determined by $S$ is therefore given by the formula

$$\langle \mathbf{p}, \mathbf{q} \rangle = (\mathbf{p})_S \cdot (\mathbf{q})_S = a_0 b_0 + a_1 b_1 + a_2 b_2$$

**Example 44**

**(For readers who have studied calculus.)**

Let $\mathbf{p} = p(x)$ and $\mathbf{q} = q(x)$ be two polynomials in $P_n$, and define

$$\langle \mathbf{p}, \mathbf{q} \rangle = \int_a^b p(x)q(x)\, dx \tag{4.13}$$

where $a$ and $b$ are any fixed real numbers such that $a < b$. We shall show that (4.13) defines an inner product on $P_n$.

(1) $\langle \mathbf{p}, \mathbf{q} \rangle = \displaystyle\int_a^b p(x)q(x)\, dx = \int_a^b q(x)p(x)\, dx = \langle \mathbf{q}, \mathbf{p} \rangle$

which proves that Axiom (1) holds.

(2) $\langle \mathbf{p} + \mathbf{q}, \mathbf{s} \rangle = \displaystyle\int_a^b (p(x) + q(x))s(x)\, dx$

$= \displaystyle\int_a^b p(x)s(x)\, dx + \int_a^b q(x)s(x)\, dx$

$= \langle \mathbf{p}, \mathbf{s} \rangle + \langle \mathbf{q}, \mathbf{s} \rangle$

which proves that Axiom (2) holds.

(3) $\langle k\mathbf{p}, \mathbf{q} \rangle = \displaystyle\int_a^b kp(x)q(x)\, dx = k\int_a^b p(x)q(x)\, dx = k\langle \mathbf{p}, \mathbf{q} \rangle$

which proves that Axiom (3) holds.

(4) If $\mathbf{p} = p(x)$ is any polynomial in $P_n$ then $p^2(x) \geq 0$ for all $x$; therefore

$$\langle \mathbf{p}, \mathbf{p} \rangle = \int_a^b p^2(x)\, dx \geq 0$$

Further, since $p^2(x) \geq 0$, $\displaystyle\int_a^b p^2(x)\, dx = 0$ if and only if $p(x) = 0$ for all $x$ satisfying $a \leq x \leq b$. Therefore, $\langle \mathbf{p}, \mathbf{p} \rangle = \displaystyle\int_a^b p^2(x)\, dx = 0$ if and only if $\mathbf{p} = \mathbf{0}$. This establishes Axiom (4).

We note that the arguments given here can also be used to show that the vector space $C[a, b]$ discussed in Example 12 is an inner product space under the inner product

$$\langle \mathbf{f}, \mathbf{g} \rangle = \int_a^b f(x)g(x)\, dx$$

If $\mathbf{u}$ and $\mathbf{v}$ are nonzero vectors in $R^3$, then $\mathbf{u} \cdot \mathbf{v} = \|\mathbf{u}\|\, \|\mathbf{v}\| \cos \theta$, where $\theta$ is the angle between $\mathbf{u}$ and $\mathbf{v}$ (Section 3.3). If we square both sides of this inequality and use the relationships $\|\mathbf{u}\|^2 = \mathbf{u} \cdot \mathbf{u}$, $\|\mathbf{v}\|^2 = \mathbf{v} \cdot \mathbf{v}$, and $\cos^2 \theta \leq 1$, we obtain the inequality

$$(\mathbf{u} \cdot \mathbf{v})^2 \leq (\mathbf{u} \cdot \mathbf{u})(\mathbf{v} \cdot \mathbf{v})$$

The following theorem shows that this inequality can be generalized to any inner product space. The resulting inequality, called the *Cauchy-Schwarz inequality*, will enable us to introduce notions of length and angle in any inner product space.

**Theorem 13.** (*Cauchy-Schwarz\* Inequality*). *If* $\mathbf{u}$ *and* $\mathbf{v}$ *are vectors in an inner product space V, then*

$$\langle \mathbf{u}, \mathbf{v} \rangle^2 \leq \langle \mathbf{u}, \mathbf{u} \rangle \langle \mathbf{v}, \mathbf{v} \rangle$$

*Proof.* We warn the reader in advance that the proof presented here depends on a clever but highly unmotivated trick. If $\mathbf{u} = \mathbf{0}$, then $\langle \mathbf{u}, \mathbf{v} \rangle = \langle \mathbf{u}, \mathbf{u} \rangle = 0$, so that the inequality clearly holds. Assume now that $\mathbf{u} \neq \mathbf{0}$. Let $a = \langle \mathbf{u}, \mathbf{u} \rangle$, $b = 2\langle \mathbf{u}, \mathbf{v} \rangle$, $c = \langle \mathbf{v}, \mathbf{v} \rangle$, and let $t$ be any real number. By the positivity axiom, the inner product of any vector with itself is always non-negative. Therefore

$$0 \leq \langle (t\mathbf{u} + \mathbf{v}), (t\mathbf{u} + \mathbf{v}) \rangle = \langle \mathbf{u}, \mathbf{u} \rangle t^2 + 2\langle \mathbf{u}, \mathbf{v} \rangle t + \langle \mathbf{v}, \mathbf{v} \rangle$$
$$= at^2 + bt + c$$

This inequality implies that the quadratic polynomial $at^2 + bt + c$ has no real roots or a repeated real root. Therefore its discriminant must satisfy $b^2 - 4ac \leq 0$. Expressing $a$, $b$, and $c$ in terms of $\mathbf{u}$ and $\mathbf{v}$ gives $4\langle \mathbf{u}, \mathbf{v} \rangle^2 - 4\langle \mathbf{u}, \mathbf{u} \rangle \langle \mathbf{v}, \mathbf{v} \rangle \leq 0$; or equivalently, $\langle \mathbf{u}, \mathbf{v} \rangle^2 \leq \langle \mathbf{u}, \mathbf{u} \rangle \langle \mathbf{v}, \mathbf{v} \rangle$. ▮

**Example 45**

If $\mathbf{u} = (u_1, u_2, \ldots, u_n)$ and $\mathbf{v} = (v_1, v_2, \ldots, v_n)$ are any two vectors in $R^n$, then the Cauchy-Schwarz inequality applied to $\mathbf{u}$ and $\mathbf{v}$ yields

$$(u_1 v_1 + u_2 v_2 + \cdots + u_n v_n)^2 \leq (u_1{}^2 + u_2{}^2 + \cdots + u_n{}^2)(v_1{}^2 + v_2{}^2 + \cdots + v_n{}^2)$$

which is called *Cauchy's inequality*.

**EXERCISE SET 4.7**

1. Compute $\langle \mathbf{u}, \mathbf{v} \rangle$ using the inner product in Example 41.
   (a) $\mathbf{u} = (2, -1)$, $\mathbf{v} = (-1, 3)$    (b) $\mathbf{u} = (0, 0)$, $\mathbf{v} = (7, 2)$
   (c) $\mathbf{u} = (3, 1)$, $\mathbf{v} = (-2, 9)$    (d) $\mathbf{u} = (4, 6)$, $\mathbf{v} = (4, 6)$

---

*\*Augustin Louis (Baron de) Cauchy (1789–1857). Sometimes called the father of modern analysis, Cauchy helped put calculus on a firm mathematical footing. A partisan of the Bourbons, he spent several years in exile for his political involvements.*

*Hermann Amandus Schwarz (1843–1921). German mathematician.*

**2.** Repeat Exercise 1 using the Euclidean inner product on $R^2$.

**3.** Compute $\langle \mathbf{u}, \mathbf{v} \rangle$ using the inner product in Example 42.

(a) $\mathbf{u} = \begin{bmatrix} 2 & -1 \\ 3 & 7 \end{bmatrix} \qquad \mathbf{v} = \begin{bmatrix} 0 & 4 \\ 2 & 2 \end{bmatrix}$

(b) $\mathbf{u} = \begin{bmatrix} 1 & 2 \\ -3 & 5 \end{bmatrix} \qquad \mathbf{v} = \begin{bmatrix} 4 & 6 \\ 0 & 8 \end{bmatrix}$

**4.** Compute $\langle \mathbf{p}, \mathbf{q} \rangle$ using the inner product in Example 43.
   (a) $\mathbf{p} = -1 + 2x + x^2 \qquad \mathbf{q} = 2 - 4x^2$
   (b) $\mathbf{p} = -3 + 2x + x^2 \qquad \mathbf{q} = 2 + 4x - 2x^2$

**5.** Let $\mathbf{u} = (u_1, u_2)$ and $\mathbf{v} = (v_1, v_2)$. Show that the following are inner products on $R^2$.
   (a) $\langle \mathbf{u}, \mathbf{v} \rangle = 6u_1v_1 + 2u_2v_2$
   (b) $\langle \mathbf{u}, \mathbf{v} \rangle = 2u_1v_1 + u_2v_1 + u_1v_2 + 2u_2v_2$

**6.** Let $\mathbf{u} = (u_1, u_2, u_3)$ and $\mathbf{v} = (v_1, v_2, v_3)$. Determine which of the following are inner products on $R^3$. For those which are not, list the axioms which do not hold.
   (a) $\langle \mathbf{u}, \mathbf{v} \rangle = u_1v_1 + u_3v_3$
   (b) $\langle \mathbf{u}, \mathbf{v} \rangle = u_1^2v_1^2 + u_2^2v_2^2 + u_3^2v_3^2$
   (c) $\langle \mathbf{u}, \mathbf{v} \rangle = 2u_1v_1 + u_2v_2 + 4u_3v_3$
   (d) $\langle \mathbf{u}, \mathbf{v} \rangle = u_1v_1 - u_2v_2 + u_3v_3$

**7.** Let $U = \begin{bmatrix} u_1 & u_2 \\ u_3 & u_4 \end{bmatrix}$ and $V = \begin{bmatrix} v_1 & v_2 \\ v_3 & v_4 \end{bmatrix}$.

Show that $\langle U, V \rangle = u_1v_1 + u_2v_3 + u_3v_2 + u_4v_4$ is an inner product on $M_{22}$.

**8.** Let $\mathbf{p} = p(x)$ and $\mathbf{q} = q(x)$ be polynomials in $P_2$. Show that $\langle \mathbf{p}, \mathbf{q} \rangle = p(0)q(0) + p(1/2)q(1/2) + p(1)q(1)$ is an inner product on $P_2$.

**9.** Consider the inner product on $R^4$ determined by the basis $\{\mathbf{v}_1, \mathbf{v}_2, \mathbf{v}_3, \mathbf{v}_4\}$, where $\mathbf{v}_1 = (1, 0, 0, 0)$, $\mathbf{v}_2 = (1, 1, 0, 0)$, $\mathbf{v}_3 = (1, 1, 1, 0)$, and $\mathbf{v}_4 = (1, 1, 1, 1)$. Let $\mathbf{x} = (x_1, x_2, x_3, x_4)$ and $\mathbf{y} = (y_1, y_2, y_3, y_4)$. Find $\langle \mathbf{x}, \mathbf{y} \rangle$.

**10.** Verify the Cauchy-Schwarz inequality for
   (a) $\mathbf{u} = (2, 1, 5)$ and $\mathbf{v} = (1, -3, 4)$, using the inner product in Example 41
   (b) $\mathbf{u} = (2, 1, 5)$ and $\mathbf{v} = (1, -3, 4)$, using the Euclidean inner product

   (c) $U = \begin{bmatrix} -1 & 2 \\ 6 & 1 \end{bmatrix}$ and $V = \begin{bmatrix} 1 & 0 \\ 3 & 3 \end{bmatrix}$

   using the inner product in Example 42
   (d) $\mathbf{p} = -1 + 2x + x^2$ and $\mathbf{q} = 2 - 4x^2$, using the inner product in Example 43.

**11.** Let $c_1$, $c_2$, and $c_3$ be positive real numbers and let $\mathbf{u} = (u_1, u_2, u_3)$ and $\mathbf{v} = (v_1, v_2, v_3)$. Show that $\langle \mathbf{u}, \mathbf{v} \rangle = c_1u_1v_1 + c_2u_2v_2 + c_3u_3v_3$ is an inner product on $R^3$.

**12.** Let $c_1, c_2, \ldots, c_n$ be positive real numbers and let $\mathbf{u} = (u_1, u_2, \ldots, u_n)$ and $\mathbf{v} = (v_1, v_2, \ldots, v_n)$. Show that $\langle \mathbf{u}, \mathbf{v} \rangle = c_1u_1v_1 + c_2u_2v_2 + \cdots + c_nu_nv_n$ is an inner product on $R^n$.

13. Prove that if $\langle \mathbf{u}, \mathbf{v} \rangle$ is any inner product, then $\langle \mathbf{0}, \mathbf{v} \rangle = \langle \mathbf{v}, \mathbf{0} \rangle = 0$.

14. Prove that if $\langle \mathbf{u}, \mathbf{v} \rangle$ is any inner product and $k$ is any scalar, then $\langle \mathbf{u}, k\mathbf{v} \rangle = k\langle \mathbf{u}, \mathbf{v} \rangle$.

15. Show that equality holds in the Cauchy-Schwarz inequality if and only if $\mathbf{u}$ and $\mathbf{v}$ are linearly dependent.

16. Let $S$ be a basis for a finite dimensional vector space $V$. Prove that the inner product on $V$ determined by $S$ satisfies the four inner product space axioms.

17. **(For readers who have studied calculus.)** Use the inner product

$$\langle \mathbf{p}, \mathbf{q} \rangle = \int_{-1}^{1} p(x)q(x)\, dx$$

to compute $\langle \mathbf{p}, \mathbf{q} \rangle$ for the vectors $\mathbf{p} = p(x)$ and $\mathbf{q} = q(x)$ in $P_3$.
(a) $\mathbf{p} = 1 - x + x^2 + 5x^3$    $\mathbf{q} = x - 3x^2$
(b) $\mathbf{p} = x - 5x^3$    $\mathbf{q} = 2 + 8x^2$

18. **(For readers who have studied calculus.)** Use the inner product

$$\langle \mathbf{f}, \mathbf{g} \rangle = \int_{0}^{1} f(x)g(x)\, dx$$

to compute $\langle \mathbf{f}, \mathbf{g} \rangle$ for the vectors $\mathbf{f} = f(x)$ and $\mathbf{g} = g(x)$ in $C[0, 1]$.
(a) $\mathbf{f} = \cos 2\pi x$    $\mathbf{g} = \sin 2\pi x$
(b) $\mathbf{f} = x$    $\mathbf{g} = e^x$

(c) $\mathbf{f} = \tan \dfrac{\pi}{4} x$    $\mathbf{g} = 1$

# 4.8 LENGTH AND ANGLE IN INNER PRODUCT SPACES

In this section we use the Cauchy-Schwarz inequality to develop notions of length, distance, and angle in general inner product spaces.

**Definition.** If $V$ is an inner product space, then the **norm** (or length) of a vector $\mathbf{u}$ is denoted by $\|\mathbf{u}\|$ and defined by

$$\|\mathbf{u}\| = \langle \mathbf{u}, \mathbf{u} \rangle^{1/2}$$

Further, the **distance** between two points (vectors) $\mathbf{u}$ and $\mathbf{v}$ is denoted by $d(\mathbf{u}, \mathbf{v})$ and defined by

$$d(\mathbf{u}, \mathbf{v}) = \|\mathbf{u} - \mathbf{v}\|$$

**Example 46**

If $\mathbf{u} = (u_1, u_2, \ldots, u_n)$ and $\mathbf{v} = (v_1, v_2, \ldots, v_n)$ are vectors in $R^n$, then

$$\|\mathbf{u}\| = \langle \mathbf{u}, \mathbf{u} \rangle^{1/2} = \sqrt{u_1^2 + u_2^2 + \cdots + u_n^2}$$

and

$$d(\mathbf{u}, \mathbf{v}) = \|\mathbf{u} - \mathbf{v}\| = \langle \mathbf{u} - \mathbf{v}, \mathbf{u} - \mathbf{v} \rangle^{1/2}$$
$$= \sqrt{(u_1 - v_1)^2 + (u_2 - v_2)^2 + \cdots + (u_n - v_n)^2}$$

Observe that these are just the formulas for the Euclidean norm and distance discussed in Section 4.1.

### Example 47

Suppose $R^2$ has the inner product $\langle \mathbf{u}, \mathbf{v} \rangle = 3u_1v_1 + 2u_2v_2$ discussed in Example 41. If $\mathbf{u} = (1, 0)$ and $\mathbf{v} = (0, 1)$ then

$$\|\mathbf{u}\| = \langle \mathbf{u}, \mathbf{u} \rangle^{1/2} = (3(1)(1) + 2(0)(0))^{1/2} = \sqrt{3}$$

and

$$d(\mathbf{u}, \mathbf{v}) = \|\mathbf{u} - \mathbf{v}\| = \langle (1, -1), (1, -1) \rangle^{1/2}$$
$$= (3(1)(1) + 2(-1)(-1))^{1/2} = \sqrt{5}$$

It is important to keep in mind that norm and distance depend on the inner product being used. If the inner product is changed, then the norms and distances between vectors change. For example if $R^2$ has the Euclidean inner product, then the norm of the vector $\mathbf{u}$ in the preceding example is 1, and the distance between $\mathbf{u}$ and $\mathbf{v}$ is $\sqrt{2}$.

The reader might, at this point, object to our use of the terms length and distance for the quantities $\langle \mathbf{u}, \mathbf{u} \rangle^{1/2}$ and $\|\mathbf{u} - \mathbf{v}\|$. Although these defining formulas arise by imitating the formulas in $R^2$ and $R^3$, the odd results obtained in Example 47 cast some doubt on the wisdom of these definitions. It does, after all, require a stretch of the imagination to state that the length of the vector $\mathbf{u} = (1, 0)$ is $\sqrt{3}$ ! We shall now give some arguments in support of these definitions.

Through the years mathematicians have isolated what are considered to be the most important properties of Euclidean length and distance in $R^2$ and $R^3$. They are listed in Figure 4.7.

| *Basic*<br>*properties of length* | *Basic*<br>*properties of distance* |
|---|---|
| $L1.$ $\|\mathbf{u}\| \geq 0$<br>$L2.$ $\|\mathbf{u}\| = 0$ if and only if $\mathbf{u} = 0$<br>$L3.$ $\|k\mathbf{u}\| = |k| \, \|\mathbf{u}\|$<br>$L4.$ $\|\mathbf{u} + \mathbf{v}\| \leq \|\mathbf{u}\| + \|\mathbf{v}\|$<br>(triangle inequality) | $D1.$ $d(\mathbf{u}, \mathbf{v}) \geq 0$<br>$D2.$ $d(\mathbf{u}, \mathbf{v}) = 0$ if and only if $\mathbf{u} = \mathbf{v}$<br>$D3.$ $d(\mathbf{u}, \mathbf{v}) = d(\mathbf{v}, \mathbf{u})$<br>$D4.$ $d(\mathbf{u}, \mathbf{v}) \leq d(\mathbf{u}, \mathbf{w}) + d(\mathbf{w}, \mathbf{v})$<br>(triangle inequality) |

**Figure 4.7**

The next theorem justifies our definitions of norm and distance in inner product spaces.

**Theorem 14.** *If $V$ is an inner product space, then the norm $\|\mathbf{u}\| = \langle \mathbf{u}, \mathbf{u} \rangle^{1/2}$ and the distance $d(\mathbf{u}, \mathbf{v}) = \|\mathbf{u} - \mathbf{v}\|$ satisfy all the properties listed in Figure 4.7.*

We shall prove property *L4* and leave proof of the remaining parts as exercises. Before beginning the proof, we rewrite the Cauchy-Schwarz inequality

$$\langle \mathbf{u}, \mathbf{v} \rangle^2 \leq \langle \mathbf{u}, \mathbf{u} \rangle \langle \mathbf{v}, \mathbf{v} \rangle$$

in a different form. Since $\|\mathbf{u}\|^2 = \langle \mathbf{u}, \mathbf{u} \rangle$ and $\|\mathbf{v}\|^2 = \langle \mathbf{v}, \mathbf{v} \rangle$ it can be written

$$\langle \mathbf{u}, \mathbf{v} \rangle^2 \leq \|\mathbf{u}\|^2 \|\mathbf{v}\|^2 \tag{4.14}$$

or, upon taking square roots, as

$$|\langle \mathbf{u}, \mathbf{v} \rangle| \leq \|\mathbf{u}\| \|\mathbf{v}\| \tag{4.15}$$

*Proof of Property L4.* By definition

$$
\begin{aligned}
\|\mathbf{u} + \mathbf{v}\|^2 &= \langle \mathbf{u} + \mathbf{v}, \mathbf{u} + \mathbf{v} \rangle \\
&= \langle \mathbf{u}, \mathbf{u} \rangle + 2\langle \mathbf{u}, \mathbf{v} \rangle + \langle \mathbf{v}, \mathbf{v} \rangle \\
&\leq \langle \mathbf{u}, \mathbf{u} \rangle + 2|\langle \mathbf{u}, \mathbf{v} \rangle| + \langle \mathbf{v}, \mathbf{v} \rangle \\
&\leq \langle \mathbf{u}, \mathbf{u} \rangle + 2\|\mathbf{u}\| \|\mathbf{v}\| + \langle \mathbf{v}, \mathbf{v} \rangle \qquad \text{(by 4.15)} \\
&= \|\mathbf{u}\|^2 + 2\|\mathbf{u}\| \|\mathbf{v}\| + \|\mathbf{v}\|^2 \\
&= (\|\mathbf{u}\| + \|\mathbf{v}\|)^2.
\end{aligned}
$$

Taking square roots gives

$$\|\mathbf{u} + \mathbf{v}\| \leq \|\mathbf{u}\| + \|\mathbf{v}\|. \ \blacksquare$$

In $R^2$ or $R^3$ the result just proved states the familiar geometric fact that the sum of the lengths of two sides of a triangle is at least as large as the length of the third side (Figure 4.8).

Suppose that $\mathbf{u}$ and $\mathbf{v}$ are nonzero vectors in an inner product space $V$. The Cauchy-Schwarz inequality as given in (4.14) can be rewritten as

$$\left( \frac{\langle \mathbf{u}, \mathbf{v} \rangle}{\|\mathbf{u}\| \|\mathbf{v}\|} \right)^2 \leq 1$$

or equivalently

$$-1 \leq \frac{\langle \mathbf{u}, \mathbf{v} \rangle}{\|\mathbf{u}\| \|\mathbf{v}\|} \leq 1$$

As a consequence of this result, there is a unique angle $\theta$ such that

$$\cos \theta = \frac{\langle \mathbf{u}, \mathbf{v} \rangle}{\|\mathbf{u}\| \|\mathbf{v}\|} \qquad \text{and} \qquad 0 \leq \theta \leq \pi \tag{4.16}$$

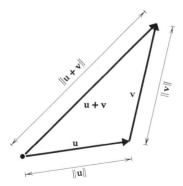

**Figure 4.8**

We define $\theta$ to be the **angle between** the vectors **u** and **v**. Observe that in $R^2$ or $R^3$ with the Euclidean inner product, (4.16) agrees with the usual formula for the cosine of the angle between two nonzero vectors (Section 3.3 of Chapter 3).

**Example 48**

Find the cosine of the angle $\theta$ between the vectors

$$\mathbf{u} = (4, 3, 1, -2) \quad \text{and} \quad \mathbf{v} = (-2, 1, 2, 3)$$

where the vector space is $R^4$ with the Euclidean inner product.

*Solution.*

$$\|\mathbf{u}\| = \sqrt{30} \quad \|\mathbf{v}\| = \sqrt{18} \quad \text{and} \quad \langle \mathbf{u}, \mathbf{v} \rangle = -9$$

so that

$$\cos \theta = -\frac{9}{\sqrt{30}\sqrt{18}} = -\frac{3}{\sqrt{30}\sqrt{2}}$$

**Example 49**

If $M_{22}$ has the inner product determined by the standard basis (Example 42), then the angle between the matrices

$$U = \begin{bmatrix} 1 & 0 \\ 1 & 1 \end{bmatrix} \quad \text{and} \quad V = \begin{bmatrix} 0 & 2 \\ 0 & 0 \end{bmatrix}$$

is $\pi/2$ since

$$\cos \theta = \frac{U \cdot V}{\|U\| \|V\|} = \frac{1(0) + 0(2) + 0(1) + 1(0)}{\|U\| \|V\|} = 0$$

If **u** and **v** are nonzero vectors such that $\langle \mathbf{u}, \mathbf{v} \rangle = 0$, then from (4.16) it follows that $\cos \theta = 0$ and $\theta = \pi/2$. This suggests the following terminology.

Figure 4.9

**Definition.** In an inner product space, two vectors **u** and **v** are called ***orthogonal*** if $\langle \mathbf{u}, \mathbf{v} \rangle = 0$. Further, if **u** is orthogonal to each vector in a set $W$, we say that **u** is ***orthogonal to W.***

We emphasize that orthogonality depends on the selection of the inner product. Two vectors can be orthogonal with respect to one inner product but not another. We conclude this section with an interesting and useful generalization of a familiar result.

**Theorem 15.** (*Generalized Theorem of Pythagoras*). *If* **u** *and* **v** *are orthogonal vectors in an inner product space, then*

$$\|\mathbf{u} + \mathbf{v}\|^2 = \|\mathbf{u}\|^2 + \|\mathbf{v}\|^2.$$

*Proof.*
$$\|\mathbf{u} + \mathbf{v}\|^2 = \langle (\mathbf{u} + \mathbf{v}), (\mathbf{u} + \mathbf{v}) \rangle = \|\mathbf{u}\|^2 + 2\langle \mathbf{u}, \mathbf{v} \rangle + \|\mathbf{v}\|^2$$
$$= \|\mathbf{u}\|^2 + \|\mathbf{v}\|^2. \ \blacksquare$$

Note that in $R^2$ or $R^3$ with the Euclidian inner product this theorem reduces to the ordinary Pythagorean theorem (Figure 4.9).

## EXERCISE SET 4.8

1. Let $R^2$ have the inner product $\langle \mathbf{u}, \mathbf{v} \rangle = 3u_1v_1 + 2u_2v_2$, where $\mathbf{u} = (u_1, u_2)$ and $\mathbf{v} = (v_1, v_2)$. Find $\|\mathbf{w}\|$ when
   (a) $\mathbf{w} = (-1, 3)$   (b) $\mathbf{w} = (6, 7)$
   (c) $\mathbf{w} = (0, 1)$   (d) $\mathbf{w} = (0, 0)$

2. Repeat Exercise 1 using the Euclidean inner product on $R^2$.

3. Let $P_2$ have the inner product in Example 43. Find $\|\mathbf{p}\|$ when
   (a) $\mathbf{p} = -1 + 2x + x^2$   (b) $\mathbf{p} = 3 - 4x^2$

4. Let $M_{22}$ have the inner product in Example 42. Find $\|A\|$ when

   (a) $A = \begin{bmatrix} -1 & 7 \\ 6 & 2 \end{bmatrix}$   (b) $A = \begin{bmatrix} 0 & 0 \\ 0 & 0 \end{bmatrix}$

5. Let $R^2$ have the inner product in Exercise 1. Find $d(\mathbf{x}, \mathbf{y})$ when
   (a) $\mathbf{x} = (-1, 2), \mathbf{y} = (2, 5)$    (b) $\mathbf{x} = (3, 9), \mathbf{y} = (3, 9)$

6. Repeat Exercise 5 using the Euclidean inner product on $R^2$.

7. Let $P_2$ have the inner product in Example 43. Find $d(\mathbf{p}, \mathbf{q})$ when

$$\mathbf{p} = 2 - x + x^2, \mathbf{q} = 1 + 5x^2$$

8. Let $M_{22}$ have the inner product in Example 42. Find $d(A, B)$ when

   (a) $A = \begin{bmatrix} 1 & 5 \\ 8 & 3 \end{bmatrix}$    $B = \begin{bmatrix} -5 & 0 \\ 7 & -3 \end{bmatrix}$

   (b) $A = \begin{bmatrix} 6 & 3 \\ 2 & 1 \end{bmatrix}$    $B = \begin{bmatrix} 6 & 3 \\ 2 & 1 \end{bmatrix}$

9. Let $R^2$, $R^3$, and $R^4$ have the Euclidean inner product. In each part find the cosine of the angle between $\mathbf{u}$ and $\mathbf{v}$.
   (a) $\mathbf{u} = (1, -3), \mathbf{v} = (2, 4)$          (b) $\mathbf{u} = (-1, 0), \mathbf{v} = (3, 8)$
   (c) $\mathbf{u} = (-1, 5, 2), \mathbf{v} = (2, 4, -9)$      (d) $\mathbf{u} = (4, 1, 8), \mathbf{v} = (1, 0, -3)$
   (e) $\mathbf{u} = (1, 0, 1, 0), \mathbf{v} = (-3, -3, -3, -3)$
   (f) $\mathbf{u} = (2, 1, 7, -1), \mathbf{v} = (4, 0, 0, 0)$

10. Let $P_2$ have the inner product in Example 43. Find the cosine of the angle between $\mathbf{p}$ and $\mathbf{q}$.
    (a) $\mathbf{p} = -1 + 5x + 2x^2$    $\mathbf{q} = 2 + 4x - 9x^2$
    (b) $\mathbf{p} = x - x^2$          $\mathbf{q} = 7 + 3x + 3x^2$

11. Let $M_{22}$ have the inner product in Example 42. Find the cosine of the angle between $A$ and $B$

    (a) $A = \begin{bmatrix} 2 & 6 \\ 1 & -3 \end{bmatrix}$    $B = \begin{bmatrix} 3 & 2 \\ 1 & 0 \end{bmatrix}$

    (b) $A = \begin{bmatrix} 2 & 4 \\ -1 & 3 \end{bmatrix}$    $B = \begin{bmatrix} -3 & 1 \\ 4 & 2 \end{bmatrix}$

12. Let $R^3$ have the Euclidean inner product. For which values of $k$ are $\mathbf{u}$ and $\mathbf{v}$ orthogonal?
    (a) $\mathbf{u} = (2, 1, 3)$    $\mathbf{v} = (1, 7, k)$
    (b) $\mathbf{u} = (k, k, 1)$    $\mathbf{v} = (k, 5, 6)$

13. Let $P_2$ have the inner product in Example 43. Show that $\mathbf{p} = 1 - x + 2x^2$ and $\mathbf{q} = 2x + x^2$ are orthogonal.

14. Let $M_{22}$ have the inner product in Example 42. Which of the following are orthogonal to

$$A = \begin{bmatrix} 2 & 1 \\ -1 & 3 \end{bmatrix}?$$

    (a) $\begin{bmatrix} -3 & 0 \\ 0 & 2 \end{bmatrix}$  (b) $\begin{bmatrix} 1 & 1 \\ 0 & -1 \end{bmatrix}$  (c) $\begin{bmatrix} 0 & 0 \\ 0 & 0 \end{bmatrix}$  (d) $\begin{bmatrix} 2 & 1 \\ 5 & 2 \end{bmatrix}$

15. Let $R^4$ have the Euclidean inner product. Find two vectors of norm 1 orthogonal to all of the vectors $\mathbf{u} = (2, 1, -4, 0), \mathbf{v} = (-1, -1, 2, 2)$, and $\mathbf{w} = (3, 2, 5, 4)$.

**16.** Let $V$ be an inner product space. Show that if $\mathbf{w}$ is orthogonal to both $\mathbf{u}_1$ and $\mathbf{u}_2$, it is orthogonal to $k_1\mathbf{u}_1 + k_2\mathbf{u}_2$ for all scalars $k_1$ and $k_2$. Interpret this result geometrically in $R^3$ with the Euclidean inner product.

**17.** Let $V$ be an inner product space. Show that if $\mathbf{w}$ is orthogonal to each of the vectors $\mathbf{u}_1, \mathbf{u}_2, \ldots, \mathbf{u}_r$, then it is orthogonal to every vector in lin $\{\mathbf{u}_1, \mathbf{u}_2, \ldots, \mathbf{u}_r\}$.

**18.** Let $V$ be an inner product space. Show that if $\mathbf{u}$ and $\mathbf{v}$ are orthogonal vectors in $V$ such that $\|\mathbf{u}\| = \|\mathbf{v}\| = 1$, then $\|\mathbf{u} - \mathbf{v}\| = \sqrt{2}$.

**19.** Let $V$ be an inner product space. Establish the identity

$$\|\mathbf{u} + \mathbf{v}\|^2 + \|\mathbf{u} - \mathbf{v}\|^2 = 2\|\mathbf{u}\|^2 + 2\|\mathbf{v}\|^2$$

for vectors in $V$.

**20.** Let $V$ be an inner product space. Establish the identity

$$\langle \mathbf{u}, \mathbf{v} \rangle = \frac{1}{4}\|\mathbf{u} + \mathbf{v}\|^2 - \frac{1}{4}\|\mathbf{u} - \mathbf{v}\|^2$$

for vectors in $V$.

**21.** Let $\{\mathbf{v}_1, \mathbf{v}_2, \ldots, \mathbf{v}_r\}$ be a basis for an inner product space $V$. Show that the zero vector is the only vector in $V$ that is orthogonal to all of the basis vectors.

**22.** Let $\mathbf{v}$ be a vector in an inner product space $V$.
(a) Show that the set of all vectors in $V$ orthogonal to $\mathbf{v}$ forms a subspace of $V$.
(b) Describe the subspace geometrically in $R^2$ and $R^3$ with the Euclidian inner product.

**23.** Prove the following generalization of Theorem 15. If $\mathbf{v}_1, \mathbf{v}_2, \ldots, \mathbf{v}_r$ are pairwise orthogonal vectors in an inner product space $V$, then

$$\|\mathbf{v}_1 + \mathbf{v}_2 + \cdots + \mathbf{v}_r\|^2 = \|\mathbf{v}_1\|^2 + \|\mathbf{v}_2\|^2 + \cdots + \|\mathbf{v}_r\|^2$$

**24.** Prove the following parts of Theorem 14.
(a) part $L1$    (b) part $L2$    (c) part $L3$
(d) part $D1$    (e) part $D2$    (f) part $D3$
(g) part $D4$

**25. (For readers who have studied calculus.)** Let $C[0, \pi]$ have the inner product

$$\langle \mathbf{f}, \mathbf{g} \rangle = \int_0^\pi f(x)g(x)\,dx$$

and let $\mathbf{f}_n = \cos nx\ (n = 0, 1, 2, \ldots)$. Show that if $k \neq l$, then $\mathbf{f}_k$ and $\mathbf{f}_l$ are orthogonal with respect to the given inner product.

## 4.9 ORTHONORMAL BASES; GRAM-SCHMIDT PROCESS

In many problems concerned with vector spaces, the selection of a basis for the space is at the discretion of the problem solver. Naturally, the best strategy is to choose the basis to simplify the solution of the problem at hand. In inner product spaces, it is often the case that the best choice is a basis in which all the vectors are orthogonal to one another. In this section we show how such bases can be constructed.

**Definition.** A set of vectors in an inner product space is called ***orthonormal*** if each vector in the set has norm 1 and any two distinct vectors in the set are orthogonal.

**Example 50**

Let

$$\mathbf{v}_1 = (0, 1, 0), \ \mathbf{v}_2 = \left(\frac{1}{\sqrt{2}}, 0, \frac{1}{\sqrt{2}}\right), \ \mathbf{v}_3 = \left(\frac{1}{\sqrt{2}}, 0, -\frac{1}{\sqrt{2}}\right).$$

The set $S = \{\mathbf{v}_1, \mathbf{v}_2, \mathbf{v}_3\}$ is orthonormal if $R^3$ has the Euclidean inner product, since

$$\langle \mathbf{v}_1, \mathbf{v}_2 \rangle = \langle \mathbf{v}_2, \mathbf{v}_3 \rangle = \langle \mathbf{v}_1, \mathbf{v}_3 \rangle = 0$$

and

$$\|\mathbf{v}_1\| = \|\mathbf{v}_2\| = \|\mathbf{v}_3\| = 1$$

**Example 51**

If $\mathbf{v}$ is a nonzero vector in an inner product space, then by property $L3$ in Theorem 14, the vector

$$\frac{1}{\|\mathbf{v}\|} \mathbf{v}$$

has norm 1, since

$$\left\|\frac{1}{\|\mathbf{v}\|} \mathbf{v}\right\| = \frac{1}{\|\mathbf{v}\|} \|\mathbf{v}\| = 1$$

This process of multiplying a nonzero vector $\mathbf{v}$ by the reciprocal of its length to obtain a vector of norm 1 is called ***normalizing*** $\mathbf{v}$.

The interest in finding orthonormal bases for inner product spaces is, in part, motivated by the next theorem that shows that the computation of coordinate vectors is exceptionally simple when the basis is orthonormal.

**Theorem 16.** *If $S = \{\mathbf{v}_1, \mathbf{v}_2, \ldots, \mathbf{v}_n\}$ is an orthonormal basis for an inner product space V, and $\mathbf{u}$ is any vector in V, then*

$$\mathbf{u} = \langle \mathbf{u}, \mathbf{v}_1 \rangle \mathbf{v}_1 + \langle \mathbf{u}, \mathbf{v}_2 \rangle \mathbf{v}_2 + \cdots + \langle \mathbf{u}, \mathbf{v}_n \rangle \mathbf{v}_n$$

*that is, the coordinate vector of $\mathbf{u}$ with respect to S is*

$$(\mathbf{u})_S = (\langle \mathbf{u}, \mathbf{v}_1 \rangle, \langle \mathbf{u}, \mathbf{v}_2 \rangle, \ldots, \langle \mathbf{u}, \mathbf{v}_n \rangle)$$

*Proof.* Since $S = \{\mathbf{v}_1, \mathbf{v}_2, \ldots, \mathbf{v}_n\}$ is a basis, a vector $\mathbf{u}$ can be expressed uniquely in the form

$$\mathbf{u} = k_1 \mathbf{v}_1 + k_2 \mathbf{v}_2 + \cdots + k_n \mathbf{v}_n$$

We shall complete the proof by showing $k_i = \langle \mathbf{u}, \mathbf{v}_i \rangle$ for $i = 1, 2, \ldots, n$. For each vector $\mathbf{v}_i$ in $S$ we have

$$\langle \mathbf{u}, \mathbf{v}_i \rangle = \langle k_1 \mathbf{v}_1 + k_2 \mathbf{v}_2 + \cdots + k_n \mathbf{v}_n, \mathbf{v}_i \rangle$$
$$= k_1 \langle \mathbf{v}_1, \mathbf{v}_i \rangle + k_2 \langle \mathbf{v}_2, \mathbf{v}_i \rangle + \cdots + k_n \langle \mathbf{v}_n, \mathbf{v}_i \rangle$$

Since $S = \{ \mathbf{v}_1, \mathbf{v}_2, \ldots, \mathbf{v}_n \}$ is an orthonormal set, we have

$$\langle \mathbf{v}_i, \mathbf{v}_i \rangle = \| \mathbf{v}_i \|^2 = 1 \qquad \text{and} \qquad \langle \mathbf{v}_i, \mathbf{v}_j \rangle = 0 \text{ if } j \neq i$$

Therefore the above equation simplifies to

$$\langle \mathbf{u}, \mathbf{v}_i \rangle = k_i \quad \blacksquare$$

**Example 52**

Let

$$\mathbf{v}_1 = (0, 1, 0), \; \mathbf{v}_2 = \left( -\frac{4}{5}, 0, \frac{3}{5} \right), \; \mathbf{v}_3 = \left( \frac{3}{5}, 0, \frac{4}{5} \right)$$

It is easy to check that $S = \{ \mathbf{v}_1, \mathbf{v}_2, \mathbf{v}_3 \}$ is an orthonormal basis for $R^3$ with the Euclidean inner product. Find the coordinate vector for $\mathbf{u} = (1, 1, 1)$ with respect to $S$.

*Solution.*

$$\langle \mathbf{u}, \mathbf{v}_1 \rangle = 1 \qquad \langle \mathbf{u}, \mathbf{v}_2 \rangle = -\frac{1}{5} \qquad \text{and} \qquad \langle \mathbf{u}, \mathbf{v}_3 \rangle = \frac{7}{5}$$

Therefore by Theorem 16

$$(\mathbf{u})_S = \left( 1, -\frac{1}{5}, \frac{7}{5} \right)$$

The usefulness of Theorem 16 should be evident from this example if it is kept in mind that for nonorthonormal bases, the computation of a coordinate vector generally involves solving a system of linear equations (Example 35).

**Theorem 17.** *If $S = \{ \mathbf{v}_1, \mathbf{v}_2, \ldots, \mathbf{v}_n \}$ is an orthogonal set of nonzero vectors in an inner product space, then $S$ is linearly independent.*

*Proof.* Assume

$$k_1 \mathbf{v}_1 + k_2 \mathbf{v}_2 + \cdots + k_n \mathbf{v}_n = \mathbf{0} \tag{4.17}$$

To demonstrate that $S = \{ \mathbf{v}_1, \mathbf{v}_2, \ldots, \mathbf{v}_n \}$ is linearly independent, we must prove that $k_1 = k_2 = \cdots = k_n = 0$.

For each $\mathbf{v}_i$ in $S$, it follows from (4.17) that

$$\langle k_1\mathbf{v}_1 + k_2\mathbf{v}_2 + \cdots + k_n\mathbf{v}_n, \mathbf{v}_i \rangle = \langle \mathbf{0}, \mathbf{v}_i \rangle = 0$$

or equivalently

$$k_1\langle \mathbf{v}_1, \mathbf{v}_i \rangle + k_2\langle \mathbf{v}_2, \mathbf{v}_i \rangle + \cdots + k_n\langle \mathbf{v}_n, \mathbf{v}_i \rangle = 0$$

From the orthogonality of $S$, $\langle \mathbf{v}_j, \mathbf{v}_i \rangle = 0$ when $j \neq i$, so that this equation reduces to

$$k_i\langle \mathbf{v}_i, \mathbf{v}_i \rangle = 0$$

Since the vectors in $S$ are assumed to be nonzero, $\langle \mathbf{v}_i, \mathbf{v}_i \rangle \neq 0$ by the positivity axiom for inner products. Therefore $k_i = 0$. Since the subscript $i$ is arbitrary, we have $k_1 = k_2 = \cdots = k_n = 0$; thus $S$ is linearly independent. ▮

**Example 53**

In Example 50 we showed that

$$\mathbf{v}_1 = (0, 1, 0), \, \mathbf{v}_2 = \left(\frac{1}{\sqrt{2}}, 0, \frac{1}{\sqrt{2}}\right), \text{ and } \mathbf{v}_3 = \left(\frac{1}{\sqrt{2}}, 0, -\frac{1}{\sqrt{2}}\right)$$

form an orthonormal set with respect to the Euclidean inner product on $R^3$. By Theorem 17, these vectors form a linearly independent set. Therefore, since $R^3$ is three dimensional, $S = \{\mathbf{v}_1, \mathbf{v}_2, \mathbf{v}_3\}$ is an orthonormal basis for $R^3$.

We now turn to the problem of constructing orthonormal bases for inner product spaces. The proof of the following preliminary result is discussed in the exercises at the end of this section.

**Theorem 18.** *Let $V$ be an inner product space and $\{\mathbf{v}_1, \mathbf{v}_2, \ldots, \mathbf{v}_r\}$ an orthonormal set of vectors in $V$. If $W$ denotes the space spanned by $\mathbf{v}_1, \mathbf{v}_2, \ldots, \mathbf{v}_r$, then every vector $\mathbf{u}$ in $V$ can be expressed uniquely in the form*

$$\mathbf{u} = \mathbf{w}_1 + \mathbf{w}_2$$

*where $\mathbf{w}_1$ is in $W$ and $\mathbf{w}_2$ is orthogonal to $W$; moreover, the vectors $\mathbf{w}_1$ and $\mathbf{w}_2$ are given by*

$$\mathbf{w}_1 = \langle \mathbf{u}, \mathbf{v}_1 \rangle \mathbf{v}_1 + \langle \mathbf{u}, \mathbf{v}_2 \rangle \mathbf{v}_2 + \cdots + \langle \mathbf{u}, \mathbf{v}_r \rangle \mathbf{v}_r \tag{4.18}$$

*and*

$$\mathbf{w}_2 = \mathbf{u} - \langle \mathbf{u}, \mathbf{v}_1 \rangle \mathbf{v}_1 - \langle \mathbf{u}, \mathbf{v}_2 \rangle \mathbf{v}_2 - \cdots - \langle \mathbf{u}, \mathbf{v}_r \rangle \mathbf{v}_r \tag{4.19}$$

*(See Figure 4.10 for an illustration in $R^3$.)*

Motivated by Figure 4.10, we call $\mathbf{w}_1$ the ***orthogonal projection*** *of* $\mathbf{u}$ on $W$ and denote it by $\text{proj}_W \mathbf{u}$. The vector $\mathbf{w}_2 = \mathbf{u} - \text{proj}_W \mathbf{u}$ is called the ***component of*** $\mathbf{u}$ ***orthogonal to*** $\mathbf{W}$.

**Figure 4.10**

**Example 54**

Let $R^3$ have the Euclidean inner product, and let $W$ be the subspace spanned by the orthonormal vectors $\mathbf{v}_1 = (0, 1, 0)$ and $\mathbf{v}_2 = (-(4/5), 0, 3/5)$. The orthogonal projection of $\mathbf{u} = (1, 1, 1)$ on $W$ is

$$\text{proj}_W \mathbf{u} = \langle \mathbf{u}, \mathbf{v}_1 \rangle \mathbf{v}_1 + \langle \mathbf{u}, \mathbf{v}_2 \rangle \mathbf{v}_2$$

$$= (1)(0, 1, 0) + \left(-\frac{1}{5}\right)\left(-\frac{4}{5}, 0, \frac{3}{5}\right)$$

$$= \left(\frac{4}{25}, 1, -\frac{3}{25}\right)$$

The component of $\mathbf{u}$ orthogonal to $W$ is

$$\mathbf{u} - \text{proj}_W \mathbf{u} = (1, 1, 1) - \left(\frac{4}{25}, 1, -\frac{3}{25}\right) = \left(\frac{21}{25}, 0, \frac{28}{25}\right)$$

Observe that $\mathbf{u} - \text{proj}_W\mathbf{u}$ is orthogonal to both $\mathbf{v}_1$ and $\mathbf{v}_2$ so that this vector is orthogonal to each vector in the space $W$ spanned by $\mathbf{v}_1$ and $\mathbf{v}_2$ as it should be.

We are now in a position to prove the main result of this section.

**Theorem 19.** *Every nonzero finite dimensional inner product space has an orthonormal basis.*

*Proof.* Let $V$ be any nonzero, $n$-dimensional inner product space, and let $S = \{\mathbf{u}_1, \mathbf{u}_2, \ldots, \mathbf{u}_n\}$ be any basis for $V$. The following sequence of steps will produce an orthonormal basis $\{\mathbf{v}_1, \mathbf{v}_2, \ldots, \mathbf{v}_n\}$ for $V$.

**Step 1.** Let $\mathbf{v}_1 = \mathbf{u}_1 / \|\mathbf{u}_1\|$. The vector $\mathbf{v}_1$ has norm 1.

**Step 2.** To construct a vector $\mathbf{v}_2$ of norm 1 that is orthogonal to $\mathbf{v}_1$, we compute the component of $\mathbf{u}_2$ orthogonal to the space $W_1$ spanned by $\mathbf{v}_1$ and then normalize it; that is,

$$\mathbf{v}_2 = \frac{\mathbf{u}_2 - \text{proj}_{W_1}\mathbf{u}_2}{\|\mathbf{u}_2 - \text{proj}_{W_1}\mathbf{u}_2\|} = \frac{\mathbf{u}_2 - \langle \mathbf{u}_2, \mathbf{v}_1 \rangle \mathbf{v}_1}{\|\mathbf{u}_2 - \langle \mathbf{u}_2, \mathbf{v}_1 \rangle \mathbf{v}_1\|}$$

(Figure 4.11).

Of course, if $\mathbf{u}_2 - \langle \mathbf{u}_2, \mathbf{v}_1 \rangle \mathbf{v}_1 = \mathbf{0}$, then we cannot carry out the normalization. But this cannot happen since we would then have

$$\mathbf{u}_2 = \langle \mathbf{u}_2, \mathbf{v}_1 \rangle \mathbf{v}_1 = \frac{\langle \mathbf{u}_2, \mathbf{v}_1 \rangle}{\|\mathbf{u}_1\|} \mathbf{u}_1$$

which says that $\mathbf{u}_2$ is a multiple of $\mathbf{u}_1$, contradicting the linear independence of the basis $S = \{\mathbf{u}_1, \mathbf{u}_2, \ldots, \mathbf{u}_n\}$.

**Step 3.** To construct a vector $\mathbf{v}_3$ of norm 1 that is orthogonal to both $\mathbf{v}_1$ and $\mathbf{v}_2$, we compute the component of $\mathbf{u}_3$ orthogonal to the space $W_2$ spanned by $\mathbf{v}_1$ and $\mathbf{v}_2$ and normalize it (Figure 4.12); that is

$$\mathbf{v}_3 = \frac{\mathbf{u}_3 - \mathrm{proj}_{W_2} \mathbf{u}_3}{\|\mathbf{u}_3 - \mathrm{proj}_{W_2} \mathbf{u}_3\|} = \frac{\mathbf{u}_3 - \langle \mathbf{u}_3, \mathbf{v}_1 \rangle \mathbf{v}_1 - \langle \mathbf{u}_3, \mathbf{v}_2 \rangle \mathbf{v}_2}{\|\mathbf{u}_3 - \langle \mathbf{u}_3, \mathbf{v}_1 \rangle \mathbf{v}_1 - \langle \mathbf{u}_3, \mathbf{v}_2 \rangle \mathbf{v}_2\|}$$

As in Step 2, the linear independence of $\{\mathbf{u}_1, \mathbf{u}_2, \ldots, \mathbf{u}_n\}$ assures that $\mathbf{u}_3 - \langle \mathbf{u}_3, \mathbf{v}_1 \rangle \mathbf{v}_1 - \langle \mathbf{u}_3, \mathbf{v}_2 \rangle \mathbf{v}_2 \neq \mathbf{0}$ so that the normalization can always be carried out. We leave the details as an exercise.

**Step 4.** To determine a vector $\mathbf{v}_4$ of norm 1 that is orthogonal to $\mathbf{v}_1$, $\mathbf{v}_2$, and $\mathbf{v}_3$, we compute the component of $\mathbf{u}_4$ orthogonal to the space $W_3$ spanned by $\mathbf{v}_1$, $\mathbf{v}_2$ and

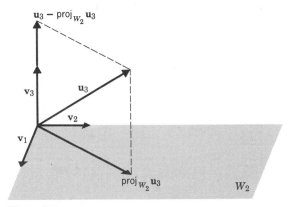

**Figure 4.12**

$v_3$ and normalize it. Thus

$$v_4 = \frac{u_4 - \text{proj}_{W_4} u_4}{\|u_4 - \text{proj}_{W_4} u_4\|} = \frac{u_4 - \langle u_4, v_1 \rangle v_1 - \langle u_4, v_2 \rangle v_2 - \langle u_4, v_3 \rangle v_3}{\|u_4 - \langle u_4, v_1 \rangle v_1 - \langle u_4, v_2 \rangle v_2 - \langle u_4, v_3 \rangle v_3\|}$$

Continuing in this way, we will obtain an orthonormal set of vectors, $\{v_1, v_2, \ldots, v_n\}$. Since $V$ is $n$-dimensional and every orthonormal set is linearly independent, $S' = \{v_1, v_2, \ldots, v_n\}$ will be an orthonormal basis for $V$. ∎

The above step by step construction for converting an arbitrary basis into an orthonormal basis is called the **Gram-Schmidt\* process.**

## Example 55

Consider the vector space $R^3$ with the Euclidean inner product. Apply the Gram-Schmidt process to transform the basis $u_1 = (1, 1, 1)$, $u_2 = (0, 1, 1)$, $u_3 = (0, 0, 1)$ into an orthonormal basis.

*Solution.*

**Step 1.**
$$v_1 = \frac{u_1}{\|u_1\|} = \frac{(1, 1, 1)}{\sqrt{3}} = \left( \frac{1}{\sqrt{3}}, \frac{1}{\sqrt{3}}, \frac{1}{\sqrt{3}} \right)$$

**Step 2.**
$$u_2 - \text{proj}_{W_1} u_2 = u_2 - \langle u_2, v_1 \rangle v_1$$

$$= (0, 1, 1) - \frac{2}{\sqrt{3}} \left( \frac{1}{\sqrt{3}}, \frac{1}{\sqrt{3}}, \frac{1}{\sqrt{3}} \right)$$

$$= \left( -\frac{2}{3}, \frac{1}{3}, \frac{1}{3} \right)$$

Therefore

$$v_2 = \frac{u_2 - \text{proj}_{W_1} u_2}{\|u_2 - \text{proj}_{W_1} u_2\|} = \frac{3}{\sqrt{6}} \left( -\frac{2}{3}, \frac{1}{3}, \frac{1}{3} \right) = \left( -\frac{2}{\sqrt{6}}, \frac{1}{\sqrt{6}}, \frac{1}{\sqrt{6}} \right)$$

**Step 3.** $\quad u_3 - \text{proj}_{W_2} u_3 = u_3 - \langle u_3, v_1 \rangle v_1 - \langle u_3, v_2 \rangle v_2$

$$= (0, 0, 1) - \frac{1}{\sqrt{3}} \left( \frac{1}{\sqrt{3}}, \frac{1}{\sqrt{3}}, \frac{1}{\sqrt{3}} \right) - \frac{1}{\sqrt{6}} \left( -\frac{2}{\sqrt{6}}, \frac{1}{\sqrt{6}}, \frac{1}{\sqrt{6}} \right)$$

$$= \left( 0, -\frac{1}{2}, \frac{1}{2} \right)$$

---

\**Jörgen Pederson Gram* (1850-1916). Danish actuary.

*Erhardt Schmidt* (1876-1959). German mathematician.

Therefore

$$\mathbf{v}_3 = \frac{\mathbf{u}_3 - \text{proj}_{W_1}\mathbf{u}_3}{\|\mathbf{u}_3 - \text{proj}_{W_2}\mathbf{u}_3\|} = \sqrt{2}\left(0, -\frac{1}{2}, \frac{1}{2}\right) = \left(0, -\frac{1}{\sqrt{2}}, \frac{1}{\sqrt{2}}\right)$$

Thus

$$\mathbf{v}_1 = \left(\frac{1}{\sqrt{3}}, \frac{1}{\sqrt{3}}, \frac{1}{\sqrt{3}}\right), \mathbf{v}_2 = \left(-\frac{2}{\sqrt{6}}, \frac{1}{\sqrt{6}}, \frac{1}{\sqrt{6}}\right), \mathbf{v}_3 = \left(0, -\frac{1}{\sqrt{2}}, \frac{1}{\sqrt{2}}\right)$$

form an orthonormal basis for $R^3$.

## EXERCISE SET 4.9

1. Let $R^2$ have the Euclidean inner product. Which of the following form orthonormal sets?

(a) $(1, 0), (0, 2)$

(b) $\left(\frac{1}{\sqrt{2}}, -\frac{1}{\sqrt{2}}\right), \left(\frac{1}{\sqrt{2}}, \frac{1}{\sqrt{2}}\right)$

(c) $\left(\frac{1}{\sqrt{2}}, \frac{1}{\sqrt{2}}\right), \left(-\frac{1}{\sqrt{2}}, -\frac{1}{\sqrt{2}}\right)$

(d) $(1, 0), (0, 0)$

2. Let $R^3$ have the Euclidean inner product. Which of the following form orthonormal sets?

(a) $\left(\frac{1}{\sqrt{2}}, 0, \frac{1}{\sqrt{2}}\right), \left(\frac{1}{\sqrt{3}}, \frac{1}{\sqrt{3}}, -\frac{1}{\sqrt{3}}\right), \left(-\frac{1}{\sqrt{2}}, 0, \frac{1}{\sqrt{2}}\right)$

(b) $\left(\frac{2}{3}, -\frac{2}{3}, \frac{1}{3}\right), \left(\frac{2}{3}, \frac{1}{3}, -\frac{2}{3}\right), \left(\frac{1}{3}, \frac{2}{3}, \frac{2}{3}\right)$

(c) $(1, 0, 0), \left(0, \frac{1}{\sqrt{2}}, \frac{1}{\sqrt{2}}\right), (0, 0, 1)$

(d) $\left(\frac{1}{\sqrt{6}}, \frac{1}{\sqrt{6}}, -\frac{2}{\sqrt{6}}\right), \left(\frac{1}{\sqrt{2}}, -\frac{1}{\sqrt{2}}, 0\right)$

3. Let $P_2$ have the inner product determined by the standard basis. Which of the following form orthonormal sets?

(a) $\frac{2}{3} - \frac{2}{3}x + \frac{1}{3}x^2, \frac{2}{3} + \frac{1}{3}x - \frac{2}{3}x^2, \frac{1}{3} + \frac{2}{3}x + \frac{2}{3}x^2$

(b) $1, \frac{1}{\sqrt{2}}x + \frac{1}{\sqrt{2}}x^2, x^2$

4. Let $M_{22}$ have the inner product in Example 42. Which of the following form orthonormal sets?

(a) $\begin{bmatrix} 1 & 0 \\ 0 & 0 \end{bmatrix}$ $\begin{bmatrix} 0 & \frac{2}{3} \\ \frac{1}{3} & -\frac{2}{3} \end{bmatrix}$ $\begin{bmatrix} 0 & \frac{2}{3} \\ -\frac{2}{3} & \frac{1}{3} \end{bmatrix}$ $\begin{bmatrix} 0 & \frac{1}{3} \\ \frac{2}{3} & \frac{2}{3} \end{bmatrix}$

(b) $\begin{bmatrix} 1 & 0 \\ 0 & 0 \end{bmatrix}$ $\begin{bmatrix} 0 & 1 \\ 0 & 0 \end{bmatrix}$ $\begin{bmatrix} 0 & 0 \\ 1 & 1 \end{bmatrix}$ $\begin{bmatrix} 0 & 0 \\ 1 & -1 \end{bmatrix}$

5. Let $\mathbf{x} = (1, -1)$ and $\mathbf{y} = \left(\dfrac{2}{\sqrt{30}}, \dfrac{3}{\sqrt{30}}\right)$.

Show that $\{\mathbf{x}, \mathbf{y}\}$ is orthonormal if $R^2$ has the inner product $\langle \mathbf{u}, \mathbf{v} \rangle = 3u_1v_1 + 2u_2v_2$, but is not orthonormal if $R^2$ has the Euclidean inner product.

6. Show that

$$\mathbf{u}_1 = \left(\frac{1}{\sqrt{2}}, 0, 0, \frac{1}{\sqrt{2}}\right), \quad \mathbf{u}_2 = \left(-\frac{1}{\sqrt{6}}, 0, \frac{2}{\sqrt{6}}, \frac{1}{\sqrt{6}}\right),$$

$$\mathbf{u}_3 = \left(\frac{2}{\sqrt{21}}, \frac{3}{\sqrt{21}}, \frac{2}{\sqrt{21}}, -\frac{2}{\sqrt{21}}\right), \text{ and } \mathbf{u}_4 = \left(-\frac{1}{\sqrt{7}}, \frac{2}{\sqrt{7}}, -\frac{1}{\sqrt{7}}, \frac{1}{\sqrt{7}}\right)$$

form an orthonormal basis for $R^4$ with the Euclidian inner product; use the method of Example 52 to find the coordinate vectors for the following vectors with respect to the basis $S = \{\mathbf{u}_1, \mathbf{u}_2, \mathbf{u}_3, \mathbf{u}_4\}$.
   (a) $(1, 1, 1, 1)$     (b) $(3, -2, 6, 1)$

7. Let $R^2$ have the Euclidean inner product. Use the Gram-Schmidt process to transform the basis $\{\mathbf{u}_1, \mathbf{u}_2\}$ into an orthonormal basis.
   (a) $\mathbf{u}_1 = (1, -3)$, $\mathbf{u}_2 = (2, 2)$     (b) $\mathbf{u}_1 = (1, 0)$, $\mathbf{u}_2 = (3, -5)$

8. Let $R^3$ have the Euclidean inner product. Use the Gram-Schmidt process to transform the basis $\{\mathbf{u}_1, \mathbf{u}_2, \mathbf{u}_3\}$ into an orthonormal basis.
   (a) $\mathbf{u}_1 = (1, 1, 1)$, $\mathbf{u}_2 = (-1, 1, 0)$, $\mathbf{u}_3 = (1, 2, 1)$
   (b) $\mathbf{u}_1 = (1, 0, 0)$, $\mathbf{u}_2 = (3, 7, -2)$, $\mathbf{u}_3 = (0, 4, 1)$

9. Let $R^4$ have the Euclidean inner product. Use the Gram-Schmidt process to transform the basis $\{\mathbf{u}_1, \mathbf{u}_2, \mathbf{u}_3, \mathbf{u}_4\}$ into an orthonormal basis.

$$\mathbf{u}_1 = (0, 2, 1, 0), \mathbf{u}_2 = (1, -1, 0, 0), \mathbf{u}_3 = (1, 2, 0, -1), \mathbf{u}_4 = (1, 0, 0, 1)$$

10. Let $R^3$ have the Euclidean inner product. Find an orthonormal basis for the subspace spanned by $(0, 1, 2)$, $(-1, 0, 1)$.

11. Let $R^3$ have the inner product $\langle u, v \rangle = u_1v_1 + 2u_2v_2 + 3u_3v_3$. Use the Gram-Schmidt process to transform
$$u_1 = (1, 1, 1) \quad u_2 = (1, 1, 0) \quad u_3 = (1, 0, 0)$$
into an orthonormal basis.

12. The subspace of $R^3$ spanned by the vectors $\mathbf{u}_1 = (4/5, 0, -(3/5))$ and $\mathbf{u}_2 = (0, 1, 0)$ is a plane passing through the origin. Express $\mathbf{w} = (1, 2, 3)$ in the form $\mathbf{w} = \mathbf{w}_1 + \mathbf{w}_2$, where $\mathbf{w}_1$ lies in the plane and $\mathbf{w}_2$ is perpendicular to the plane.

13. Repeat Exercise 12 with $\mathbf{u}_1 = (1, 1, 1)$ and $\mathbf{u}_2 = (2, 0, -1)$.

14. Let $R^4$ have the Euclidean inner product. Express $\mathbf{w} = (-1, 2, 6, 0)$ in the form $\mathbf{w} = \mathbf{w}_1 + \mathbf{w}_2$, where $\mathbf{w}_1$ is in the space $W$ spanned by $\mathbf{u}_1 = (-1, 0, 1, 2)$ and $\mathbf{u}_2 = (0, 1, 0, 1)$, and $\mathbf{w}_2$ is orthogonal to $W$.

15. Let $\{\mathbf{v}_1, \mathbf{v}_2, \mathbf{v}_3\}$ be an orthonormal basis for an inner product space $V$. Show that if $\mathbf{w}$ is a vector in $V$, then $\|\mathbf{w}\|^2 = \langle \mathbf{w}, \mathbf{v}_1 \rangle^2 + \langle \mathbf{w}, \mathbf{v}_2 \rangle^2 + \langle \mathbf{w}, \mathbf{v}_3 \rangle^2$.

**16.** Let $\{\mathbf{v}_1, \mathbf{v}_2, \ldots, \mathbf{v}_n\}$ be an orthonormal basis for an inner product space $V$. Show that if $\mathbf{w}$ is a vector in $V$, then

$$\|\mathbf{w}\|^2 = \langle \mathbf{w}, \mathbf{v}_1 \rangle^2 + \langle \mathbf{w}, \mathbf{v}_2 \rangle^2 + \cdots + \langle \mathbf{w}, \mathbf{v}_n \rangle^2$$

**17.** In Step 3 of the proof of Theorem 19, it was stated that, "the linear independence of $\{\mathbf{u}_1, \mathbf{u}_2, \ldots, \mathbf{u}_n\}$ assures that

$$\mathbf{u}_3 - \langle \mathbf{u}_3, \mathbf{v}_1 \rangle \mathbf{v}_1 - \langle \mathbf{u}_3, \mathbf{v}_2 \rangle \mathbf{v}_2 \neq \mathbf{0}.\text{''}$$

Prove this statement.

**18.** Prove Theorem 18.

(***Hint.*** Show first that the vector $\mathbf{w}_1$ in (4.18) lies in $W$, the vector $\mathbf{w}_2$ in (4.19) is orthogonal to $W$, and that $\mathbf{u} = \mathbf{w}_1 + \mathbf{w}_2$. Next show that if $\mathbf{u} = \mathbf{w}_1' + \mathbf{w}_2'$, where $\mathbf{w}_1'$ is in $W$ and $\mathbf{w}_2'$ is orthogonal to $W$, then $\mathbf{w}_1 = \mathbf{w}_1'$ and $\mathbf{w}_2 = \mathbf{w}_2'$.)

**19. (For readers who have studied calculus.)** Let the vector space $P_2$ have the inner product

$$\langle \mathbf{p}, \mathbf{q} \rangle = \int_{-1}^{1} p(x)q(x)\, dx$$

Apply the Gram-Schmidt process to transform the standard basis $S = \{1, x, x^2\}$ into an orthonormal basis. (The polynomials in the resulting basis are called the first three ***normalized Legendre polynomials.***)

**20. (For readers who have studied calculus.)** Use Theorem 16 to express the following as linear combinations of the first three normalized Legendre polynomials (Exercise 19).
(a) $1 + x + 4x^2$   (b) $2 - 7x^2$   (c) $4 + 3x$

**21. (For readers who have studied calculus.)** Let $P_2$ have the inner product

$$\langle \mathbf{p}, \mathbf{q} \rangle = \int_{0}^{1} p(x)q(x)\, dx$$

Apply the Gram-Schmidt process to transform the standard basis $S = \{1, x, x^2\}$ into an orthonormal basis.

# Chapter 5 | *Linear Transformations*

## 5.1 INTRODUCTION TO LINEAR TRANSFORMATIONS

In this section we begin the study of vector-valued functions of a vector variable. That is, functions having the form $\mathbf{w} = F(\mathbf{v})$, where the independent variable $\mathbf{v}$ and the dependent variable $\mathbf{w}$ are both vectors. We shall concentrate on a special class of vector functions called linear transformations. These have many important applications in physics, engineering, social sciences, and various branches of mathematics.

If $V$ and $W$ are vector spaces and $F$ is a function that associates a unique vector in $W$ with each vector in $V$, we say $F$ *maps* $V$ into $W$, and write $F: V \to W$. Further, if $F$ associates the vector $\mathbf{w}$ with the vector $\mathbf{v}$, we write $\mathbf{w} = F(\mathbf{v})$ and say that $\mathbf{w}$ is the *image* of $\mathbf{v}$ under $F$.

To illustrate, if $\mathbf{v} = (x, y)$ is a vector in $R^2$, then the formula

$$F(\mathbf{v}) = (x, x + y, x - y) \tag{5.1}$$

defines a function that maps $R^2$ into $R^3$. In particular, if $\mathbf{v} = (1, 1)$, then the image of $\mathbf{v}$ under $F$ is $F(\mathbf{v}) = (1, 2, 0)$.

**Definition:** If $F: V \to W$ is a function from the vector space $V$ into the vector space $W$, then $F$ is called a *linear transformation* if

(i) $F(\mathbf{u} + \mathbf{v}) = F(\mathbf{u}) + F(\mathbf{v})$ for all vectors $\mathbf{u}$ and $\mathbf{v}$ in $V$.
(ii) $F(k\mathbf{u}) = kF(\mathbf{u})$ for all vectors $\mathbf{u}$ in $V$ and all scalars $k$.

To illustrate, let $F: R^2 \to R^3$ be the function defined by (5.1). If $\mathbf{u} = (x_1, y_1)$ and $\mathbf{v} = (x_2, y_2)$, then $\mathbf{u} + \mathbf{v} = (x_1 + x_2, y_1 + y_2)$, so that

$$
\begin{aligned}
F(\mathbf{u} + \mathbf{v}) &= (x_1 + x_2, (x_1 + x_2) + (y_1 + y_2), (x_1 + x_2) - (y_1 + y_2)) \\
&= (x_1, x_1 + y_1, x_1 - y_1) + (x_2, x_2 + y_2, x_2 - y_2) \\
&= F(\mathbf{u}) + F(\mathbf{v})
\end{aligned}
$$

Also, if $k$ is a scalar, $k\mathbf{u} = (kx_1, ky_1)$, so that

$$F(k\mathbf{u}) = (kx_1, kx_1 + ky_1, kx_1 - ky_1)$$
$$= k(x_1, x_1 + y_1, x_1 - y_1)$$
$$= kF(\mathbf{u})$$

Thus $F$ is a linear transformation.

If $F: V \to W$ is a linear transformation, then for any $\mathbf{v}_1$ and $\mathbf{v}_2$ in $V$ and any scalars $k_1$ and $k_2$, we have

$$F(k_1\mathbf{v}_1 + k_2\mathbf{v}_2) = F(k_1\mathbf{v}_1) + F(k_2\mathbf{v}_2) = k_1F(\mathbf{v}_1) + k_2F(\mathbf{v}_2)$$

Similarly, if $\mathbf{v}_1, \mathbf{v}_2, \ldots, \mathbf{v}_r$ are vectors in $V$ and $k_1, k_2, \ldots, k_r$ are scalars, then

$$F(k_1\mathbf{v}_1 + k_2\mathbf{v}_2 + \cdots + k_r\mathbf{v}_r) = k_1F(\mathbf{v}_1) + k_2F(\mathbf{v}_2) + \cdots + k_rF(\mathbf{v}_r)$$

We now give some further examples of linear transformations.

### Example 1

Let $A$ be a fixed $m \times n$ matrix. If we use matrix notation for vectors in $R^m$ and $R^n$, then we can define a function $T: R^n \to R^m$ by

$$T(\mathbf{x}) = A\mathbf{x}$$

Observe that if $\mathbf{x}$ is an $n \times 1$ matrix, then the product $A\mathbf{x}$ is an $m \times 1$ matrix; thus $T$ maps $R^n$ into $R^m$. Moreover, $T$ is linear; to see this, let $\mathbf{u}$ and $\mathbf{v}$ be $n \times 1$ matrices and let $k$ be a scalar. Using properties of matrix multiplication, we obtain

$$A(\mathbf{u} + \mathbf{v}) = A\mathbf{u} + A\mathbf{v} \qquad \text{and} \qquad A(k\mathbf{u}) = k(A\mathbf{u})$$

or equivalently

$$T(\mathbf{u} + \mathbf{v}) = T(\mathbf{u}) + T(\mathbf{v}) \qquad \text{and} \qquad T(k\mathbf{u}) = kT(\mathbf{u})$$

We shall call the linear transformation in this example *multiplication by A*. Linear transformations of this kind are called *matrix transformations.*

### Example 2

As a special case of the previous example, let $\theta$ be a fixed angle, and let $T: R^2 \to R^2$ be multiplication by the matrix

$$A = \begin{bmatrix} \cos\theta & -\sin\theta \\ \sin\theta & \cos\theta \end{bmatrix}$$

If $\mathbf{v}$ is the vector

$$\mathbf{v} = \begin{bmatrix} x \\ y \end{bmatrix}$$

then

$$T(\mathbf{v}) = A\mathbf{v} = \begin{bmatrix} \cos\theta & -\sin\theta \\ \sin\theta & \cos\theta \end{bmatrix}\begin{bmatrix} x \\ y \end{bmatrix} = \begin{bmatrix} x\cos\theta - y\sin\theta \\ x\sin\theta + y\cos\theta \end{bmatrix}$$

Geometrically, $T(\mathbf{v})$ is the vector that results if $\mathbf{v}$ is rotated through an angle $\theta$. To see this, let $\phi$ be the angle between $\mathbf{v}$ and the positive $x$ axis, and let

$$\mathbf{v}' = \begin{bmatrix} x' \\ y' \end{bmatrix}$$

be the vector that results when $\mathbf{v}$ is rotated through an angle $\theta$ (Figure 5.1). We shall show $\mathbf{v}' = T(\mathbf{v})$.

If $r$ denotes the length of $\mathbf{v}$, then

$$x = r \cos \phi \qquad y = r \sin \phi$$

Similarly, since $\mathbf{v}'$ has the same length as $\mathbf{v}$, we have

$$x' = r \cos (\theta + \phi) \qquad y' = r \sin (\theta + \phi)$$

Therefore

$$\mathbf{v}' = \begin{bmatrix} x' \\ y' \end{bmatrix} = \begin{bmatrix} r \cos (\theta + \phi) \\ r \sin (\theta + \phi) \end{bmatrix}$$

$$= \begin{bmatrix} r \cos \theta \cos \phi - r \sin \theta \sin \phi \\ r \sin \theta \cos \phi + r \cos \theta \sin \phi \end{bmatrix}$$

$$= \begin{bmatrix} x \cos \theta - y \sin \theta \\ x \sin \theta + y \cos \theta \end{bmatrix}$$

$$= \begin{bmatrix} \cos \theta & -\sin \theta \\ \sin \theta & \cos \theta \end{bmatrix} \begin{bmatrix} x \\ y \end{bmatrix}$$

$$= A\mathbf{v} = T(\mathbf{v})$$

The linear transformation in this example is called the ***rotation of $R^2$ through the angle $\theta$***.

### Example 3

Let $V$ and $W$ be any two vector spaces. The linear transformation $T: V \rightarrow W$ such that $T(\mathbf{v}) = \mathbf{0}$ for every $\mathbf{v}$ in $V$ is called the ***zero transformation.*** To see that $T$ is linear, observe that

$$T(\mathbf{u} + \mathbf{v}) = \mathbf{0}, \; T(\mathbf{u}) = \mathbf{0}, \; T(\mathbf{v}) = \mathbf{0} \qquad \text{and} \qquad T(k\mathbf{u}) = \mathbf{0}$$

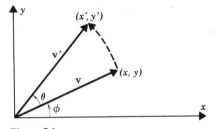

**Figure 5.1**

Therefore

$$T(\mathbf{u} + \mathbf{v}) = T(\mathbf{u}) + T(\mathbf{v}) \quad \text{and} \quad T(k\mathbf{u}) = kT(\mathbf{u})$$

**Example 4**

Let $V$ be any vector space. The mapping $T: V \to V$ defined by $T(\mathbf{v}) = \mathbf{v}$ is called the *identity transformation* on $V$. The verification that $T$ is linear is left as an exercise.

If, as in Examples 2 and 4, $T: V \to V$ is a linear transformation from a vector space $V$ into itself, then $T$ is called a *linear operator* on $V$.

**Example 5**

Let $V$ be any vector space and $k$ any fixed scalar. We leave it as an exercise to check that the function $T: V \to V$ defined by

$$T(\mathbf{v}) = k\mathbf{v}$$

is a linear operator on $V$. If $k > 1$, $T$ is called a *dilation* of $V$ and if $0 < k < 1$, then $T$ is called a *contraction* of $V$. Geometrically, a dilation "stretches" each vector in $V$ by a factor of $k$, and a contraction of $V$ "compresses" each vector by a factor of $k$ (Figure 5.2).

**Example 6**

Let $V$ be an inner product space, and suppose $W$ is a finite dimensional subspace of $V$ having

$$S = \{\mathbf{w}_1, \mathbf{w}_2, \ldots, \mathbf{w}_r\}$$

as an orthonormal basis. Let $T: V \to W$ be the function that maps a vector $\mathbf{v}$ in $V$ into its orthogonal projection on $W$ (Section 4.9); that is

$$T(\mathbf{v}) = \langle \mathbf{v}, \mathbf{w}_1 \rangle \mathbf{w}_1 + \langle \mathbf{v}, \mathbf{w}_2 \rangle \mathbf{w}_2 + \cdots + \langle \mathbf{v}, \mathbf{w}_r \rangle \mathbf{w}_r$$

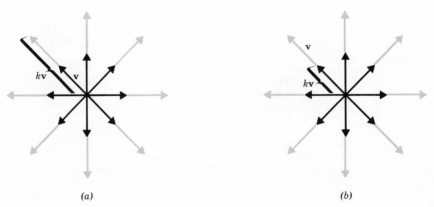

(a)                                               (b)

**Figure 5.2** (a) Dilation of $V$. (b) Contraction of $V$.

**Figure 5.3**

(See Figure 5.3.)

The mapping $T$ is called the ***orthogonal projection of V onto W;*** its linearity follows from the basic properties of the inner product. For example

$$T(\mathbf{u} + \mathbf{v}) = \langle \mathbf{u} + \mathbf{v}, \mathbf{w}_1 \rangle \mathbf{w}_1 + \langle \mathbf{u} + \mathbf{v}, \mathbf{w}_2 \rangle \mathbf{w}_2 + \cdots + \langle \mathbf{u} + \mathbf{v}, \mathbf{w}_r \rangle \mathbf{w}_r$$
$$= \langle \mathbf{u}, \mathbf{w}_1 \rangle \mathbf{w}_1 + \langle \mathbf{u}, \mathbf{w}_2 \rangle \mathbf{w}_2 + \cdots + \langle \mathbf{u}, \mathbf{w}_r \rangle \mathbf{w}_r$$
$$+ \langle \mathbf{v}, \mathbf{w}_1 \rangle \mathbf{w}_1 + \langle \mathbf{v}, \mathbf{w}_2 \rangle \mathbf{w}_2 + \cdots + \langle \mathbf{v}, \mathbf{w}_r \rangle \mathbf{w}_r$$
$$= T(\mathbf{u}) + T(\mathbf{v})$$

Similarly, $T(k\mathbf{u}) = kT(\mathbf{u})$.

**Example 7**

As a special case of the previous example, let $V = R^3$ have the Euclidean inner product. The vectors $\mathbf{w}_1 = (1, 0, 0)$ and $\mathbf{w}_2 = (0, 1, 0)$ form an orthonormal basis for the $xy$-plane. Thus, if $\mathbf{v} = (x, y, z)$ is any vector in $R^3$, the orthogonal projection of $R^3$ onto the $xy$-plane is given by

$$T(\mathbf{v}) = \langle \mathbf{v}, \mathbf{w}_1 \rangle \mathbf{w}_1 + \langle \mathbf{v}, \mathbf{w}_2 \rangle \mathbf{w}_2$$
$$= x(1, 0, 0) + y(0, 1, 0)$$
$$= (x, y, 0)$$

(Figure 5.4).

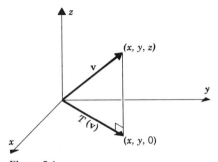

**Figure 5.4**

### Example 8

Let $V$ be an $n$-dimensional vector space and $S$ a fixed basis for $V$. Let $T: V \to R^n$ be the function that maps a vector $\mathbf{v}$ in $V$ into its coordinate vector with respect to $S$; that is

$$T(\mathbf{v}) = (\mathbf{v})_S$$

In the previous chapter we observed that

$$(\mathbf{u} + \mathbf{v})_S = (\mathbf{u})_S + (\mathbf{v})_S$$

and

$$(k\mathbf{u})_S = k(\mathbf{u})_S$$

In terms of $T$ these relationships become

$$T(\mathbf{u} + \mathbf{v}) = T(\mathbf{u}) + T(\mathbf{v})$$

and

$$T(k\mathbf{u}) = kT(\mathbf{u})$$

Thus $T$ is a linear transformation from $V$ into $R^n$.

### Example 9

**(For readers who have studied calculus.)**

Let $V = C[0, 1]$ be the vector space of all real-valued functions continuous on the interval $0 \leq x \leq 1$, and let $W$ be the subspace of $C[0, 1]$ consisting of all functions with continuous first derivatives on the interval $0 \leq x \leq 1$.

Let $D: W \to V$ be the transformation that maps $\mathbf{f}$ into its derivative; that is

$$D(\mathbf{f}) = \mathbf{f}'$$

From the properties of differentiation, we have

$$D(\mathbf{f} + \mathbf{g}) = D(\mathbf{f}) + D(\mathbf{g})$$

and

$$D(k\mathbf{f}) = kD(\mathbf{f})$$

Thus $D$ is a linear transformation.

### Example 10

**(For readers who have studied calculus.)**

Let $V = C[0, 1]$ be as in the previous example, and let $J: V \to R$ be defined by

$$J(\mathbf{f}) = \int_0^1 f(x)\, dx$$

For example, if $f(x) = x^2$, then

$$J(\mathbf{f}) = \int_0^1 x^2\, dx = \frac{1}{3}$$

Since

$$\int_0^1 (f(x) + g(x)) \, dx = \int_0^1 f(x) \, dx + \int_0^1 g(x) \, dx$$

and

$$\int_0^1 kf(x) \, dx = k \int_0^1 f(x) \, dx$$

for any constant $k$, it follows that

$$J(\mathbf{f} + \mathbf{g}) = J(\mathbf{f}) + J(\mathbf{g})$$
$$J(k\mathbf{f}) = kJ(\mathbf{f})$$

Thus $J$ is a linear transformation.

## EXERCISE SET 5.1

In Exercises 1 to 8 a formula is given for a function $F: R^2 \to R^2$. In each exercise determine if $F$ is linear.

**1.** $F(x, y) = (2x, y)$      **2.** $F(x, y) = (x^2, y)$

**3.** $F(x, y) = (y, x)$      **4.** $F(x, y) = (0, y)$

**5.** $F(x, y) = (x, y + 1)$      **6.** $F(x, y) = (2x + y, x - y)$

**7.** $F(x, y) = (y, y)$      **8.** $F(x, y) = (\sqrt[3]{x}, \sqrt[3]{y})$

In Exercises 9 to 12, a formula is given for a function $F: R^3 \to R^2$. In each exercise determine if $F$ is linear.

**9.** $F(x, y, z) = (x, x + y + z)$      **10.** $F(x, y, z) = (0, 0)$

**11.** $F(x, y, z) = (1, 1)$      **12.** $F(x, y, z) = (2x + y, 3y - 4z)$

In Exercises 13 to 16, a formula is given for a function $F: M_{22} \to R$. In each exercise determine if $F$ is linear.

**13.** $F\left(\begin{bmatrix} a & b \\ c & d \end{bmatrix}\right) = a + d$      **14.** $F\left(\begin{bmatrix} a & b \\ c & d \end{bmatrix}\right) = \det \begin{bmatrix} a & b \\ c & d \end{bmatrix}$

**15.** $F\left(\begin{bmatrix} a & b \\ c & d \end{bmatrix}\right) = 2a + 3b + c - d$      **16.** $F\left(\begin{bmatrix} a & b \\ c & d \end{bmatrix}\right) = a^2 + b^2$

In Exercises 17 to 20, a formula is given for a function $F: P_2 \to P_2$. In each exercise determine if $F$ is linear.

**17.** $F(a_0 + a_1 x + a_2 x^2) = a_0 + (a_1 + a_2)x + (2a_0 - 3a_1)x^2$

**18.** $F(a_0 + a_1 x + a_2 x^2) = a_0 + a_1(x + 1) + a_2(x + 1)^2$

**19.** $F(a_0 + a_1 x + a_2 x^2) = 0$

**20.** $F(a_0 + a_1 x + a_2 x^2) = (a_0 + 1) + a_1 x + a_2 x^2$

**21.** Let $F: R^2 \to R^2$ be the function that maps each point in the plane into its reflection about the $y$-axis. Find a formula for $F$, and show that it is a linear operator on $R^2$.

**22.** Let $B$ be a fixed $2 \times 3$ matrix. Show that the function $T: M_{22} \to M_{23}$ defined by $T(A) = AB$ is a linear transformation.

**23.** Let $T: R^3 \to R^2$ be a matrix transformation, and suppose

$$T\left(\begin{bmatrix} 1 \\ 0 \\ 0 \end{bmatrix}\right) = \begin{bmatrix} 1 \\ 1 \end{bmatrix}, \quad T\left(\begin{bmatrix} 0 \\ 1 \\ 0 \end{bmatrix}\right) = \begin{bmatrix} 3 \\ 0 \end{bmatrix}, \text{ and } T\left(\begin{bmatrix} 0 \\ 0 \\ 1 \end{bmatrix}\right) = \begin{bmatrix} 4 \\ -7 \end{bmatrix}.$$

(a) Find the matrix.

(b) Find $T\left(\begin{bmatrix} 1 \\ 3 \\ 8 \end{bmatrix}\right)$

(c) Find $T\left(\begin{bmatrix} x \\ y \\ z \end{bmatrix}\right)$

**24.** Let $T: R^3 \to W$ be the orthogonal projection of $R^3$ onto the $xz$-plane $W$.
   (a) Find $T(2, 7, -1)$.
   (b) Find a formula for $T$.

**25.** Let $T: R^3 \to W$ be the orthogonal projection of $R^3$ onto the plane $W$ having the equation $x + y + z = 0$.
   (a) Find $T(3, 8, 4)$.
   (b) Find a formula for $T$.

**26.** In each part let $T: R^2 \to R^2$ be the linear operator that rotates each vector in the plane through the angle $\theta$. Find $T(-1, 2)$ and $T(x, y)$ when

   (a) $\theta = \dfrac{\pi}{4}$    (b) $\theta = \pi$    (c) $\theta = \dfrac{\pi}{6}$    (d) $\theta = -\dfrac{\pi}{3}$

**27.** Prove that if $T: V \to W$ is a linear transformation, then $T(\mathbf{u} - \mathbf{v}) = T(\mathbf{u}) - T(\mathbf{v})$ for all vectors $\mathbf{u}$ and $\mathbf{v}$ in $V$.

**28.** Let $\{\mathbf{v}_1, \mathbf{v}_2, \ldots, \mathbf{v}_n\}$ be a basis for a vector space $V$ and let $T: V \to W$ be a linear transformation. Show that if $T(\mathbf{v}_1) = T(\mathbf{v}_2) = \cdots = T(\mathbf{v}_n) = \mathbf{0}$, then $T$ is the zero transformation.

**29.** Let $\{\mathbf{v}_1, \mathbf{v}_2, \ldots, \mathbf{v}_n\}$ be a basis for a vector space $V$ and let $T: V \to V$ be a linear operator. Show that if $T(\mathbf{v}_1) = \mathbf{v}_1, T(\mathbf{v}_2) = \mathbf{v}_2, \ldots, T(\mathbf{v}_n) = \mathbf{v}_n$, then $T$ is the identity transformation on $V$.

**30.** In each part let $T: R^3 \to R^3$ be multiplication by the given matrix $A$. Describe the transformation $T$ geometrically.

   (a) $A = \begin{bmatrix} \cos\theta & -\sin\theta & 0 \\ \sin\theta & \cos\theta & 0 \\ 0 & 0 & 1 \end{bmatrix}$    (b) $A = \begin{bmatrix} \cos\theta & 0 & -\sin\theta \\ 0 & 1 & 0 \\ \sin\theta & 0 & \cos\theta \end{bmatrix}$

   (c) $A = \begin{bmatrix} 1 & 0 & 0 \\ 0 & \cos\theta & -\sin\theta \\ 0 & \sin\theta & \cos\theta \end{bmatrix}$

**31.** Let $\mathbf{v}_0$ be a fixed vector in an inner product space $V$. Show that the function $F: V \to R$ defined by $F(\mathbf{v}) = \langle \mathbf{v}, \mathbf{v}_0 \rangle$ is a linear transformation.

## 5.2 PROPERTIES OF LINEAR TRANSFORMATIONS; KERNEL AND RANGE

In this section we develop some basic properties of linear transformations. In particular, we show that once the images of the basis vectors under a linear transformation are known, it is possible to find the images of the remaining vectors in the space.

**Theorem 1.** *If $T: V \to W$ is a linear transformation, then:*

(a) $T(\mathbf{0}) = \mathbf{0}$

(b) $T(-\mathbf{v}) = -\mathbf{v}$ *for all* $\mathbf{v}$ *in* $V$

(c) $T(\mathbf{v} - \mathbf{w}) = T(\mathbf{v}) - T(\mathbf{w})$ *for all* $\mathbf{v}$ *and* $\mathbf{w}$ *in* $V$

*Proof.* Let $\mathbf{v}$ be any vector in $V$. Since $0\mathbf{v} = \mathbf{0}$ we have

$$T(\mathbf{0}) = T(0\mathbf{v}) = 0T(\mathbf{v}) = \mathbf{0}$$

which proves (a).

Also, $T(-\mathbf{v}) = T((-1)\mathbf{v}) = (-1)T(\mathbf{v}) = -T(\mathbf{v})$, which proves (b).

Finally, $\mathbf{v} - \mathbf{w} = \mathbf{v} + (-1)\mathbf{w}$; thus

$$\begin{aligned} T(\mathbf{v} - \mathbf{w}) &= T(\mathbf{v} + (-1)\mathbf{w}) \\ &= T(\mathbf{v}) + (-1)T(\mathbf{w}) \\ &= T(\mathbf{v}) - T(\mathbf{w}) \end{aligned}$$

**Definition.** If $T: V \to W$ is a linear transformation, then the set of vectors in $V$ that $T$ maps into $\mathbf{0}$ is called the **kernel** (or **nullspace**) of $T$; it is denoted by $\ker(T)$. The set of all vectors in $W$ that are images under $T$ of at least one vector in $V$ is called the **range** of $T$; it is denoted by $R(T)$.

### Example 11

Let $T: V \to W$ be the zero transformation. Since $T$ maps every vector into $\mathbf{0}$, $\ker(T) = V$. Since $\mathbf{0}$ is the only possible image under $T$, $R(T)$ consists only of the zero vector.

### Example 12

Let $T: R^n \to R^m$ be multiplication by

$$A = \begin{bmatrix} a_{11} & a_{12} & \cdots & a_{1n} \\ a_{21} & a_{22} & \cdots & a_{2n} \\ \vdots & \vdots & & \vdots \\ a_{m1} & a_{m2} & \cdots & a_{mn} \end{bmatrix}$$

The kernel of $T$ consists of all

$$\mathbf{x} = \begin{bmatrix} x_1 \\ x_2 \\ \vdots \\ x_n \end{bmatrix}$$

that are solution vectors of the homogeneous system

$$A \begin{bmatrix} x_1 \\ x_2 \\ \vdots \\ x_n \end{bmatrix} = \begin{bmatrix} 0 \\ 0 \\ \vdots \\ 0 \end{bmatrix}$$

The range of $T$ consists of vectors

$$\mathbf{b} = \begin{bmatrix} b_1 \\ b_2 \\ \vdots \\ b_m \end{bmatrix}$$

such that the system

$$A \begin{bmatrix} x_1 \\ x_2 \\ \vdots \\ x_n \end{bmatrix} = \begin{bmatrix} b_1 \\ b_2 \\ \vdots \\ b_m \end{bmatrix}$$

has a solution.

**Theorem 2.** *If $T: V \to W$ is a linear transformation then:*

(a) *The kernel of $T$ is a subspace of $V$.*
(b) *The range of $T$ is a subspace of $W$.*

*Proof.*

(a) To show that $\ker(T)$ is a subspace, we must show it is closed under addition and scalar multiplication. Let $\mathbf{v}_1$ and $\mathbf{v}_2$ be vectors in $\ker(T)$, and let $k$ be any scalar. Then

$$T(\mathbf{v}_1 + \mathbf{v}_2) = T(\mathbf{v}_1) + T(\mathbf{v}_2)$$
$$= \mathbf{0} + \mathbf{0} = \mathbf{0}$$

so that $\mathbf{v}_1 + \mathbf{v}_2$ is in $\ker(T)$. Also

$$T(k\mathbf{v}_1) = kT(\mathbf{v}_1) = k\mathbf{0} = \mathbf{0}$$

so that $k\mathbf{v}_1$ is in $\ker(T)$.

(*b*) Let $\mathbf{w}_1$ and $\mathbf{w}_2$ be vectors in the range of $T$. To prove this part we must show that $\mathbf{w}_1 + \mathbf{w}_2$ and $k\mathbf{w}_1$ are in the range of $T$ for any scalar $k$; that is, we must find vectors $\mathbf{a}$ and $\mathbf{b}$ in $V$ such that $T(\mathbf{a}) = \mathbf{w}_1 + \mathbf{w}_2$ and $T(\mathbf{b}) = k\mathbf{w}_1$.

Since $\mathbf{w}_1$ and $\mathbf{w}_2$ are in the range of $T$, there are vectors $\mathbf{a}_1$ and $\mathbf{a}_2$ in $V$ such that $T(\mathbf{a}_1) = \mathbf{w}_1$ and $T(\mathbf{a}_2) = \mathbf{w}_2$. Let $\mathbf{a} = \mathbf{a}_1 + \mathbf{a}_2$ and $\mathbf{b} = k\mathbf{a}_1$. Then

$$T(\mathbf{a}) = T(\mathbf{a}_1 + \mathbf{a}_2) = T(\mathbf{a}_1) + T(\mathbf{a}_2) = \mathbf{w}_1 + \mathbf{w}_2$$

also

$$T(\mathbf{b}) = T(k\mathbf{a}_1) = kT(\mathbf{a}_1) = k\mathbf{w}_1$$

which completes the proof. ▮

**Definition.** If $T: V \rightarrow W$ is a linear transformation, then the dimension of the range of $T$ is called the ***rank of T*** and the dimension of the kernel is called the ***nullity of T***.

**Example 13**

Let $T: R^2 \rightarrow R^2$ be the rotation of $R^2$ through the angle $\pi/4$. It is geometrically obvious that the range of $T$ is all of $R^2$ and the kernel of $T$ is $\{\mathbf{0}\}$. Therefore, $T$ has rank $= 2$ and nullity $= 0$.

The next theorem shows that two linear transformations that "agree" on the basis vectors of a finite dimensional vector space $V$, must "agree" everywhere on $V$.

**Theorem 3.** *Let* $\{\mathbf{v}_1, \mathbf{v}_2, \ldots, \mathbf{v}_n\}$ *be a basis for a finite dimensional vector space* $V$, *and let* $F: V \rightarrow W$ *and* $T: V \rightarrow W$ *be linear transformations. If*

$$T(\mathbf{v}_1) = F(\mathbf{v}_1), \; T(\mathbf{v}_2) = F(\mathbf{v}_2), \ldots, T(\mathbf{v}_n) = F(\mathbf{v}_n)$$

*then* $T(\mathbf{v}) = F(\mathbf{v})$ *for all* $\mathbf{v}$ *in* $V$.

*Proof.* Each vector $\mathbf{v}$ in $V$ can be written in the form

$$\mathbf{v} = k_1\mathbf{v}_1 + k_2\mathbf{v}_2 + \cdots + k_n\mathbf{v}_n$$

Therefore

$$
\begin{aligned}
T(\mathbf{v}) &= T(k_1\mathbf{v}_1 + k_2\mathbf{v}_2 + \cdots + k_n\mathbf{v}_n) \\
&= k_1 T(\mathbf{v}_1) + k_2 T(\mathbf{v}_2) + \cdots + k_n T(\mathbf{v}_n) \\
&= k_1 F(\mathbf{v}_1) + k_2 F(\mathbf{v}_2) + \cdots + k_n F(\mathbf{v}_n) \\
&= F(k_1\mathbf{v}_1 + k_2\mathbf{v}_2 + \cdots + k_n\mathbf{v}_n) \\
&= F(\mathbf{v})
\end{aligned}
$$
▮

If $\{\mathbf{v}_1, \mathbf{v}_2, \ldots, \mathbf{v}_n\}$ is a basis for a vector space $V$, and if $\mathbf{w}_1, \mathbf{w}_2, \ldots, \mathbf{w}_n$ are arbitrary vectors (not necessarily distinct) in a vector space $W$, we shall show it is always possible to construct a linear transformation $T: V \rightarrow W$ such that

$$T(\mathbf{v}_1) = \mathbf{w}_1, \; T(\mathbf{v}_2) = \mathbf{w}_2, \ldots, T(\mathbf{v}_n) = \mathbf{w}_n$$

In other words, we can always find a linear transformation that maps the basis vectors of $V$ into any prescribed vectors in $W$.

From Theorem 3 there cannot be more than one such linear transformation. To see that there is one, let $\mathbf{x}$ be any vector in $V$. By Theorem 12 in Section 4.6, $\mathbf{x}$ can be expressed uniquely in the form

$$\mathbf{x} = c_1\mathbf{v}_1 + c_2\mathbf{v}_2 + \cdots + c_n\mathbf{v}_n \tag{5.2}$$

Define $T: V \to W$ by

$$T(\mathbf{x}) = c_1\mathbf{w}_1 + c_2\mathbf{w}_2 + \cdots + c_n\mathbf{w}_n \tag{5.3}$$

The verification that $T$ is linear is left as an exercise. To see that $T$ maps $\mathbf{v}_1, \mathbf{v}_2, \ldots, \mathbf{v}_n$ into the prescribed images, let $\mathbf{x} = \mathbf{v}_1$; from (5.2)

$$c_1 = 1, c_2 = 0, \ldots, c_n = 0$$

Substituting these values in (5.3), we obtain

$$T(\mathbf{v}_1) = \mathbf{w}_1$$

Similarly, $T(\mathbf{v}_2) = \mathbf{w}_2, \ldots, T(\mathbf{v}_n) = \mathbf{w}_n$.

**Example 14**

Consider the basis $S = \{\mathbf{v}_1, \mathbf{v}_2, \mathbf{v}_3\}$ for $R^3$, where $\mathbf{v}_1 = (1, 1, 1)$, $\mathbf{v}_2 = (1, 1, 0)$, $\mathbf{v}_3 = (1, 0, 0)$. Find a linear transformation $T: R^3 \to R^2$ such that $T(\mathbf{v}_1) = (1, 0)$, $T(\mathbf{v}_2) = (2, -1)$, and $T(\mathbf{v}_3) = (4, 3)$, and compute $T(\mathbf{x})$ for $\mathbf{x} = (2, -3, 5)$.

*Solution.* If $\mathbf{x} = (x_1, x_2, x_3)$ is any vector in $R^3$, then the expression for $\mathbf{x}$ as a linear combination of the vectors in $S$ is

$$\mathbf{x} = x_3\mathbf{v}_1 + (x_2 - x_3)\mathbf{v}_2 + (x_1 - x_2)\mathbf{v}_3$$

By (5.3) with

$$\mathbf{w}_1 = T(\mathbf{v}_1) = (1, 0), \mathbf{w}_2 = T(\mathbf{v}_2) = (2, -1), \text{ and } \mathbf{w}_3 = T(\mathbf{v}_3) = (4, 3)$$

we obtain

$$\begin{aligned} T(\mathbf{x}) &= x_3\mathbf{w}_1 + (x_2 - x_3)\mathbf{w}_2 + (x_1 - x_2)\mathbf{w}_3 \\ &= x_3(1, 0) + (x_2 - x_3)(2, -1) + (x_1 - x_2)(4, 3) \\ &= (4x_1 - 2x_2 - x_3, 3x_1 - 4x_2 + x_3) \end{aligned}$$

In particular, if $\mathbf{x} = (2, -3, 5)$ we obtain $T(\mathbf{x}) = (9, 23)$.

**EXERCISE SET 5.2**

**1.** Let $T: R^2 \to R^2$ be multiplication by

$$\begin{bmatrix} 2 & -1 \\ -8 & 4 \end{bmatrix}$$

Which of the following are in $R(T)$?

(a) $\begin{bmatrix} 1 \\ -4 \end{bmatrix}$  (b) $\begin{bmatrix} 5 \\ 0 \end{bmatrix}$  (c) $\begin{bmatrix} -3 \\ 12 \end{bmatrix}$

2. Let $T:R^2 \rightarrow R^2$ be the linear transformation in Exercise 1. Which of the following are in ker($T$)?

(a) $\begin{bmatrix} 5 \\ 10 \end{bmatrix}$  (b) $\begin{bmatrix} 3 \\ 2 \end{bmatrix}$  (c) $\begin{bmatrix} 1 \\ 1 \end{bmatrix}$

3. Let $T:R^4 \rightarrow R^3$ be multiplication by

$$\begin{bmatrix} 4 & 1 & -2 & -3 \\ 2 & 1 & 1 & -4 \\ 6 & 0 & -9 & 9 \end{bmatrix}$$

Which of the following are in $R(T)$?

(a) $\begin{bmatrix} 0 \\ 0 \\ 6 \end{bmatrix}$  (b) $\begin{bmatrix} 1 \\ 3 \\ 0 \end{bmatrix}$  (c) $\begin{bmatrix} 2 \\ 4 \\ 1 \end{bmatrix}$

4. Let $T:R^4 \rightarrow R^3$ be the linear transformation in Exercise 3. Which of the following are in ker($T$)?

(a) $\begin{bmatrix} 3 \\ -8 \\ 2 \\ 0 \end{bmatrix}$  (b) $\begin{bmatrix} 0 \\ 0 \\ 0 \\ 1 \end{bmatrix}$  (c) $\begin{bmatrix} 0 \\ -4 \\ 1 \\ 0 \end{bmatrix}$

5. Let $T:P_2 \rightarrow P_3$ be the linear transformation defined by $T(p(x)) = xp(x)$. Which of the following are in ker($T$)?
   (a) $x^2$  (b) 0  (c) $1 + x$

6. Let $T:P_2 \rightarrow P_3$ be the linear transformation in Exercise 5. Which of the following are in $R(T)$?
   (a) $x + x^2$  (b) $1 + x$  (c) $3 - x^2$

7. Let $V$ be any vector space, and let $T:V \rightarrow V$ be defined by $T(v) = 3v$.
   (a) What is the kernel of $T$?
   (b) What is the range of $T$?

8. Find the rank and nullity of the linear transformation in Exercise 1.

9. Find the rank and nullity of the linear transformation in Exercise 5.

10. Let $V$ be an $n$-dimensional vector space. Find the rank and nullity of the linear transformation $T:V \rightarrow V$ defined by
    (a) $T(\mathbf{x}) = \mathbf{x}$  (b) $T(\mathbf{x}) = \mathbf{0}$  (c) $T(\mathbf{x}) = 3\mathbf{x}$

11. Consider the basis $S = \{v_1, v_2, v_3\}$ for $R^3$, where $v_1 = (1, 2, 3)$, $v_2 = (2, 5, 3)$, and $v_3 = (1, 0, 10)$. Find a formula for the linear transformation $T:R^3 \rightarrow R^2$ for which $T(v_1) = (1, 0)$, $T(v_2) = (1, 0)$, and $T(v_3) = (0, 1)$. Compute $T(1, 1, 1)$.

12. Find the linear transformation $T:P_2 \rightarrow P_2$ for which $T(1) = 1 + x$, $T(x) = 3 - x^2$, and $T(x^2) = 4 + 2x - 3x^2$. Compute $T(2 - 2x + 3x^2)$.

13. Let $T: R^3 \to V$ be a linear transformation from $R^3$ to any vector space. Show that the kernel of $T$ is a line through the origin, a plane through the origin, the origin only, or all of $R^3$.

14. Let $T: V \to R^3$ be a linear transformation from any vector space to $R^3$. Show that the range of $T$ is a line through the origin, a plane through the origin, the origin only, or all of $R^3$.

15. Let $T: R^3 \to R^3$ be multiplication by

$$
\begin{bmatrix}
1 & 3 & 4 \\
3 & 4 & 7 \\
-2 & 2 & 0
\end{bmatrix}
$$

   (a) Show that the range of $T$ is a plane through the origin and find an equation for it.
   (b) Show that the kernel of $T$ is a line through the origin and find parametric equations for it.

16. Prove that the function defined by (5.3) is a linear transformation.

17. **(For readers who have studied calculus.)** Let $D: P_3 \to P_2$ be the differentiation transformation $D(\mathbf{p}) = \mathbf{p}'$. Describe the kernel of $D$.

18. **(For readers who have studied calculus.)** Let $J: P_1 \to R$ be the integration transformation $J(\mathbf{p}) = \int_{-1}^{1} p(x)\, dx$. Describe the kernel of $J$.


## 5.3 MATRIX TRANSFORMATIONS IN DETAIL; DIMENSION THEOREM

In this section we shall develop more results about matrix transformations. In particular, we shall show that the rank and nullity of a matrix transformation can be obtained by inspection from any row-echelon form of the matrix. We shall also establish a basic relationship between the rank and nullity of a linear transformation.

**Definition.** The *column vectors* of a matrix

$$
A = \begin{bmatrix}
a_{11} & a_{12} & \cdots & a_{1n} \\
a_{21} & a_{22} & \cdots & a_{2n} \\
\vdots & \vdots & & \vdots \\
a_{m1} & a_{m2} & \cdots & a_{mn}
\end{bmatrix}
$$

are the vectors

$$
\mathbf{c}_1 = \begin{bmatrix} a_{11} \\ a_{21} \\ \vdots \\ a_{m1} \end{bmatrix}, \quad
\mathbf{c}_2 = \begin{bmatrix} a_{12} \\ a_{22} \\ \vdots \\ a_{m2} \end{bmatrix}, \quad \ldots, \quad
\mathbf{c}_n = \begin{bmatrix} a_{1n} \\ a_{2n} \\ \vdots \\ a_{mn} \end{bmatrix}
$$

formed from the columns of $A$. The subspace of $R^m$ spanned by the column vectors is called the **column space** of $A$.

**Example 15**

The column vectors of

$$A = \begin{bmatrix} 2 & 1 & 0 \\ 3 & -1 & 4 \end{bmatrix}$$

are

$$\mathbf{c}_1 = \begin{bmatrix} 2 \\ 3 \end{bmatrix} \qquad \mathbf{c}_2 = \begin{bmatrix} 1 \\ -1 \end{bmatrix} \qquad \text{and} \qquad \mathbf{c}_3 = \begin{bmatrix} 0 \\ 4 \end{bmatrix}$$

Let $T: R^n \to R^m$ be multiplication by an $m \times n$ matrix $A$. A vector

$$\mathbf{b} = \begin{bmatrix} b_1 \\ b_2 \\ \vdots \\ b_m \end{bmatrix}$$

will be in the range of $T$ if and only if $T(\mathbf{x}) = \mathbf{b}$ for some vector $\mathbf{x}$ in $R^n$, or equivalently, if and only if the system of equations

$$A\mathbf{x} = \begin{bmatrix} a_{11} & a_{12} & \cdots & a_{1n} \\ a_{21} & a_{22} & \cdots & a_{2n} \\ \vdots & \vdots & & \vdots \\ a_{m1} & a_{m2} & \cdots & a_{mn} \end{bmatrix} \begin{bmatrix} x_1 \\ x_2 \\ \vdots \\ x_n \end{bmatrix} = \begin{bmatrix} b_1 \\ b_2 \\ \vdots \\ b_m \end{bmatrix} \qquad (5.4)$$

is consistent. By multiplying out the matrices on the left side, this system can be rewritten as

$$\begin{bmatrix} a_{11}x_1 + a_{12}x_2 + \cdots + a_{1n}x_n \\ a_{21}x_1 + a_{22}x_2 + \cdots + a_{2n}x_n \\ \vdots \\ a_{m1}x_1 + a_{m2}x_2 + \cdots + a_{mn}x_n \end{bmatrix} = \begin{bmatrix} b_1 \\ b_2 \\ \vdots \\ b_m \end{bmatrix}$$

or

$$x_1 \begin{bmatrix} a_{11} \\ a_{21} \\ \vdots \\ a_{m1} \end{bmatrix} + x_2 \begin{bmatrix} a_{12} \\ a_{22} \\ \vdots \\ a_{m2} \end{bmatrix} + \cdots + x_n \begin{bmatrix} a_{1n} \\ a_{2n} \\ \vdots \\ a_{mn} \end{bmatrix} = \begin{bmatrix} b_1 \\ b_2 \\ \vdots \\ b_m \end{bmatrix}$$

Since the matrices on the left side form the columns of $A$, it follows that (5.4) is consistent if and only if $\mathbf{b}$ is a linear combination of the column vectors from $A$. Thus we have the following useful theorem.

**Theorem 4.** *If $T: R^n \to R^m$ is multiplication by an $m \times n$ matrix $A$, then the range of $T$ is the column space of $A$.*

**Example 16**

Let $T: R^4 \to R^5$ be multiplication by

$$A = \begin{bmatrix} 1 & 2 & 0 & 2 \\ -2 & -5 & 5 & 6 \\ 0 & -3 & 15 & 18 \\ 0 & -2 & 10 & 8 \\ 3 & 6 & 0 & 6 \end{bmatrix}$$

Find a basis for the range of $T$.

*Solution.* By Theorem 4, the range of $T$ is the column space of $A$. The column space of $A$ is the row space of

$$A^t = \begin{bmatrix} 1 & -2 & 0 & 0 & 3 \\ 2 & -5 & -3 & -2 & 6 \\ 0 & 5 & 15 & 10 & 0 \\ 2 & 6 & 18 & 8 & 6 \end{bmatrix}$$

From Example 34 in Section 4.6

$$R = \begin{bmatrix} 1 & -2 & 0 & 0 & 3 \\ 0 & 1 & 3 & 2 & 0 \\ 0 & 0 & 1 & 1 & 0 \\ 0 & 0 & 0 & 0 & 0 \end{bmatrix}$$

is a row-echelon form of $A^t$. By Theorem 11 in Section 4.6, the nonzero rows of $R$ form a basis for the row space of $A^t$. Thus the nonzero columns of

$$R^t = \begin{bmatrix} 1 & 0 & 0 & 0 \\ -2 & 1 & 0 & 0 \\ 0 & 3 & 1 & 0 \\ 0 & 2 & 1 & 0 \\ 3 & 0 & 0 & 0 \end{bmatrix}$$

form a basis for the column space of $A$. Therefore

$$\begin{bmatrix} 1 \\ -2 \\ 0 \\ 0 \\ 3 \end{bmatrix}, \begin{bmatrix} 0 \\ 1 \\ 3 \\ 2 \\ 0 \end{bmatrix}, \text{ and } \begin{bmatrix} 0 \\ 0 \\ 1 \\ 1 \\ 0 \end{bmatrix}$$

form a basis for the range of $T$.

The next theorem will enable us to determine the rank of a matrix transformation by inspecting any row-echelon form of the matrix. The proof is deferred to the end of this section.

**Theorem 5.** *If A is any matrix, then the row space and column space of A have the same dimension.*

As an application of this theorem, let $T: R^n \to R^m$ be multiplication by an $m \times n$ matrix $A$. Since the range of $T$ is the column space of $A$, we have

$$\begin{aligned}
\dim(\text{range of } T) &= \dim(\text{column space of } A) \\
&= \dim(\text{row space of } A) \\
&= \text{number of nonzero rows in any row-echelon form of } A \text{ (Theorem} \\
&\quad \text{11 in Section 4.6)}
\end{aligned}$$

Thus *the rank of a matrix transformation is the number of nonzero rows in any row-echelon form of the matrix.*

**Example 17**

Let $T: R^4 \to R^5$ be multiplication by the matrix $A$ in Example 16. The reader can check that

$$\begin{bmatrix}
1 & 2 & 0 & 2 \\
0 & 1 & -5 & -10 \\
0 & 0 & 0 & 1 \\
0 & 0 & 0 & 0 \\
0 & 0 & 0 & 0
\end{bmatrix}$$

is a row-echelon form of $A$. Since there are three nonzero rows, rank of $T = 3$. This agrees with the results obtained in Example 16.

**Example 18**

Let $T: R^4 \to R^5$ be multiplication by the matrix $A$ in Example 16. Find the nullity of $T$ and a basis for the kernel of $T$.

*Solution.* From Example 12 the kernel of $T$ is the solution space of

$$\begin{bmatrix}
1 & 2 & 0 & 2 \\
-2 & -5 & 5 & 6 \\
0 & -3 & 15 & 18 \\
0 & -2 & 10 & 8 \\
3 & 6 & 0 & 6
\end{bmatrix}
\begin{bmatrix} x_1 \\ x_2 \\ x_3 \\ x_4 \end{bmatrix} =
\begin{bmatrix} 0 \\ 0 \\ 0 \\ 0 \\ 0 \end{bmatrix}.$$

Solving this system, we obtain

$$x_1 = -10t \qquad x_2 = 5t \qquad x_3 = t \qquad x_4 = 0$$

that is,

$$\begin{bmatrix} x_1 \\ x_2 \\ x_3 \\ x_4 \end{bmatrix} = \begin{bmatrix} -10t \\ 5t \\ t \\ 0 \end{bmatrix} = t \begin{bmatrix} -10 \\ 5 \\ 1 \\ 0 \end{bmatrix}$$

Thus

$$\begin{bmatrix} -10 \\ 5 \\ 1 \\ 0 \end{bmatrix}$$

is a basis for the kernel, and nullity of $T = 1$.

Our next theorem establishes a relationship between the range and kernel of a linear transformation defined on a finite dimensional vector space. We shall defer the proof to the end of the section.

**Theorem 6.** (*Dimension Theorem*). *If* $T: V \rightarrow W$ *is a linear transformation from an n-dimensional vector space* $V$ *to a vector space* $W$, *then*

$$(rank\ of\ T) + (nullity\ of\ T) = n$$

**Example 19**

Let $T: R^4 \rightarrow R^5$ be multiplication by the matrix $A$ in Example 16. From the dimension theorem, the rank and nullity of $T$ should satisfy

$$(\text{rank of } T) + (\text{nullity of } T) = 4$$

This agrees with the results obtained in Examples 17 and 18, where we showed rank of $T = 3$ and nullity of $T = 1$.

Let $T: R^n \rightarrow R^m$ be multiplication by an $m \times n$ matrix $A$. From the dimension theorem

$$\begin{aligned} (\text{nullity of } T) &= n - (\text{rank of } T) \\ &= (\text{number of columns of } A) - (\text{rank of } T) \end{aligned}$$

Thus *the nullity of a matrix transformation is the number of columns in the matrix minus the rank of the transformation.*

**Example 20**

Let $T: R^5 \rightarrow R^4$ be multiplication by

$$A = \begin{bmatrix} 1 & 1 & 2 & 0 & 1 \\ 2 & 2 & 5 & 0 & 3 \\ 0 & 0 & 0 & 1 & 0 \\ 0 & 0 & 0 & 0 & 0 \end{bmatrix}$$

Find the rank and nullity of $T$.

*Solution.* Reducing $A$ to row-echelon form we obtain

$$R = \begin{bmatrix} 1 & 1 & 2 & 0 & 1 \\ 0 & 0 & 1 & 0 & 1 \\ 0 & 0 & 0 & 1 & 0 \\ 0 & 0 & 0 & 0 & 0 \end{bmatrix}$$

Since this matrix has three nonzero rows, rank of $T = 3$. Also, since $A$ has five columns, nullity of $T = 5 - 3 = 2$.

Some writers refer to the rank and nullity of a matrix. The **rank** and **nullity of an $m \times n$ matrix** $A$ are defined to be the rank and nullity of $T: R^n \rightarrow R^m$, where $T$ is multiplication by $A$. Thus, if $A$ is the matrix in Example 20, rank of $A = 3$ and nullity of $A = 2$.

We conclude this section by using the Dimension Theorem to help add two more statements to those given in the corollary of Theorem 12 in Section 1.8 and in Theorem 6 of Section 2.3.

**Theorem 7.** *If $A$ is an $n \times n$ matrix, then the following statements are equivalent.*

(*a*) *$A$ is invertible.*
(*b*) *$Ax = 0$ has only the trivial solution.*
(*c*) *$A$ is row equivalent to $I_n$.*
(*d*) *$Ax = b$ is consistent for every $n \times 1$ matrix $b$.*
(*e*) *$det(A) \neq 0$.*
(*f*) *The row vectors of $A$ are linearly independent.*
(*g*) *The column vectors of $A$ are linearly independent.*

*Proof.* We shall show that (c), (f), and (g) are equivalent by proving the sequence of implications $(c) \Rightarrow (f) \Rightarrow (g) \Rightarrow (c)$. This will complete the proof since we already know that (c) is equivalent to (a), (b), (d), and (e).

$(c) \Rightarrow (f)$ Since $A$ is row equivalent to $I_n$, and $I_n$ has $n$ nonzero rows, the row space of $A$ is $n$-dimensional by Theorem 11 in Section 4.6. Since the row vectors of $A$ span the row space, the row vectors of $A$ are linearly independent by Theorem 9 in Section 4.5.

$(f) \Rightarrow (g)$ Assume the row vectors of $A$ are linearly independent. Thus the row space of $A$ is $n$-dimensional. By Theorem 5 the column space of $A$ is also $n$-dimensional. Since the column vectors of $A$ span the column space, the column vectors of $A$ are linearly independent by Theorem 9 in Section 4.6.

$(g) \Rightarrow (c)$. Assume the column vectors of $A$ are linearly independent. Thus the column space of $A$ is $n$-dimensional. If $T: R^n \rightarrow R^n$ is multiplication by $A$, then the range of $T$ is $n$-dimensional by Theorem 4. By the dimension theorem, the kernel of $T$ is zero dimensional, that is, $\ker(T) = \{\mathbf{0}\}$.

Equivalently,

$$A\mathbf{x} = \mathbf{0}$$

has only the trivial solution. From the equivalence of (b) and (c), $A$ is row equivalent to $I_n$. ▨

**Optional**

---

**Proof of Theorem 5.** Denote the row vectors of

$$A = \begin{bmatrix} a_{11} & a_{12} & \cdots & a_{1n} \\ a_{21} & a_{22} & \cdots & a_{2n} \\ \vdots & \vdots & & \vdots \\ a_{m1} & a_{m2} & \cdots & a_{mn} \end{bmatrix}$$

by

$$\mathbf{r}_1, \mathbf{r}_2, \ldots, \mathbf{r}_m$$

Suppose the row space of $A$ has dimension $k$ and that $S = \{\mathbf{b}_1, \mathbf{b}_2, \ldots, \mathbf{b}_k\}$ is a basis for the row space, where $\mathbf{b}_i = (b_{i1}, b_{i2}, \ldots, b_{in})$. Since $S$ is a basis for the row space, each row vector is expressible as a linear combination of $\mathbf{b}_1, \mathbf{b}_2, \ldots, \mathbf{b}_k$; thus

$$\begin{aligned} \mathbf{r}_1 &= c_{11}\mathbf{b}_1 + c_{12}\mathbf{b}_2 + \cdots + c_{1k}\mathbf{b}_k \\ \mathbf{r}_2 &= c_{21}\mathbf{b}_1 + c_{22}\mathbf{b}_2 + \cdots + c_{2k}\mathbf{b}_k \\ \vdots \quad & \qquad \vdots \qquad\quad \vdots \qquad\qquad\quad \vdots \\ \mathbf{r}_m &= c_{m1}\mathbf{b}_1 + c_{m2}\mathbf{b}_2 + \cdots + c_{mk}\mathbf{b}_k \end{aligned} \tag{5.5}$$

Since two vectors in $R^n$ are equal if and only if corresponding components are equal, we can equate the $j$th component on each side of (5.5) to obtain

$$\begin{aligned} a_{1j} &= c_{11}b_{1j} + c_{12}b_{2j} + \cdots + c_{1k}b_{kj} \\ a_{2j} &= c_{21}b_{1j} + c_{22}b_{2j} + \cdots + c_{2k}b_{kj} \\ \vdots \quad & \qquad \vdots \qquad\quad \vdots \qquad\qquad\quad \vdots \\ a_{mj} &= c_{m1}b_{1j} + c_{m2}b_{2j} + \cdots + c_{mk}b_{kj} \end{aligned}$$

or equivalently

$$\begin{bmatrix} a_{1j} \\ a_{2j} \\ \vdots \\ a_{mj} \end{bmatrix} = b_{1j}\begin{bmatrix} c_{11} \\ c_{21} \\ \vdots \\ c_{m1} \end{bmatrix} + b_{2j}\begin{bmatrix} c_{12} \\ c_{22} \\ \vdots \\ c_{m2} \end{bmatrix} + \cdots + b_{kj}\begin{bmatrix} c_{1k} \\ c_{2k} \\ \vdots \\ c_{mk} \end{bmatrix} \tag{5.6}$$

The left side of this equation is the $j$th column vector of $A$, and $j = 1, 2, \ldots, n$ is arbitrary; therefore each column vector of $A$ lies in the space spanned by the $k$ vectors on the right side of (5.6). Thus the column space of $A$ has dimension $\leq k$. Since

$$k = \dim(\text{row space of } A)$$

we have

$$\dim(\text{column space of } A) \leq \dim(\text{row space of } A). \qquad (5.7)$$

Since the matrix $A$ is completely arbitrary, this same conclusion applies to the transpose of $A$, that is

$$\dim(\text{column space of } A^t) \leq \dim(\text{row space of } A^t) \qquad (5.8)$$

But transposing a matrix converts columns to rows and rows to columns so that

$$\text{column space of } A^t = \text{row space of } A$$

and

$$\text{row space of } A^t = \text{column space of } A$$

Thus (5.8) can be rewritten as

$$\dim(\text{row space of } A) \leq \dim(\text{column space of } A).$$

From this result and (5.7) we conclude that

$$\dim(\text{row space of } A) = \dim(\text{column space of } A). \quad \blacksquare$$

**Optional**

---

**Proof of Theorem 6.** We must show that

$$\dim(R(T)) + \dim(\ker(T)) = n$$

We shall give the proof for the case where $1 \leq \dim(\ker(T)) < n$. The cases $\dim(\ker(T)) = 0$ and $\dim(\ker(T)) = n$ are left as exercises. Assume $\dim(\ker(T)) = r$, and let $v_1, \ldots, v_r$ be a basis for the kernel. Since $\{v_1, \ldots, v_r\}$ is linearly independent, part (c) of Theorem 9 in Chapter 4 states that there are $n - r$ vectors, $v_{r+1}, \ldots, v_n$, such that $\{v_1, \ldots, v_r, v_{r+1}, \ldots, v_n\}$ is a basis for $V$. To complete the proof, we shall show that the $n - r$ vectors in the set $S = \{T(v_{r+1}), \ldots, T(v_n)\}$ form a basis for the range of $T$. It will then follow that

$$\dim(R(T)) + \dim(\ker(T)) = (n - r) + r = n$$

First we show that $S$ spans the range of $T$. If $b$ is any vector in the range of $T$, then $b = T(v)$ for some vector $v$ in $V$. Since $\{v_1, \ldots, v_r, v_{r+1}, \ldots, v_n\}$ is a basis for $V$, $v$ can be written in the form

$$v = c_1 v_1 + \cdots + c_r v_r + c_{r+1} v_{r+1} + \cdots + c_n v_n$$

Since $v_1, \ldots, v_r$ lie in the kernel of $T$, $T(v_1) = \cdots = T(v_r) = 0$, so that

$$b = T(v) = c_{r+1} T(v_{r+1}) + \cdots + c_n T(v_n)$$

Thus $S$ spans the range of $T$.

Finally, we show that $S$ is a linearly independent set and consequently forms a basis for the range of $T$. Suppose some linear combination of the vectors in $S$ is zero, that is

$$k_{r+1}T(\mathbf{v}_{r+1}) + \cdots + k_nT(\mathbf{v}_n) = \mathbf{0} \tag{5.9}$$

We must show $k_{r+1} = \cdots = k_n = 0$. Since $T$ is linear, (5.9) can be rewritten as

$$T(k_{r+1}\mathbf{v}_{r+1} + \cdots + k_n\mathbf{v}_n) = \mathbf{0}$$

which says that $k_{r+1}\mathbf{v}_{r+1} + \cdots + k_n\mathbf{v}_n$ is in the kernel of $T$. This vector can therefore be written as a linear combination of the basis vectors $\{\mathbf{v}_1, \ldots, \mathbf{v}_r\}$ say,

$$k_{r+1}\mathbf{v}_{r+1} + \cdots + k_n\mathbf{v}_n = k_1\mathbf{v}_1 + \cdots + k_r\mathbf{v}_r.$$

Thus

$$k_1\mathbf{v}_1 + \cdots + k_r\mathbf{v}_r - k_{r+1}\mathbf{v}_{r+1} - \cdots - k_n\mathbf{v}_n = \mathbf{0}$$

Since $\{\mathbf{v}_1, \ldots, \mathbf{v}_n\}$ is linearly independent, all the $k$'s are zero; in particular, $k_{r+1} = \cdots = k_n = 0$, which completes the proof. ∎

## EXERCISE SET 5.3

1. List the column vectors of the matrix

$$\begin{bmatrix} 2 & -1 & 0 & 1 \\ 3 & 5 & 7 & -1 \\ 1 & 4 & 2 & 7 \end{bmatrix}$$

2. In each part the matrix $A$ is in row-echelon form. By inspection find the rank and nullity of $T: R^m \rightarrow R^n$, where $T$ is multiplication by $A$.

(a) $A = \begin{bmatrix} 1 & 0 & 0 & 4 \\ 0 & 1 & 0 & 3 \\ 0 & 0 & 1 & 2 \end{bmatrix}$

(b) $A = \begin{bmatrix} 1 & 0 & 2 & 3 & 2 \\ 0 & 1 & 0 & -1 & 4 \\ 0 & 0 & 1 & 1 & 2 \end{bmatrix}$

(c) $A = \begin{bmatrix} 1 & 5 & 3 & 0 & 5 & -1 \\ 0 & 0 & 1 & 0 & 3 & 1 \\ 0 & 0 & 0 & 1 & 4 & 2 \\ 0 & 0 & 0 & 0 & 0 & 0 \end{bmatrix}$

(d) $A = \begin{bmatrix} 1 & 3 & 6 & 4 & 2 & 0 & 0 \\ 0 & 0 & 1 & 2 & 0 & 0 & 0 \\ 0 & 0 & 0 & 0 & 0 & 0 & 0 \\ 0 & 0 & 0 & 0 & 0 & 0 & 0 \end{bmatrix}$

In Exercises 3–6 find the rank and nullity of the matrix.

3. $\begin{bmatrix} 1 & -3 \\ 2 & -6 \end{bmatrix}$

4. $\begin{bmatrix} 1 & 2 & -1 \\ 2 & 4 & 6 \\ 0 & 0 & -8 \end{bmatrix}$

5. $\begin{bmatrix} -3 & 5 & 1 & 2 \\ 7 & 2 & 0 & -4 \\ -8 & 3 & 1 & 6 \end{bmatrix}$

6. $\begin{bmatrix} 1 & -3 & 2 & 1 \\ 0 & 3 & 6 & -2 \\ 2 & -3 & -2 & 4 \\ 3 & -3 & 6 & 3 \\ 5 & -3 & 10 & 5 \end{bmatrix}$

**7.** Verify that the row space and column space have the same dimension (as guaranteed by Theorem 5).

(a) $\begin{bmatrix} 2 & 0 & 2 & 2 \\ 3 & -4 & -1 & -9 \\ 1 & 2 & 3 & 7 \\ -3 & 1 & -2 & 0 \end{bmatrix}$   (b) $\begin{bmatrix} 2 & 3 & 5 & 7 & 4 \\ -1 & 2 & 1 & 0 & -2 \\ 4 & 1 & 5 & 9 & 8 \end{bmatrix}$

**8.** In each part use the given information to find the nullity of $T$.
   (a) $T: R^5 \to R^7$ has rank 3.        (b) $T: P_4 \to P_3$ has rank 1.
   (c) The range of $T: R^6 \to R^3$ is $R^3$.   (d) $T: M_{22} \to M_{22}$ has rank 3.

**9.** Let $A$ be a $7 \times 6$ matrix such that $A\mathbf{x} = \mathbf{0}$ has only the trivial solution, and let $T: R^6 \to R^7$ be multiplication by $A$. Find the rank and nullity of $T$.

**10.** Let $A$ be a $5 \times 7$ matrix with rank 4.
   (a) What is the dimension of the solution space of $A\mathbf{x} = \mathbf{0}$?
   (b) Is $A\mathbf{x} = \mathbf{b}$ consistent for all vectors $\mathbf{b}$ in $R^5$? Explain.

In Exercises 11–14, let $T$ be multiplication by the given matrix. Find:
   (a) A basis for the range of $T$.
   (b) A basis for the kernel of $T$.
   (c) The rank and nullity of $T$.

**11.** $\begin{bmatrix} 1 & -1 & 3 \\ 5 & 6 & -4 \\ 7 & 4 & 2 \end{bmatrix}$   **12.** $\begin{bmatrix} 2 & 0 & -1 \\ 4 & 0 & -2 \\ 0 & 0 & 0 \end{bmatrix}$

**13.** $\begin{bmatrix} 4 & 1 & 5 & 2 \\ 1 & 2 & 3 & 0 \end{bmatrix}$   **14.** $\begin{bmatrix} 1 & 4 & 5 & 0 & 9 \\ 3 & -2 & 1 & 0 & -1 \\ -1 & 0 & -1 & 0 & -1 \\ 2 & 3 & 5 & 1 & 8 \end{bmatrix}$

**15.** Prove that a system of equations $A\mathbf{x} = \mathbf{b}$ is consistent if and only if the rank of its augmented matrix is equal to the rank of $A$.

**16.** Prove the dimension theorem in the cases:
   (a) $\dim(\ker(T)) = 0$
   (b) $\dim(\ker(T)) = n$

**17.** Let $T: V \to V$ be a linear operator on a finite dimensional vector space $V$. Prove that $R(T) = V$ if and only if $\ker(T) = \{\mathbf{0}\}$.

## 5.4 MATRICES OF LINEAR TRANSFORMATIONS

In this section we show that every linear transformation on a finite dimensional vector space can be regarded as a matrix transformation. Our work here will enable us to exploit our knowledge of matrix transformations to study more general linear transformations.

We shall show first that every linear transformation from $R^n$ to $R^m$ is a matrix transformation. More precisely, we shall show that if $T: R^n \to R^m$ is any linear transformation, then we can find an $m \times n$ matrix $A$ such that $T$ is multiplication by $A$. To see this, let

$$\mathbf{e}_1, \mathbf{e}_2, \ldots, \mathbf{e}_n$$

be the standard basis for $R^n$, and let $A$ be the $m \times n$ matrix having

$$T(\mathbf{e}_1), T(\mathbf{e}_2), \ldots, T(\mathbf{e}_n)$$

as its column vectors. (We shall assume in this section that all vectors are expressed in matrix notation.) For example, if $T: R^2 \to R^2$ is given by

$$T\left(\begin{bmatrix} x_1 \\ x_2 \end{bmatrix}\right) = \begin{bmatrix} x_1 + 2x_2 \\ x_1 - x_2 \end{bmatrix}$$

then

$$T(\mathbf{e}_1) = T\left(\begin{bmatrix} 1 \\ 0 \end{bmatrix}\right) = \begin{bmatrix} 1 \\ 1 \end{bmatrix} \quad \text{and} \quad T(\mathbf{e}_2) = T\left(\begin{bmatrix} 0 \\ 1 \end{bmatrix}\right) = \begin{bmatrix} 2 \\ -1 \end{bmatrix}$$

$$A = \begin{bmatrix} 1 & 2 \\ 1 & -1 \end{bmatrix}$$
$$\quad\uparrow\quad\uparrow$$
$$T(\mathbf{e}_1)\ T(\mathbf{e}_2)$$

More generally, if

$$T(\mathbf{e}_1) = \begin{bmatrix} a_{11} \\ a_{21} \\ \vdots \\ a_{m1} \end{bmatrix}, T(\mathbf{e}_2) = \begin{bmatrix} a_{12} \\ a_{22} \\ \vdots \\ a_{m2} \end{bmatrix}, \ldots, T(\mathbf{e}_n) = \begin{bmatrix} a_{1n} \\ a_{2n} \\ \vdots \\ a_{mn} \end{bmatrix}$$

then

$$A = \begin{bmatrix} a_{11} & a_{12} & \cdots & a_{1n} \\ a_{21} & a_{22} & & a_{2n} \\ \vdots & \vdots & & \vdots \\ a_{m1} & a_{m2} & \cdots & a_{mn} \end{bmatrix} \tag{5.10}$$
$$\quad\uparrow\qquad\uparrow\qquad\quad\uparrow$$
$$T(\mathbf{e}_1)\quad T(\mathbf{e}_2)\quad \cdots \quad T(\mathbf{e}_n)$$

We shall show that the linear transformation $T: R^n \to R^m$ is multiplication by $A$. To see this, observe first that

$$\mathbf{x} = \begin{bmatrix} x_1 \\ x_2 \\ \vdots \\ x_n \end{bmatrix} = x_1\mathbf{e}_1 + x_2\mathbf{e}_2 + \cdots + x_n\mathbf{e}_n$$

Therefore, by the linearity of $T$,

$$T(\mathbf{x}) = x_1 T(\mathbf{e}_1) + x_2 T(\mathbf{e}_2) + \cdots + x_n T(\mathbf{e}_n) \tag{5.11}$$

On the other hand

$$
A\mathbf{x} =
\begin{bmatrix}
a_{11} & a_{12} & \cdots & a_{1n} \\
a_{21} & a_{22} & \cdots & a_{2n} \\
\vdots & \vdots & & \vdots \\
a_{m1} & a_{m2} & \cdots & a_{mn}
\end{bmatrix}
\begin{bmatrix}
x_1 \\ x_2 \\ \vdots \\ x_n
\end{bmatrix}
=
\begin{bmatrix}
a_{11}x_1 + a_{12}x_2 + \cdots + a_{1n}x_n \\
a_{21}x_1 + a_{22}x_2 + \cdots + a_{2n}x_n \\
\vdots \\
a_{m1}x_1 + a_{m2}x_2 + \cdots + a_{mn}x_n
\end{bmatrix}
$$

$$
= x_1
\begin{bmatrix} a_{11} \\ a_{21} \\ \vdots \\ a_{m1} \end{bmatrix}
+ x_2
\begin{bmatrix} a_{12} \\ a_{22} \\ \vdots \\ a_{m2} \end{bmatrix}
+ \cdots + x_n
\begin{bmatrix} a_{1n} \\ a_{2n} \\ \vdots \\ a_{mn} \end{bmatrix}
$$

$$
= x_1 T(\mathbf{e}_1) + x_2 T(\mathbf{e}_2) + \cdots + x_n T(\mathbf{e}_n) \tag{5.12}
$$

Comparing (5.11) and (5.12) yields $T(\mathbf{x}) = A\mathbf{x}$, that is, $T$ is multiplication by $A$. We shall refer to the matrix $A$ in (5.10) as the ***standard matrix for T.***

**Example 21**

Find the standard matrix for the transformation $T: R^3 \to R^4$ defined by

$$
T\left( \begin{bmatrix} x_1 \\ x_2 \\ x_3 \end{bmatrix} \right) =
\begin{bmatrix} x_1 + x_2 \\ x_1 - x_2 \\ x_3 \\ x_1 \end{bmatrix}
$$

*Solution.*

$$
T(\mathbf{e}_1) = T\left( \begin{bmatrix} 1 \\ 0 \\ 0 \end{bmatrix} \right) = \begin{bmatrix} 1 \\ 1 \\ 0 \\ 1 \end{bmatrix}
\qquad
T(\mathbf{e}_2) = T\left( \begin{bmatrix} 0 \\ 1 \\ 0 \end{bmatrix} \right) = \begin{bmatrix} 1 \\ -1 \\ 0 \\ 0 \end{bmatrix}
\qquad
T(\mathbf{e}_3) = T\left( \begin{bmatrix} 0 \\ 0 \\ 1 \end{bmatrix} \right) = \begin{bmatrix} 0 \\ 0 \\ 1 \\ 0 \end{bmatrix}
$$

Using $T(\mathbf{e}_1)$, $T(\mathbf{e}_2)$, and $T(\mathbf{e}_3)$ as column vectors, we obtain

$$
A =
\begin{bmatrix}
1 & 1 & 0 \\
1 & -1 & 0 \\
0 & 0 & 1 \\
1 & 0 & 0
\end{bmatrix}
$$

As a check, observe that

$$
A \begin{bmatrix} x_1 \\ x_2 \\ x_3 \end{bmatrix} =
\begin{bmatrix} x_1 + x_2 \\ x_1 - x_2 \\ x_3 \\ x_1 \end{bmatrix}
$$

which agrees with the given formula for $T$.

**Figure 5.5**

We show next that if $V$ and $W$ are any finite dimensional vector spaces (not necessarily $R^n$ and $R^m$), then with a little ingenuity, any linear transformation $T: V \rightarrow W$ can be regarded as a matrix transformation. The basic idea is as follows. Let $B$ and $B'$ be bases for $V$ and $W$ respectively and suppose that somehow we can find a matrix $A$ such that

$$A[\mathbf{x}]_B = [T(\mathbf{x})]_{B'} \tag{5.13}$$

for each $\mathbf{x}$ in $V$, where $[\mathbf{x}]_B$ and $[T(\mathbf{x})]_{B'}$ are the coordinate matrices for $\mathbf{x}$ and $T(\mathbf{x})$. In other words, multiplication by $A$ maps the coordinate matrix for $\mathbf{x}$ into the coordinate matrix for $T(\mathbf{x})$. Then, as shown in Figure 5.5, $T(\mathbf{x})$ can be computed in three steps by the following indirect procedure.

**1.** Compute the coordinate matrix $[\mathbf{x}]_B$.
**2.** Multiply $[\mathbf{x}]_B$ on the left by $A$ to produce $[T(\mathbf{x})]_{B'}$.
**3.** Reconstruct $T(\mathbf{x})$ from its coordinate matrix $[T(\mathbf{x})]_{B'}$.

The indirect procedure is well suited for computer computation. Further, since the matrix $A$ depends on the bases $B$ and $B'$, it is often possible to simplify various computations by judiciously choosing the bases $B$ and $B'$ so that $A$ has lots of zero entries.

We turn now to the problem of finding a matrix $A$ satisfying (5.13). Suppose $V$ is an $n$-dimensional space with basis $B = \{\mathbf{u}_1, \mathbf{u}_2, \ldots, \mathbf{u}_n\}$ and $W$ is an $m$-dimensional space with basis $B' = \{\mathbf{v}_1, \mathbf{v}_2, \ldots, \mathbf{v}_m\}$. We are looking for an $m \times n$ matrix

$$A = \begin{bmatrix} a_{11} & a_{12} & \cdots & a_{1n} \\ a_{21} & a_{22} & \cdots & a_{2n} \\ \vdots & \vdots & & \vdots \\ a_{m1} & a_{m2} & \cdots & a_{mn} \end{bmatrix}$$

such that (5.13) holds for all vectors $\mathbf{x}$ in $V$. In particular, when $\mathbf{x}$ is the basis vector $\mathbf{u}_1$ we want

$$A[\mathbf{u}_1]_B = [T(\mathbf{u}_1)]_{B'} \tag{5.14}$$

But

$$[\mathbf{u}_1]_B = \begin{bmatrix} 1 \\ 0 \\ 0 \\ \vdots \\ 0 \end{bmatrix}$$

so that

$$A[\mathbf{u}_1]_B = \begin{bmatrix} a_{11} & a_{12} & \cdots & a_{1n} \\ a_{21} & a_{22} & \cdots & a_{2n} \\ \vdots & \vdots & & \vdots \\ a_{m1} & a_{m2} & \cdots & a_{mn} \end{bmatrix} \begin{bmatrix} 1 \\ 0 \\ 0 \\ \vdots \\ 0 \end{bmatrix} = \begin{bmatrix} a_{11} \\ a_{21} \\ \vdots \\ a_{m1} \end{bmatrix}$$

Thus (5.14) implies that

$$\begin{bmatrix} a_{11} \\ a_{21} \\ \vdots \\ a_{m1} \end{bmatrix} = [T(\mathbf{u}_1)]_{B'}$$

That is, the first column of $A$ is the coordinate matrix for the vector $T(\mathbf{u}_1)$ with respect to the basis $B'$. Similarly, if we let $\mathbf{x} = \mathbf{u}_2$ in (5.13) we obtain

$$A[\mathbf{u}_2]_B = [T(\mathbf{u}_2)]_{B'} \tag{5.15}$$

But

$$[\mathbf{u}_2]_B = \begin{bmatrix} 0 \\ 1 \\ 0 \\ \vdots \\ 0 \end{bmatrix}$$

so that

$$A[\mathbf{u}_2]_B = \begin{bmatrix} a_{11} & a_{12} & \cdots & a_{1n} \\ a_{21} & a_{22} & \cdots & a_{2n} \\ \vdots & \vdots & & \vdots \\ a_{m1} & a_{m2} & \cdots & a_{mn} \end{bmatrix} \begin{bmatrix} 0 \\ 1 \\ 0 \\ \vdots \\ 0 \end{bmatrix} = \begin{bmatrix} a_{12} \\ a_{22} \\ \vdots \\ a_{m2} \end{bmatrix}$$

Thus

$$\begin{bmatrix} a_{12} \\ a_{22} \\ \vdots \\ a_{m2} \end{bmatrix} = [T(\mathbf{u}_2)]_{B'}$$

That is, the second column of $A$ is the coordinate matrix for the vector $T(\mathbf{u}_2)$ with respect to the basis $B'$. Continuing in this way, we find that *the jth column of A is the coordinate matrix for the vector $T(\mathbf{u}_j)$ with respect to B'*. The unique matrix $A$ obtained in this way is called the ***matrix of T with respect to the bases B*** and ***B'***.

**Example 22**

Let $T: R^3 \to R^4$ be the linear transformation defined by

$$T\left(\begin{bmatrix} x_1 \\ x_2 \\ x_3 \end{bmatrix}\right) = \begin{bmatrix} x_1 + x_2 \\ x_1 - x_2 \\ x_3 \\ x_1 \end{bmatrix}$$

Find the matrix of $T$ with respect to the bases $B = \{\mathbf{u}_1, \mathbf{u}_2, \mathbf{u}_3\}$ and $B' = \{\mathbf{v}_1, \mathbf{v}_2, \mathbf{v}_3, \mathbf{v}_4\}$ for $R^3$ and $R^4$, where

$$\mathbf{u}_1 = \begin{bmatrix} 2 \\ 0 \\ 0 \end{bmatrix} \quad \mathbf{u}_2 = \begin{bmatrix} 0 \\ 3 \\ 0 \end{bmatrix} \quad \mathbf{u}_3 = \begin{bmatrix} 0 \\ 0 \\ 4 \end{bmatrix} \quad \mathbf{v}_1 = \begin{bmatrix} 1 \\ 0 \\ 0 \\ 0 \end{bmatrix} \quad \mathbf{v}_2 = \begin{bmatrix} 1 \\ 1 \\ 0 \\ 0 \end{bmatrix} \quad \mathbf{v}_3 = \begin{bmatrix} 1 \\ 1 \\ 1 \\ 0 \end{bmatrix} \quad \mathbf{v}_4 = \begin{bmatrix} 1 \\ 1 \\ 1 \\ 1 \end{bmatrix}$$

*Solution.* Let $A$ be the desired matrix. From the formula for $T$

$$T(\mathbf{u}_1) = \begin{bmatrix} 2 \\ 2 \\ 0 \\ 2 \end{bmatrix} \quad T(\mathbf{u}_2) = \begin{bmatrix} 3 \\ -3 \\ 0 \\ 0 \end{bmatrix} \quad \text{and } T(\mathbf{u}_3) = \begin{bmatrix} 0 \\ 0 \\ 4 \\ 0 \end{bmatrix}$$

The first column of $A$ is $[T(\mathbf{u}_1)]_{B'}$. To determine this coordinate matrix, we must express $T(\mathbf{u}_1)$ in terms of the basis vectors in $B'$; that is, we must find $c_1, c_2, c_3, c_4$ such that

$$T(\mathbf{u}_1) = c_1\mathbf{v}_1 + c_2\mathbf{v}_2 + c_3\mathbf{v}_3 + c_4\mathbf{v}_4$$

or

$$\begin{bmatrix} 2 \\ 2 \\ 0 \\ 2 \end{bmatrix} = c_1 \begin{bmatrix} 1 \\ 0 \\ 0 \\ 0 \end{bmatrix} + c_2 \begin{bmatrix} 1 \\ 1 \\ 0 \\ 0 \end{bmatrix} + c_3 \begin{bmatrix} 1 \\ 1 \\ 1 \\ 0 \end{bmatrix} + c_4 \begin{bmatrix} 1 \\ 1 \\ 1 \\ 1 \end{bmatrix}$$

Equating corresponding components gives

$$\begin{aligned} c_1 + c_2 + c_3 + c_4 &= 2 \\ c_2 + c_3 + c_4 &= 2 \\ c_3 + c_4 &= 0 \\ c_4 &= 2 \end{aligned}$$

Solving this system we obtain

$$c_1 = 0 \quad c_2 = 2 \quad c_3 = -2 \quad c_4 = 2$$

thus

$$[T(\mathbf{u}_1)]_{B'} = \begin{bmatrix} 0 \\ 2 \\ -2 \\ 2 \end{bmatrix}$$

which is the first column of $A$. Similar computations give

$$[T(\mathbf{u}_2)]_{B'} = \begin{bmatrix} 6 \\ -3 \\ 0 \\ 0 \end{bmatrix} \qquad [T(\mathbf{u}_3)]_{B'} = \begin{bmatrix} 0 \\ -4 \\ 4 \\ 0 \end{bmatrix}$$

so that

$$A = \begin{bmatrix} 0 & 6 & 0 \\ 2 & -3 & -4 \\ -2 & 0 & 4 \\ 2 & 0 & 0 \end{bmatrix}$$

We emphasize that the matrix of a linear transformation depends on the bases chosen. To illustrate, in Examples 21 and 22 we have two different matrices of the same linear transformation.

**Example 23**

Let $T: R^3 \to R^4$, $B$, and $B'$ be as in Example 22, and let

$$\mathbf{x} = \begin{bmatrix} 4 \\ -6 \\ 4 \end{bmatrix}$$

Use the matrix of $T$ with respect to $B$ and $B'$ to compute $T(\mathbf{x})$ by the indirect procedure in Figure 5.5.

*Solution.* The coordinate matrix of $\mathbf{x}$ with respect to $B$ is

$$[\mathbf{x}]_B = \begin{bmatrix} 2 \\ -2 \\ 1 \end{bmatrix}$$

Therefore

$$[T(\mathbf{x})]_{B'} = A[\mathbf{x}]_B = \begin{bmatrix} 0 & 6 & 0 \\ 2 & -3 & -4 \\ -2 & 0 & 4 \\ 2 & 0 & 0 \end{bmatrix} \begin{bmatrix} 2 \\ -2 \\ 1 \end{bmatrix} = \begin{bmatrix} -12 \\ 6 \\ 0 \\ 4 \end{bmatrix}$$

Thus

$$T(\mathbf{x}) = -12\mathbf{v}_1 + 6\mathbf{v}_2 + 0\mathbf{v}_3 + 4\mathbf{v}_4 = \begin{bmatrix} -2 \\ 10 \\ 4 \\ 4 \end{bmatrix}$$

The reader should check to see that this result agrees with that obtained by directly applying the formula for $T$.

**Example 24**

Let $T: P_1 \rightarrow P_2$ be the linear transformation defined by

$$T(p(x)) = xp(x)$$

Find the matrix of $T$ with respect to the bases

$$B = \{\mathbf{q}_1, \mathbf{q}_2\} \quad \text{and} \quad B' = \{\mathbf{p}_1, \mathbf{p}_2, \mathbf{p}_3\}$$

where

$$\mathbf{q}_1 = 1 \quad \mathbf{q}_2 = x$$

and

$$\mathbf{p}_1 = 1 + 2x + x^2 \quad \mathbf{p}_2 = 2 + 9x \quad \mathbf{p}_3 = 3 + 3x + 4x^2$$

*Solution.* Let $A$ be the matrix of $T$ with respect to the bases $B$ and $B'$. From the definition of $T$

$$T(\mathbf{q}_1) = xq_1(x) = x$$
$$T(\mathbf{q}_2) = xq_2(x) = x^2$$

The reader can check that the coordinate matrices for $T(\mathbf{q}_1)$ and $T(\mathbf{q}_2)$ are

$$[T(\mathbf{q}_1)]_{B'} = \begin{bmatrix} 8 \\ -1 \\ -2 \end{bmatrix} \quad \text{and} \quad [T(\mathbf{q}_2)]_{B'} = \begin{bmatrix} 21 \\ -3 \\ -5 \end{bmatrix}$$

Therefore

$$A = \begin{bmatrix} 8 & 21 \\ -1 & -3 \\ -2 & -5 \end{bmatrix}$$

In the special case where $V = W$ (so that $T: V \rightarrow V$ is a linear operator) it is usual to take $B = B'$ when constructing a matrix of $T$. The resulting matrix is called the ***matrix of T with respect to the basis B.***

**Example 25**

If $B = \{\mathbf{u}_1, \mathbf{u}_2, \ldots, \mathbf{u}_n\}$ is any basis for a finite dimensional vector space $V$ and $I: V \rightarrow V$ is the identity operator on $V$, then $I(\mathbf{u}_1) = \mathbf{u}_1$, $I(\mathbf{u}_2) = \mathbf{u}_2, \ldots, I(\mathbf{u}_n) = \mathbf{u}_n$. Therefore

$$[I(\mathbf{u}_1)]_B = \begin{bmatrix} 1 \\ 0 \\ 0 \\ \vdots \\ 0 \end{bmatrix}, \quad [I(\mathbf{u}_2)]_B = \begin{bmatrix} 0 \\ 1 \\ 0 \\ \vdots \\ 0 \end{bmatrix}, \ldots, [I(\mathbf{u}_n)]_B = \begin{bmatrix} 0 \\ 0 \\ 0 \\ \vdots \\ 1 \end{bmatrix}$$

thus

$$[I]_B = \begin{bmatrix} 1 & 0 & \cdots & 0 \\ 0 & 1 & \cdots & 0 \\ 0 & 0 & \cdots & 0 \\ \vdots & \vdots & & \vdots \\ 0 & 0 & \cdots & 1 \end{bmatrix}$$

Consequently, the matrix of the identity operator with respect to any basis is the $n \times n$ identity matrix.

A square matrix in which all the *non*diagonal entries are zeros is called a ***diagonal matrix.*** Diagonal matrices have many desirable properties. For example, the $k$th power of a diagonal matrix

$$D = \begin{bmatrix} d_1 & 0 & \cdots & 0 \\ 0 & d_2 & \cdots & 0 \\ \vdots & \vdots & & \vdots \\ 0 & 0 & \cdots & d_n \end{bmatrix}$$

is

$$D^k = \begin{bmatrix} d_1{}^k & 0 & \cdots & 0 \\ 0 & d_2{}^k & \cdots & 0 \\ \vdots & \vdots & & \vdots \\ 0 & 0 & \cdots & d_n{}^k \end{bmatrix}$$

Thus, to raise a diagonal matrix to the $k$th power, we need only raise each diagonal entry to the $k$th power. For a nondiagonal matrix there is much more computation involved in obtaining the $k$th power. Diagonal matrices also have other useful properties.

The next example illustrates that sometimes the right choice of basis can lead to a diagonal matrix of a linear operator. In the next chapter we consider the problem of finding such bases.

**Example 26**

Let $T: R^2 \to R^2$ be the linear operator defined by

$$T\left(\begin{bmatrix} x_1 \\ x_2 \end{bmatrix}\right) = \begin{bmatrix} x_1 + x_2 \\ -2x_1 + 4x_2 \end{bmatrix}$$

Find the matrix of $T$ with respect to the basis $B = \{\mathbf{u}_1, \mathbf{u}_2\}$ where

$$\mathbf{u}_1 = \begin{bmatrix} 1 \\ 1 \end{bmatrix} \quad \text{and} \quad \mathbf{u}_2 = \begin{bmatrix} 1 \\ 2 \end{bmatrix}$$

*Solution.*

From the definition of $T$

$$T(\mathbf{u}_1) = \begin{bmatrix} 2 \\ 2 \end{bmatrix} = 2\mathbf{u}_1 \quad \text{and} \quad T(\mathbf{u}_2) = \begin{bmatrix} 3 \\ 6 \end{bmatrix} = 3\mathbf{u}_2$$

Therefore

$$[T(\mathbf{u}_1)]_B = \begin{bmatrix} 2 \\ 0 \end{bmatrix} \quad \text{and} \quad [T(\mathbf{u}_2)]_B = \begin{bmatrix} 0 \\ 3 \end{bmatrix}$$

Consequently, the matrix of $T$ with respect to $B$ is

$$A = \begin{bmatrix} 2 & 0 \\ 0 & 3 \end{bmatrix}$$

## EXERCISE SET 5.4

**1.** Find the standard matrix of each of the following linear operators.

(a) $T\left(\begin{bmatrix} x_1 \\ x_2 \end{bmatrix}\right) = \begin{bmatrix} 2x_1 - x_2 \\ x_1 + x_2 \end{bmatrix}$      (b) $T\left(\begin{bmatrix} x_1 \\ x_2 \end{bmatrix}\right) = \begin{bmatrix} x_1 \\ x_2 \end{bmatrix}$

(c) $T\left(\begin{bmatrix} x_1 \\ x_2 \\ x_3 \end{bmatrix}\right) = \begin{bmatrix} x_1 + 2x_2 + x_3 \\ x_1 + 5x_2 \\ x_3 \end{bmatrix}$      (d) $T\left(\begin{bmatrix} x_1 \\ x_2 \\ x_3 \end{bmatrix}\right) = \begin{bmatrix} 4x_1 \\ 7x_2 \\ -8x_3 \end{bmatrix}$

**2.** Find the standard matrix of each of the following linear transformations.

(a) $T\left(\begin{bmatrix} x_1 \\ x_2 \end{bmatrix}\right) = \begin{bmatrix} x_2 \\ -x_1 \\ x_1 + 3x_2 \\ x_1 - x_2 \end{bmatrix}$      (b) $T\left(\begin{bmatrix} x_1 \\ x_2 \\ x_3 \\ x_4 \end{bmatrix}\right) = \begin{bmatrix} 7x_1 + 2x_2 - x_3 + x_4 \\ x_2 + x_3 \\ -x_1 \end{bmatrix}$

(c) $T\left(\begin{bmatrix} x_1 \\ x_2 \\ x_3 \end{bmatrix}\right) = \begin{bmatrix} 0 \\ 0 \\ 0 \\ 0 \\ 0 \end{bmatrix}$      (d) $T\left(\begin{bmatrix} x_1 \\ x_2 \\ x_3 \\ x_4 \end{bmatrix}\right) = \begin{bmatrix} x_4 \\ x_1 \\ x_3 \\ x_2 \\ x_1 - x_3 \end{bmatrix}$

**3.** Let $T: P_2 \to P_1$ be the linear transformation defined by

$$T(a_0 + a_1 x + a_2 x^2) = (a_0 + a_1) - (2a_1 + 3a_2)x$$

Find the matrix of $T$ with respect to the standard bases for $P_2$ and $P_1$.

**4.** Let $T: M_{22} \to M_{22}$ be the linear operator defined by $T(A) = A^t$. Find the matrix of $T$ with respect to the basis in Example 26 of Section 4.5.

**5.** Let $R^3$ have the Euclidean inner product, and let $T: R^3 \to R^3$ be the orthogonal projection on the $yz$-plane. Find the standard matrix for $T$.

**6.** Let $T: R^2 \to R^3$ be defined by

$$T\left(\begin{bmatrix} x_1 \\ x_2 \end{bmatrix}\right) = \begin{bmatrix} x_1 + 2x_2 \\ -x_1 \\ 0 \end{bmatrix}$$

(a) Find the matrix of $T$ with respect to the bases $B = \{\mathbf{u}_1, \mathbf{u}_2\}$ and $B' = \{\mathbf{v}_1, \mathbf{v}_2, \mathbf{v}_3\}$, where

$$\mathbf{u}_1 = \begin{bmatrix} 1 \\ 3 \end{bmatrix} \qquad \mathbf{u}_2 = \begin{bmatrix} -2 \\ 4 \end{bmatrix} \qquad \mathbf{v}_1 = \begin{bmatrix} 1 \\ 1 \\ 1 \end{bmatrix} \qquad \mathbf{v}_2 = \begin{bmatrix} 2 \\ 2 \\ 0 \end{bmatrix} \qquad \mathbf{v}_3 = \begin{bmatrix} 3 \\ 0 \\ 0 \end{bmatrix}$$

(b) Use the matrix obtained in (a) to compute

$$T\left(\begin{bmatrix} 8 \\ 3 \end{bmatrix}\right)$$

**7.** Let $T: R^3 \to R^3$ be defined by

$$T\left(\begin{bmatrix} x_1 \\ x_2 \\ x_3 \end{bmatrix}\right) = \begin{bmatrix} x_1 - x_2 \\ x_2 - x_1 \\ x_1 - x_3 \end{bmatrix}$$

(a) Find the matrix of $T$ with respect to the basis $B = \{\mathbf{v}_1, \mathbf{v}_2, \mathbf{v}_3\}$, where

$$\mathbf{v}_1 = \begin{bmatrix} 1 \\ 0 \\ 1 \end{bmatrix} \qquad \mathbf{v}_2 = \begin{bmatrix} 0 \\ 1 \\ 1 \end{bmatrix} \qquad \mathbf{v}_3 = \begin{bmatrix} 1 \\ 1 \\ 0 \end{bmatrix}$$

(b) Use the matrix obtained in (a) to compute

$$T\left(\begin{bmatrix} 2 \\ 0 \\ 0 \end{bmatrix}\right)$$

**8.** Let $T: P_2 \to P_4$ be the linear transformation defined by $T(p(x)) = x^2 p(x)$.
(a) Find the matrix of $T$ with respect to the bases $B = \{\mathbf{p}_1, \mathbf{p}_2, \mathbf{p}_3\}$ and $B'$, where $\mathbf{p}_1 = 1 + x^2$, $\mathbf{p}_2 = 1 + 2x + 3x^2$, $\mathbf{p}_3 = 4 + 5x + x^2$, and $B'$ is the standard basis for $P_4$.
(b) Use the matrix obtained in (a) to compute $T(-3 + 5x - 2x^2)$.

**9.** Let $\mathbf{v}_1 = \begin{bmatrix} 1 \\ 3 \end{bmatrix}$ and $\mathbf{v}_2 = \begin{bmatrix} -1 \\ 4 \end{bmatrix}$, and let

$$A = \begin{bmatrix} 1 & 3 \\ -2 & 5 \end{bmatrix}$$

be the matrix of $T: R^2 \to R^2$ with respect to the basis $B = \{\mathbf{v}_1, \mathbf{v}_2\}$.
(a) Find $[T(\mathbf{v}_1)]_B$ and $[T(\mathbf{v}_2)]_B$.
(b) Find $T(\mathbf{v}_1)$ and $T(\mathbf{v}_2)$.

(c) Find $T\left(\begin{bmatrix} 1 \\ 1 \end{bmatrix}\right)$.

**10.** Let $A = \begin{bmatrix} 3 & -2 & 1 & 0 \\ 1 & 6 & 2 & 1 \\ -3 & 0 & 7 & 1 \end{bmatrix}$ be the matrix of $T: R^4 \to R^3$ with respect to the bases $B = \{v_1, v_2, v_3, v_4\}$ and $B' = \{w_1, w_2, w_3\}$, where

$$v_1 = \begin{bmatrix} 0 \\ 1 \\ 1 \\ 1 \end{bmatrix} \quad v_2 = \begin{bmatrix} 2 \\ 1 \\ -1 \\ -1 \end{bmatrix} \quad v_3 = \begin{bmatrix} 1 \\ 4 \\ -1 \\ 2 \end{bmatrix} \quad v_4 = \begin{bmatrix} 6 \\ 9 \\ 4 \\ 2 \end{bmatrix}$$

$$w_1 = \begin{bmatrix} 0 \\ 8 \\ 8 \end{bmatrix} \quad w_2 = \begin{bmatrix} -7 \\ 8 \\ 1 \end{bmatrix} \quad w_3 = \begin{bmatrix} -6 \\ 9 \\ 1 \end{bmatrix}$$

(a) Find $[T(v_1)]_{B'}$, $[T(v_2)]_{B'}$, $[T(v_3)]_{B'}$, and $[T(v_4)]_{B'}$.

(b) Find $T(v_1)$, $T(v_2)$, $T(v_3)$, and $T(v_4)$.

(c) Find $T \begin{bmatrix} 2 \\ 2 \\ 0 \\ 0 \end{bmatrix}$

11. Let $A = \begin{bmatrix} 1 & 3 & -1 \\ 2 & 0 & 5 \\ 6 & -2 & 4 \end{bmatrix}$ be the matrix of $T: P_2 \to P_2$ with respect to the basis $B = \{v_1, v_2, v_3\}$, where $v_1 = 3x + 3x^2$, $v_2 = -1 + 3x + 2x^2$, $v_3 = 3 + 7x + 2x^2$.

(a) Find $[T(v_1)]_B$, $[T(v_2)]_B$, and $[T(v_3)]_B$.

(b) Find $T(v_1)$, $T(v_2)$, and $T(v_3)$.

(c) Find $T(1 + x^2)$.

12. Show that if $T: V \to W$ is the zero transformation (Example 3), then the matrix of $T$ with respect to any bases for $V$ and $W$ is a zero matrix.

13. Show that if $T: V \to V$ is a contraction or a dilation of $V$ (Example 5), then the matrix of $T$ with respect to any basis for $V$ is a diagonal matrix.

14. Let $B = \{v_1, v_2, v_3, v_4\}$ be a basis for a vector space $V$. Find the matrix with respect to $B$ of the linear operator $T: V \to V$ defined by $T(v_1) = v_2$, $T(v_2) = v_3$, $T(v_3) = T(v_4)$, $T(v_4) = v_1$.

15. (**For readers who have studied calculus.**) Let $D: P_2 \to P_2$ be the differentiation operator $D(p) = p'$. In parts (a) and (b) find the matrix of $D$ with respect to the basis $B = \{p_1, p_2, p_3\}$.

(a) $p_1 = 1$, $p_2 = x$, $p_3 = x^2$

(b) $p_1 = 2$, $p_2 = 2 - 3x$, $p_3 = 2 - 3x + 8x^2$

(c) Use the matrix in part (a) to compute $D(6 - 6x + 24x^2)$.

(d) Repeat the directions of part (c) for the matrix in part (b).

16. (**For readers who have studied calculus.**) In each part, $B = (f_1, f_2, f_3)$ is a basis for a subspace $V$ of the vector space of real-valued functions defined on the real line. Find the matrix with respect to $B$ of the differentiation operator $D: V \to V$.

(a) $f_1 = 1$, $f_2 = \sin x$, $f_3 = \cos x$

(b) $f_1 = 1$, $f_2 = e^x$, $f_3 = e^{2x}$

(c) $f_1 = e^{2x}$, $f_2 = xe^{2x}$, $f_3 = x^2 e^{2x}$

## 5.5 CHANGE OF BASIS

In some problems it is necessary to change the basis being used for a vector space. Since the coordinate matrix of a vector depends on the basis, this matrix changes when the basis changes. In this section we show how the new coordinate matrices are related to the old ones when a new basis is introduced.

**Definition.** If $B = \{\mathbf{u}_1, \mathbf{u}_2, \ldots, \mathbf{u}_n\}$ and $B' = \{\mathbf{u}_1', \mathbf{u}_2', \ldots, \mathbf{u}_n'\}$ are bases for a vector space $V$, then the ***transition matrix from $B$ to $B'$*** is the $n \times n$ matrix

$$P = \begin{bmatrix} a_{11} & a_{12} & \cdots & a_{1n} \\ a_{21} & a_{22} & \cdots & a_{2n} \\ \vdots & \vdots & & \vdots \\ a_{n1} & a_{n2} & \cdots & a_{nn} \end{bmatrix}$$

where

$$\begin{bmatrix} a_{11} \\ a_{21} \\ \vdots \\ a_{n1} \end{bmatrix} = [\mathbf{u}_1]_{B'}, \quad \begin{bmatrix} a_{12} \\ a_{22} \\ \vdots \\ a_{n2} \end{bmatrix} = [\mathbf{u}_2]_{B'}, \ldots, \quad \begin{bmatrix} a_{1n} \\ a_{2n} \\ \vdots \\ a_{nn} \end{bmatrix} = [\mathbf{u}_n]_{B'}$$

To express this idea another way, think of the transition as going from an "old" basis to a "new" basis; then *the jth column of the transition matrix is the coordinate matrix of the jth vector in the old basis with respect to the new basis.*

### Example 27

Consider the bases $B = \{\mathbf{e}_1, \mathbf{e}_2\}$ and $B' = \{\mathbf{v}_1, \mathbf{v}_2\}$ for $R^2$ where

$$\mathbf{e}_1 = \begin{bmatrix} 1 \\ 0 \end{bmatrix} \qquad \mathbf{e}_2 = \begin{bmatrix} 0 \\ 1 \end{bmatrix}$$

and

$$\mathbf{v}_1 = \begin{bmatrix} 1 \\ 1 \end{bmatrix} \qquad \mathbf{v}_2 = \begin{bmatrix} 2 \\ 1 \end{bmatrix}$$

(i) Find the transition matrix from $B$ to $B'$.
(ii) Find the transition matrix from $B'$ to $B$.

*Solution.*
  (i) Since the transition is from $B$ to $B'$, $B$ is the old basis and $B'$ is the new basis, If we let $P$ denote the transition matrix, then the first column of $P$ is $[\mathbf{e}_1]_{B'}$ and the second is $[\mathbf{e}_2]_{B'}$.
  It is easy to verify that

$$\mathbf{e}_1 = -\mathbf{v}_1 + \mathbf{v}_2$$

and

$$\mathbf{e}_2 = 2\mathbf{v}_1 - \mathbf{v}_2$$

Therefore

$$[\mathbf{e}_1]_{B'} = \begin{bmatrix} -1 \\ 1 \end{bmatrix} \qquad [\mathbf{e}_2]_{B'} = \begin{bmatrix} 2 \\ -1 \end{bmatrix}$$

Consequently

$$P = \begin{bmatrix} -1 & 2 \\ 1 & -1 \end{bmatrix}$$

(ii) Since the transition is from $B'$ to $B$, the old basis is $B'$ and the new one is $B$. If we let $Q$ denote the transition matrix, then the first column of $Q$ is $[\mathbf{v}_1]_B$ and the second is $[\mathbf{v}_2]_B$. But

$$[\mathbf{v}_1]_B = \begin{bmatrix} 1 \\ 1 \end{bmatrix} \quad \text{and} \quad [\mathbf{v}_2]_B = \begin{bmatrix} 2 \\ 1 \end{bmatrix}$$

so that

$$Q = \begin{bmatrix} 1 & 2 \\ 1 & 1 \end{bmatrix}$$

**Example 28**

Consider the bases $B = \{1, x, x^2\}$ and $B' = \{1 + 2x + x^2, 2 + 9x, 3 + 3x + 4x^2\}$ for the vector space $P_2$. Find the transition matrix from $B$ to $B'$.

*Solution.* Let $\mathbf{p}_1 = 1$, $\mathbf{p}_2 = x$, $\mathbf{p}_3 = x^2$. The coordinate matrices of these vectors with respect to $B'$ are

$$[\mathbf{p}_1]_{B'} = \begin{bmatrix} -36 \\ 5 \\ 9 \end{bmatrix} \qquad [\mathbf{p}_2]_{B'} = \begin{bmatrix} 8 \\ -1 \\ -2 \end{bmatrix} \qquad [\mathbf{p}_3]_{B'} = \begin{bmatrix} 21 \\ -3 \\ -5 \end{bmatrix}$$

The transition matrix is therefore

$$P = \begin{bmatrix} -36 & 8 & 21 \\ 5 & -1 & -3 \\ 9 & -2 & -5 \end{bmatrix}$$

The next example illustrates how transition matrices relate the old and new coordinate matrices of a vector when a change of basis is made.

**Example 29**

Let $B = \{\mathbf{u}_1, \mathbf{u}_2\}$ be a basis for a two-dimensional vector space $V$ and let

$$P = \begin{bmatrix} a_{11} & a_{12} \\ a_{21} & a_{22} \end{bmatrix}$$

be the transition matrix from $B$ to a new basis $B' = \{\mathbf{u}_1', \mathbf{u}_2'\}$. Thus

$$\mathbf{u}_1 = a_{11}\mathbf{u}_1' + a_{21}\mathbf{u}_2'$$
$$\mathbf{u}_2 = a_{12}\mathbf{u}_1' + a_{22}\mathbf{u}_2'$$

Suppose that $\mathbf{x}$ is any vector in $V$; we shall show *the new coordinate matrix* $[\mathbf{x}]_{B'}$ *can be obtained by multiplying the old coordinate matrix* $[\mathbf{x}]_B$ *on the left by the transition matrix* $P$; *that is* $[\mathbf{x}]_{B'} = P[\mathbf{x}]_B$. If

$$[\mathbf{x}]_B = \begin{bmatrix} c_1 \\ c_2 \end{bmatrix}$$

if the coordinate matrix of $\mathbf{x}$ with respect to $B$ then

$$\mathbf{x} = c_1\mathbf{u}_1 + c_2\mathbf{u}_2 = c_1(a_{11}\mathbf{u}_1' + a_{21}\mathbf{u}_2') + c_2(a_{12}\mathbf{u}_1' + a_{22}\mathbf{u}_2')$$
$$= (c_1a_{11} + c_2a_{12})\mathbf{u}_1' + (c_1a_{21} + c_2a_{22})\mathbf{u}_2'$$

This expresses $\mathbf{x}$ in terms of the $B'$ basis so that

$$[\mathbf{x}]_{B'} = \begin{bmatrix} c_1a_{11} + c_2a_{12} \\ c_1a_{21} + c_2a_{22} \end{bmatrix} = \begin{bmatrix} a_{11} & a_{12} \\ a_{21} & a_{22} \end{bmatrix} \begin{bmatrix} c_1 \\ c_2 \end{bmatrix} = P[\mathbf{x}]_B$$

Although the results in this example were obtained only for two-dimensional vector spaces, the same procedure can be used to establish the following general theorem.

**Theorem 8.** *Let $P$ be the transition matrix from a basis $B$ to a basis $B'$ for a finite dimensional vector space $V$. If $\mathbf{x}$ is any vector in $V$, then the coordinate matrices $[\mathbf{x}]_{B'}$ and $[\mathbf{x}]_B$ are related by the equation*

$$[\mathbf{x}]_{B'} = P[\mathbf{x}]_B \tag{5.16}$$

**Example 30**

In Example 27 we found the transition matrix $P$ from $B = \{\mathbf{e}_1, \mathbf{e}_2\}$ to $B' = \{\mathbf{v}_1, \mathbf{v}_2\}$ where

$$\mathbf{e}_1 = \begin{bmatrix} 1 \\ 0 \end{bmatrix} \quad \mathbf{e}_2 = \begin{bmatrix} 0 \\ 1 \end{bmatrix} \quad \mathbf{v}_1 = \begin{bmatrix} 1 \\ 1 \end{bmatrix} \quad \mathbf{v}_2 = \begin{bmatrix} 2 \\ 1 \end{bmatrix}$$

Use Theorem 8 to find $[\mathbf{x}]_{B'}$, if

$$\mathbf{x} = \begin{bmatrix} 7 \\ 2 \end{bmatrix}$$

*Solution.* By inspection

$$[\mathbf{x}]_B = \begin{bmatrix} 7 \\ 2 \end{bmatrix}$$

Using (5.16) and the transition matrix computed in Example 27, we obtain

$$[\mathbf{x}]_{B'} = \begin{bmatrix} -1 & 2 \\ 1 & -1 \end{bmatrix} \begin{bmatrix} 7 \\ 2 \end{bmatrix} = \begin{bmatrix} -3 \\ 5 \end{bmatrix}$$

The reader may wish to check this result by verifying that $\mathbf{x} = -3\mathbf{v}_1 + 5\mathbf{v}_2$.

In Example 27 we computed the transition matrix $P$ from $B$ to $B'$ and the transition matrix $Q$ from $B'$ to $B$. A simple check will show that $Q = P^{-1}$. Our next theorem shows that this relationship is not accidental.

**Theorem 9.** *If $P$ is the transition matrix from a basis $B$ to a basis $B'$ for a finite dimensional vector space $V$, then:*

(a) *$P$ is invertible.*
(b) *$P^{-1}$ is the transition matrix from $B'$ to $B$.*

*Proof.* Let $Q$ be the transition matrix from $B'$ to $B$. We shall show that $QP = I$ and thus conclude that $Q = P^{-1}$ to complete the proof.

Assume that $B = \{u_1, u_2, \ldots, u_n\}$ and suppose

$$QP = \begin{bmatrix} c_{11} & c_{12} & \cdots & c_{1n} \\ c_{21} & c_{22} & \cdots & c_{2n} \\ \vdots & \vdots & & \vdots \\ c_{n1} & c_{n2} & \cdots & c_{nn} \end{bmatrix}$$

By Theorem 8

$$[x]_{B'} = P[x]_B$$

and

$$[x]_B = Q[x]_{B'}$$

for all $x$ in $V$. Multiplying the top equation through by $Q$ and substituting from second equation gives

$$[x]_B = QP[x]_B \qquad (5.17)$$

for all $x$ in $V$. Letting $x = u_1$ in (5.17) gives

$$\begin{bmatrix} 1 \\ 0 \\ 0 \\ \vdots \\ 0 \end{bmatrix} = \begin{bmatrix} c_{11} & c_{12} & \cdots & c_{1n} \\ c_{21} & c_{22} & \cdots & c_{2n} \\ \vdots & \vdots & & \vdots \\ c_{n1} & c_{n2} & \cdots & c_{nn} \end{bmatrix} \begin{bmatrix} 1 \\ 0 \\ 0 \\ \vdots \\ 0 \end{bmatrix}$$

or

$$\begin{bmatrix} 1 \\ 0 \\ 0 \\ \vdots \\ 0 \end{bmatrix} = \begin{bmatrix} c_{11} \\ c_{21} \\ \vdots \\ c_{n1} \end{bmatrix}$$

Similarly, successively substituting $x = u_2, \ldots, u_n$ in (5.17) yields

$$\begin{bmatrix} c_{12} \\ c_{22} \\ \vdots \\ c_{n2} \end{bmatrix} = \begin{bmatrix} 0 \\ 1 \\ 0 \\ \vdots \\ 0 \end{bmatrix}, \dots, \begin{bmatrix} c_{1n} \\ c_{2n} \\ \vdots \\ c_{nn} \end{bmatrix} = \begin{bmatrix} 0 \\ 0 \\ \vdots \\ 1 \end{bmatrix}$$

Therefore $QP = I$. ∎

The next theorem shows that if $P$ is the transition matrix from one *orthonormal* basis for an inner product space to another *orthonormal* basis, then the inverse of $P$ is particularly easy to find.

**Theorem 10.** *If $P$ is the transition matrix from one orthonormal basis to another orthonormal basis in a finite dimensional inner product space, then*

$$P^{-1} = P^t$$

(We shall omit the proof.)

**Definition.** A square matrix $A$ is called **orthogonal** if $A^{-1} = A^t$.

**Example 31**

Consider the orthonormal bases $B = \{\mathbf{v}_1, \mathbf{v}_2, \mathbf{v}_3\}$ and $B' = \{\mathbf{u}_1, \mathbf{u}_2, \mathbf{u}_3\}$ for $R^3$, where

$$\mathbf{v}_1 = \begin{bmatrix} \dfrac{1}{\sqrt{2}} \\ 0 \\ \dfrac{1}{\sqrt{2}} \end{bmatrix} \quad \mathbf{v}_2 = \begin{bmatrix} \dfrac{1}{\sqrt{2}} \\ 0 \\ -\dfrac{1}{\sqrt{2}} \end{bmatrix} \quad \mathbf{v}_3 = \begin{bmatrix} 0 \\ 1 \\ 0 \end{bmatrix} \quad \mathbf{u}_1 = \begin{bmatrix} 1 \\ 0 \\ 0 \end{bmatrix} \quad \mathbf{u}_2 = \begin{bmatrix} 0 \\ 1 \\ 0 \end{bmatrix} \quad \mathbf{u}_3 = \begin{bmatrix} 0 \\ 0 \\ 1 \end{bmatrix}$$

Since

$$\mathbf{v}_1 = \frac{1}{\sqrt{2}} \mathbf{u}_1 + 0\mathbf{u}_2 + \frac{1}{\sqrt{2}} \mathbf{u}_3$$

$$\mathbf{v}_2 = \frac{1}{\sqrt{2}} \mathbf{u}_1 + 0\mathbf{u}_2 - \frac{1}{\sqrt{2}} \mathbf{u}_3$$

$$\mathbf{v}_3 = 0\mathbf{u}_1 + \mathbf{u}_2 + 0\mathbf{u}_3$$

the transition matrix from $B$ to $B'$ is

$$P = \begin{bmatrix} \dfrac{1}{\sqrt{2}} & \dfrac{1}{\sqrt{2}} & 0 \\ 0 & 0 & 1 \\ \dfrac{1}{\sqrt{2}} & -\dfrac{1}{\sqrt{2}} & 0 \end{bmatrix} \qquad (5.18)$$

The reader can verify that $P$ is an orthogonal matrix (as guaranteed by Theorem 10) by showing that $PP^t = P^tP = I$.

**REMARK.** We leave it as an exercise to prove the equivalence of the following statements.

(i)  $A$ is an orthogonal $n \times n$ matrix.
(ii)  The row vectors in $A$ form an orthonormal set in $R^n$.
(iii)  The column vectors in $A$ form an orthonormal set in $R^n$.

**Example 32**

The row vectors of the orthogonal matrix (5.18) in Example 31 are

$$\mathbf{r}_1 = \left(\frac{1}{\sqrt{2}}, \frac{1}{\sqrt{2}}, 0\right) \qquad \mathbf{r}_2 = (0, 0, 1) \qquad \mathbf{r}_3 = \left(\frac{1}{\sqrt{2}}, -\frac{1}{\sqrt{2}}, 0\right)$$

Since

$$\|\mathbf{r}_1\| = \|\mathbf{r}_2\| = \|\mathbf{r}_3\| = 1$$

and

$$\mathbf{r}_1 \cdot \mathbf{r}_2 = \mathbf{r}_1 \cdot \mathbf{r}_3 = \mathbf{r}_2 \cdot \mathbf{r}_3 = 0$$

these row vectors form an orthonormal set in $R^3$. Similarly, the column vectors form an orthonormal set in $R^3$.

We conclude this section with a geometric interpretation in $R^2$ and $R^3$ of some of the results of this section. Let

$$\mathbf{e}_1 = \begin{bmatrix} 1 \\ 0 \end{bmatrix} \quad \text{and} \quad \mathbf{e}_2 = \begin{bmatrix} 0 \\ 1 \end{bmatrix}$$

be the unit vectors along the positive coordinate axes of a rectangular $xy$-coordinate system in $R^2$. Let

$$\mathbf{v}_1 = \begin{bmatrix} a_1 \\ b_1 \end{bmatrix} \quad \text{and} \quad \mathbf{v}_2 = \begin{bmatrix} a_2 \\ b_2 \end{bmatrix}$$

be an *orthonormal* set of vectors, and introduce an $x'y'$-coordinate system having the same origin as the $xy$-system and such that the positive $x'$ and $y'$ axes are in the directions of $\mathbf{v}_1$ and $\mathbf{v}_2$, respectively (Figure 5.6).

Since

$$\mathbf{v}_1 = a_1\mathbf{e}_1 + b_1\mathbf{e}_2$$

and

$$\mathbf{v}_2 = a_2\mathbf{e}_1 + b_2\mathbf{e}_2$$

the transition matrix from the basis $\{\mathbf{v}_1, \mathbf{v}_2\}$ to the basis $\{\mathbf{e}_1, \mathbf{e}_2\}$ is

$$P = \begin{bmatrix} a_1 & a_2 \\ b_1 & b_2 \end{bmatrix}$$

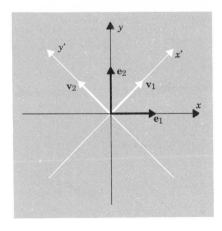

**Figure 5.6**

Thus, from Theorem 8, the $xy$-coordinates and the $x'y'$-coordinates of a point in the plane are related by

$$\begin{bmatrix} x \\ y \end{bmatrix} = \begin{bmatrix} a_1 & a_2 \\ b_1 & b_2 \end{bmatrix} \begin{bmatrix} x' \\ y' \end{bmatrix} \tag{5.19}$$

(Note that $\mathbf{v}_1$ and $\mathbf{v}_2$ are the column vectors of $P$.)

By Theorem 10 the transition matrix $P$ is orthogonal. Thus (5.19) is called an *orthogonal coordinate transformation in $R^2$*.

**Example 33**

Consider the orthogonal coordinate transformation

$$\begin{bmatrix} x \\ y \end{bmatrix} = \begin{bmatrix} \dfrac{1}{\sqrt{2}} & -\dfrac{1}{\sqrt{2}} \\[2mm] \dfrac{1}{\sqrt{2}} & \dfrac{1}{\sqrt{2}} \end{bmatrix} \begin{bmatrix} x' \\ y' \end{bmatrix} \tag{5.20}$$

The positive $x'$ and $y'$ axes are in the directions of

$$\mathbf{v}_1 = \begin{bmatrix} \dfrac{1}{\sqrt{2}} \\[2mm] \dfrac{1}{\sqrt{2}} \end{bmatrix} \quad \text{and} \quad \mathbf{v}_2 = \begin{bmatrix} -\dfrac{1}{\sqrt{2}} \\[2mm] \dfrac{1}{\sqrt{2}} \end{bmatrix}$$

respectively (Figure 5.7).

Let $Q$ be the point with $xy$-coordinates $(3, -1)$. To find the $x'y'$-coordinates of $Q$, we substitute $x = 3$, $y = -1$ into (5.20) and solve for $x'$ and $y'$. This yields

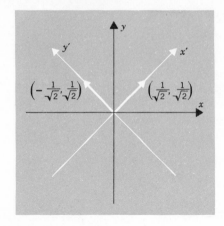

**Figure 5.7**

$$\begin{bmatrix} 3 \\ -1 \end{bmatrix} = \begin{bmatrix} \dfrac{1}{\sqrt{2}} & -\dfrac{1}{\sqrt{2}} \\ \dfrac{1}{\sqrt{2}} & \dfrac{1}{\sqrt{2}} \end{bmatrix} \begin{bmatrix} x' \\ y' \end{bmatrix}$$

or

$$\begin{bmatrix} x' \\ y' \end{bmatrix} = \begin{bmatrix} \dfrac{1}{\sqrt{2}} & -\dfrac{1}{\sqrt{2}} \\ \dfrac{1}{\sqrt{2}} & \dfrac{1}{\sqrt{2}} \end{bmatrix}^{-1} \begin{bmatrix} 3 \\ -1 \end{bmatrix}$$

Using the orthogonality, we obtain

$$\begin{bmatrix} x' \\ y' \end{bmatrix} = \begin{bmatrix} \dfrac{1}{\sqrt{2}} & -\dfrac{1}{\sqrt{2}} \\ \dfrac{1}{\sqrt{2}} & \dfrac{1}{\sqrt{2}} \end{bmatrix}^{t} \begin{bmatrix} 3 \\ -1 \end{bmatrix} = \begin{bmatrix} \dfrac{1}{\sqrt{2}} & \dfrac{1}{\sqrt{2}} \\ -\dfrac{1}{\sqrt{2}} & \dfrac{1}{\sqrt{2}} \end{bmatrix} \begin{bmatrix} 3 \\ -1 \end{bmatrix} = \begin{bmatrix} \dfrac{2}{\sqrt{2}} \\ -\dfrac{4}{\sqrt{2}} \end{bmatrix}$$

Thus the $x'y'$-coordinates of $Q$ are $\left( \dfrac{2}{\sqrt{2}}, -\dfrac{4}{\sqrt{2}} \right)$.

An analog of (5.19) holds in $R^3$. Let a rectangular $xyz$-coordinate system in $R^3$ be given. If an $x'y'z'$-coordinate system is introduced with the same origin as the $xyz$-system and is such that the positive $x'$, $y'$, and $z'$ axes are in the directions of the orthonormal vectors

$$\mathbf{v}_1 = \begin{bmatrix} a_1 \\ b_1 \\ c_1 \end{bmatrix} \qquad \mathbf{v}_2 = \begin{bmatrix} a_2 \\ b_2 \\ c_2 \end{bmatrix} \qquad \mathbf{v}_3 = \begin{bmatrix} a_3 \\ b_3 \\ c_3 \end{bmatrix}$$

then the $xyz$-coordinates and the $x'y'z'$-coordinates of a point are related by

$$\begin{bmatrix} x \\ y \\ z \end{bmatrix} = \begin{bmatrix} a_1 & a_2 & a_3 \\ b_1 & b_2 & b_3 \\ c_1 & c_2 & c_3 \end{bmatrix} \begin{bmatrix} x' \\ y' \\ z' \end{bmatrix} \tag{5.21}$$

Since

$$P = \begin{bmatrix} a_1 & a_2 & a_3 \\ b_1 & b_2 & b_3 \\ c_1 & c_2 & c_3 \end{bmatrix}$$

is orthogonal, (5.21) is called an ***orthogonal coordinate transformation in*** $R^3$.

The determinant of an orthogonal matrix is easily seen to be 1 or $-1$ (Exercise 22). If

$$\mathbf{x} = P\mathbf{x}'$$

is an orthogonal coordinate transformation in $R^2$ or $R^3$ and $\det(P) = 1$, it can be shown that the coordinate axes of either system can be obtained by rotating the coordinate axes of the other system. We shall omit the details.

**Example 34**

Since

$$\det \begin{bmatrix} \dfrac{1}{\sqrt{2}} & -\dfrac{1}{\sqrt{2}} \\ \dfrac{1}{\sqrt{2}} & \dfrac{1}{\sqrt{2}} \end{bmatrix} = 1$$

the orthogonal coordinate transformation in Example 33 is a rotation of axes (Figure 5.7).

**EXERCISE SET 5.5**

1. Consider the bases $B = \{\mathbf{u}_1, \mathbf{u}_2\}$ and $B' = \{\mathbf{v}_1, \mathbf{v}_2\}$ for $R^2$, where

$$\mathbf{u}_1 = \begin{bmatrix} 1 \\ 0 \end{bmatrix} \qquad \mathbf{u}_2 = \begin{bmatrix} 0 \\ 1 \end{bmatrix} \qquad \mathbf{v}_1 = \begin{bmatrix} 2 \\ 1 \end{bmatrix} \quad \text{and} \quad \mathbf{v}_2 = \begin{bmatrix} -3 \\ 4 \end{bmatrix}$$

   (a) Find the transition matrix from $B$ to $B'$.
   (b) Compute the coordinate matrix $[\mathbf{w}]_B$, where

$$\mathbf{w} = \begin{bmatrix} 3 \\ -5 \end{bmatrix}$$

   and use Theorem 8 to compute $[\mathbf{w}]_{B'}$.
   (c) Check your work by computing $[\mathbf{w}]_{B'}$ directly.
   (d) Find the transition matrix from $B'$ to $B$.

2. Repeat the directions of Exercise 1 with

   (a) $\mathbf{u}_1 = \begin{bmatrix} 2 \\ 2 \end{bmatrix}$   $\mathbf{u}_2 = \begin{bmatrix} 4 \\ -1 \end{bmatrix}$   $\mathbf{v}_1 = \begin{bmatrix} 1 \\ 3 \end{bmatrix}$   $\mathbf{v}_2 = \begin{bmatrix} -1 \\ -1 \end{bmatrix}$

3. Consider the bases $B = \{\mathbf{u}_1, \mathbf{u}_2, \mathbf{u}_3\}$ and $B' = \{\mathbf{v}_1, \mathbf{v}_2, \mathbf{v}_3\}$ for $R^3$, where

   $$\mathbf{u}_1 = \begin{bmatrix} -3 \\ 0 \\ -3 \end{bmatrix} \quad \mathbf{u}_2 = \begin{bmatrix} -3 \\ 2 \\ -1 \end{bmatrix} \quad \mathbf{u}_3 = \begin{bmatrix} 1 \\ 6 \\ -1 \end{bmatrix} \quad \mathbf{v}_1 = \begin{bmatrix} -6 \\ -6 \\ 0 \end{bmatrix} \quad \mathbf{v}_2 = \begin{bmatrix} -2 \\ -6 \\ 4 \end{bmatrix} \quad \text{and} \quad \mathbf{v}_3 = \begin{bmatrix} -2 \\ -3 \\ 7 \end{bmatrix}$$

   (a) Find the transition matrix from $B$ to $B'$.
   (b) Compute the coordinate matrix $[\mathbf{w}]_B$, where

   $$\mathbf{w} = \begin{bmatrix} -5 \\ 8 \\ -5 \end{bmatrix}$$

   and use Theorem 8 to compute $[\mathbf{w}]_{B'}$.
   (c) Check your work by computing $[\mathbf{w}]_{B'}$ directly.

4. Repeat the directions of Exercise 3 with

   $$\mathbf{u}_1 = \begin{bmatrix} 2 \\ 1 \\ 1 \end{bmatrix} \quad \mathbf{u}_2 = \begin{bmatrix} 2 \\ -1 \\ 1 \end{bmatrix} \quad \mathbf{u}_3 = \begin{bmatrix} 1 \\ 2 \\ 1 \end{bmatrix} \quad \mathbf{v}_1 = \begin{bmatrix} 3 \\ 1 \\ -5 \end{bmatrix} \quad \mathbf{v}_2 = \begin{bmatrix} 1 \\ 1 \\ -3 \end{bmatrix} \quad \mathbf{v}_3 = \begin{bmatrix} -1 \\ 0 \\ 2 \end{bmatrix}$$

5. Consider the bases $B = \{\mathbf{p}_1, \mathbf{p}_2\}$ and $B' = \{\mathbf{q}_1, \mathbf{q}_2\}$ for $P_1$, where $\mathbf{p}_1 = 6 + 3x$, $\mathbf{p}_2 = 10 + 2x$, $\mathbf{q}_1 = 2$, $\mathbf{q}_2 = 3 + 2x$.

   (a) Find the transition matrix from $B$ to $B'$.
   (b) Compute the coordinate matrix $[\mathbf{p}]_B$, where $\mathbf{p} = -4 + x$, and use Theorem 8 to compute $[\mathbf{p}]_{B'}$.
   (c) Check your work by computing $[\mathbf{p}]_{B'}$ directly.
   (d) Find the transition matrix from $B'$ to $B$.

6. Let $V$ be the space spanned by $\mathbf{f}_1 = \sin x$ and $\mathbf{f}_2 = \cos x$.
   (a) Show that $\mathbf{g}_1 = 2 \sin x + \cos x$ and $\mathbf{g}_2 = 3 \cos x$ form a basis for $V$.
   (b) Find the transition matrix from $B = \{\mathbf{f}_1, \mathbf{f}_2\}$ to $B' = \{\mathbf{g}_1, \mathbf{g}_2\}$.
   (c) Compute the coordinate matrix $[\mathbf{h}]_B$, where $\mathbf{h} = 2 \sin x - 5 \cos x$, and use Theorem 8 to obtain $[\mathbf{h}]_{B'}$.
   (d) Check your work by computing $[\mathbf{h}]_{B'}$ directly.
   (e) Find the transition matrix from $B'$ to $B$.

7. Use the remark preceding Example 32 to determine which of the following are orthogonal.

   (a) $\begin{bmatrix} 1 & 0 \\ 0 & 1 \end{bmatrix}$   (b) $\begin{bmatrix} \dfrac{1}{\sqrt{2}} & -\dfrac{1}{\sqrt{2}} \\ \dfrac{1}{\sqrt{2}} & \dfrac{1}{\sqrt{2}} \end{bmatrix}$

(c) $\begin{bmatrix} 0 & 1 & \dfrac{1}{\sqrt{2}} \\ 1 & 0 & 0 \\ 0 & 0 & \dfrac{1}{\sqrt{2}} \end{bmatrix}$

(d) $\begin{bmatrix} -\dfrac{1}{\sqrt{2}} & \dfrac{1}{\sqrt{6}} & \dfrac{1}{\sqrt{3}} \\ 0 & -\dfrac{2}{\sqrt{6}} & \dfrac{1}{\sqrt{3}} \\ \dfrac{1}{\sqrt{2}} & \dfrac{1}{\sqrt{6}} & \dfrac{1}{\sqrt{3}} \end{bmatrix}$

(e) $\begin{bmatrix} \dfrac{1}{2} & \dfrac{1}{2} & \dfrac{1}{2} & \dfrac{1}{2} \\ \dfrac{1}{2} & -\dfrac{5}{6} & \dfrac{1}{6} & \dfrac{1}{6} \\ \dfrac{1}{2} & \dfrac{1}{6} & \dfrac{1}{6} & -\dfrac{5}{6} \\ \dfrac{1}{2} & \dfrac{1}{6} & -\dfrac{5}{6} & \dfrac{1}{6} \end{bmatrix}$

(f) $\begin{bmatrix} 1 & 0 & 0 & 0 \\ 0 & \dfrac{1}{\sqrt{3}} & -\dfrac{1}{2} & 0 \\ 0 & \dfrac{1}{\sqrt{3}} & 0 & 1 \\ 0 & \dfrac{1}{\sqrt{3}} & \dfrac{1}{2} & 0 \end{bmatrix}$

**8.** Find the inverses of those matrices in Exercise 7 that are orthogonal.

**9.** Show that the following are orthogonal matrices for every value of $\theta$.

(a) $\begin{bmatrix} \cos\theta & -\sin\theta \\ \sin\theta & \cos\theta \end{bmatrix}$

(b) $\begin{bmatrix} \cos\theta & -\sin\theta & 0 \\ \sin\theta & \cos\theta & 0 \\ 0 & 0 & 1 \end{bmatrix}$

**10.** Find the inverses of each of the matrices in Exercise 9.

**11.** Consider the orthogonal coordinate transformation

$$\begin{bmatrix} x \\ y \end{bmatrix} = \begin{bmatrix} -\dfrac{3}{5} & -\dfrac{4}{5} \\ \dfrac{4}{5} & -\dfrac{3}{5} \end{bmatrix} \begin{bmatrix} x' \\ y' \end{bmatrix}$$

Find $(x', y')$ for the points with the following $(x, y)$ coordinates.

(a) $(2, -1)$    (b) $(4, 2)$    (c) $(-7, -8)$    (d) $(0, 0)$

**12.** Sketch the $xy$-axes and the $x'y'$-axes for the coordinate transformation in Exercise 11.

**13.** For which of the following is $\mathbf{x} = P\mathbf{x}'$ a rotation?

(a) $P = \begin{bmatrix} \dfrac{1}{\sqrt{2}} & \dfrac{1}{\sqrt{2}} \\ -\dfrac{1}{\sqrt{2}} & \dfrac{1}{\sqrt{2}} \end{bmatrix}$

(b) $P = \begin{bmatrix} -\dfrac{1}{\sqrt{2}} & \dfrac{1}{\sqrt{2}} \\ \dfrac{1}{\sqrt{2}} & \dfrac{1}{\sqrt{2}} \end{bmatrix}$

(c) $P = \begin{bmatrix} \dfrac{3}{5} & \dfrac{4}{5} \\ -\dfrac{4}{5} & \dfrac{3}{5} \end{bmatrix}$

(d) $P = \begin{bmatrix} -\dfrac{3}{5} & -\dfrac{4}{5} \\ -\dfrac{4}{5} & \dfrac{3}{5} \end{bmatrix}$

**14.** Sketch the $xy$-axes and the $x'y'$-axes for the coordinate transformations in Exercise 13.

**15.** Consider the orthogonal coordinate transformation

$$\begin{bmatrix} x \\ y \\ z \end{bmatrix} = \begin{bmatrix} \frac{4}{5} & -\frac{3}{5} & 0 \\ \frac{3}{5} & \frac{4}{5} & 0 \\ 0 & 0 & 1 \end{bmatrix} \begin{bmatrix} x' \\ y' \\ z' \end{bmatrix}$$

Find $(x', y', z')$ for the points with the following $(x, y, z)$ coordinates.

(a) $(3, 0, -7)$   (b) $(1, 2, 6)$   (c) $(-9, -2, -3)$   (d) $(0, 0, 0)$

**16.** Sketch the $xyz$-axes and the $x'y'z'$-axes for the coordinate transformation in Exercise 15.

**17.** For which of the following is $\mathbf{x} = P\mathbf{x}'$ a rotation?

(a) $P = \begin{bmatrix} \frac{4}{5} & 0 & -\frac{3}{5} \\ \frac{3}{5} & 0 & \frac{4}{5} \\ 0 & 1 & 0 \end{bmatrix}$   (b) $P = \begin{bmatrix} \frac{6}{7} & \frac{2}{7} & \frac{3}{7} \\ \frac{2}{7} & \frac{3}{7} & -\frac{6}{7} \\ -\frac{3}{7} & \frac{6}{7} & \frac{2}{7} \end{bmatrix}$

**18.** Sketch the $xyz$-axes and the $x'y'z'$-axes for the coordinate transformations in Exercise 17.

**19.** Show that if $A$ is an orthogonal matrix, then $A^t$ is also orthogonal.

**20.** Prove that an $n \times n$ matrix is orthogonal if and only if its rows form an orthonormal set in $R^n$.

**21.** Use Exercises 19 and 20 to show that an $n \times n$ matrix is orthogonal if and only if its columns form an orthonormal set in $R^n$.

**22.** Prove that if $P$ is an orthogonal matrix, then $\det(P) = 1$ or $-1$.

## 5.6 SIMILARITY

Since the matrix of a linear transformation depends on the bases used for the vector spaces, the matrix changes when the bases are changed. In this section we show how the old and new matrices of a linear operator are related when a change of basis is made.

The main objective of this section is to establish the following result.

**Theorem 11.** *Let $T: V \rightarrow V$ be a linear operator on a finite dimensional vector space V. If A is the matrix of T with respect to a basis B, and A' is the matrix of T with respect to a basis B', then*

$$A' = P^{-1}AP \tag{5.22}$$

*where P is the transition matrix from B' to B.*

To establish this theorem, it will be convenient to describe the relationship

$$A\mathbf{u} = \mathbf{v}$$

pictorially by writing

$$\mathbf{u} \xrightarrow{\ A\ } \mathbf{v}$$

Since $A$ is the matrix of $T$ with respect to $B$, and $A'$ is the matrix of $T$ with respect to $B'$, the following relationships hold for all $\mathbf{x}$ in $V$.

$$A[\mathbf{x}]_B = [T(\mathbf{x})]_B$$

and

$$A'[\mathbf{x}]_{B'} = [T(\mathbf{x})]_{B'}$$

These can be written

$$[\mathbf{x}]_B \xrightarrow{\ A\ } [T(\mathbf{x})]_B$$

and                                                                                                   (5.23)

$$[\mathbf{x}]_{B'} \xrightarrow{\ A'\ } [T(\mathbf{x})]_{B'}$$

To see how the matrices $A$ and $A'$ are related, let $P$ be the transition matrix from the $B'$ basis to the $B$ basis. From Theorems 8 and 9 we know that

$$P[\mathbf{x}]_{B'} = [\mathbf{x}]_B$$

and

$$P^{-1}[T(\mathbf{x})]_B = [T(\mathbf{x})]_{B'}$$

which can be written as $\qquad [\mathbf{x}]_{B'} \xrightarrow{\ P\ } [\mathbf{x}]_B$

and $\qquad\qquad\qquad [T(\mathbf{x})]_B \xrightarrow{\ P^{-1}\ } [T(\mathbf{x})]_{B'}$                    (5.24)

For compactness, relationships (5.23) and (5.24) can be linked together in a single figure as follows:

$$
\begin{array}{ccc}
[\mathbf{x}]_B & \xrightarrow{\ A\ } & [T(\mathbf{x})]_B \\
\Big\uparrow{\scriptstyle P} & & \Big\downarrow{\scriptstyle P^{-1}} \\
[\mathbf{x}]_{B'} & \xrightarrow{\ A'\ } & [T(\mathbf{x})]_{B'}
\end{array}
$$

This figure illustrates that there are two ways to obtain the matrix $[T(\mathbf{x})]_{B'}$ from the matrix $[\mathbf{x}]_{B'}$. We can take the bottom path across the figure, that is

$$A'[\mathbf{x}]_{B'} = [T(\mathbf{x})]_{B'}$$                                                     (5.25)

or we can go up the left side, across the top, and down the right side, that is

$$P^{-1}AP[\mathbf{x}]_{B'} = [T(\mathbf{x})]_{B'} \qquad (5.26)$$

It follows from (5.25) and (5.26) that

$$P^{-1}AP[\mathbf{x}]_{B'} = A'[\mathbf{x}]_{B'} \qquad (5.27)$$

for all $\mathbf{x}$ in $V$. It follows from (5.27) and part (b) of Exercise 11 that

$$P^{-1}AP = A'$$

This proves Theorem 11. ▮

**Example 35**

Let $T: R^2 \to R^2$ be defined by

$$T\left(\begin{bmatrix} x_1 \\ x_2 \end{bmatrix}\right) = \begin{bmatrix} x_1 + x_2 \\ -2x_1 + 4x_2 \end{bmatrix}$$

and let

$$\mathbf{u}_1 = \begin{bmatrix} 1 \\ 1 \end{bmatrix} \quad \text{and} \quad \mathbf{u}_2 = \begin{bmatrix} 1 \\ 2 \end{bmatrix}$$

From Example 26 the matrix for $T$ with respect to $B = \{\mathbf{u}_1, \mathbf{u}_2\}$ is

$$A = \begin{bmatrix} 2 & 0 \\ 0 & 3 \end{bmatrix}$$

Use Theorem 11 to find the matrix for $T$ with respect to the basis $B' = \{\mathbf{e}_1, \mathbf{e}_2\}$ where

$$\mathbf{e}_1 = \begin{bmatrix} 1 \\ 0 \end{bmatrix} \quad \mathbf{e}_2 = \begin{bmatrix} 0 \\ 1 \end{bmatrix}$$

*Solution.* Since

$$\mathbf{e}_1 = 2\mathbf{u}_1 - \mathbf{u}_2$$

and

$$\mathbf{e}_2 = -\mathbf{u}_1 + \mathbf{u}_2$$

the transition matrix from $B'$ to $B$ is

$$P = \begin{bmatrix} 2 & -1 \\ -1 & 1 \end{bmatrix}$$

Let $A'$ be the matrix of $T$ with respect to $B'$. Then

$$A' = P^{-1}AP = \begin{bmatrix} 1 & 1 \\ 1 & 2 \end{bmatrix}\begin{bmatrix} 2 & 0 \\ 0 & 3 \end{bmatrix}\begin{bmatrix} 2 & -1 \\ -1 & 1 \end{bmatrix} = \begin{bmatrix} 1 & 1 \\ -2 & 4 \end{bmatrix}$$

**Warning.** *In applying Theorem 11, it is easy to forget whether P is the transition matrix from B to B' (incorrect) or from B' to B (correct). It may help to call B the old basis, B' the new basis, A the old matrix, and A' the new matrix. Since P is the transition* *matrix from B' to B, $P^{-1}$ is the transition matrix from B to B'. Thus (5.22) can be expressed as:*

$$new\ matrix = P^{-1}\ (old\ matrix)\ P$$

*where $P^{-1}$ is the transition matrix from the old basis to the new basis.*

Theorem 11 motivates the following definition.

**Definition.** If $A$ and $B$ are square matrices, we say that **B is similar to A** if there is an invertible matrix $P$ such that $B = P^{-1}AP$.

Note that the equation $B = P^{-1}AP$ can be rewritten as

$$A = PBP^{-1} \quad \text{or} \quad A = (P^{-1})^{-1}BP^{-1}$$

Letting $Q = P^{-1}$ yields

$$A = Q^{-1}BQ$$

which says that $A$ is similar to $B$. Therefore, $B$ is similar to $A$ if and only if $A$ is similar to $B$; consequently, we shall usually say simply that **A and B are similar.**

In this terminology, Theorem 11 asserts that *two matrices that represent the same linear operator $T: V \to V$ with respect to different bases are similar.*

## EXERCISE SET 5.6

In Exercises 1–7 find the matrix of $T$ with respect to $B$, and use Theorem 11 to compute the matrix of $T$ with respect to $B'$.

**1.** $T: R^2 \to R^2$ is defined by

$$T\left(\begin{bmatrix} x_1 \\ x_2 \end{bmatrix}\right) = \begin{bmatrix} x_1 - 2x_2 \\ -x_2 \end{bmatrix}$$

$B$ and $B'$ are the bases in Exercise 1 of Section 5.5.

**2.** $T: R^2 \to R^2$ is defined by

$$T\left(\begin{bmatrix} x_1 \\ x_2 \end{bmatrix}\right) = \begin{bmatrix} x_1 + 7x_2 \\ 3x_1 - 4x_2 \end{bmatrix}$$

$B$ and $B'$ are the bases in Exercise 2 of Section 5.5.

**3.** $T: R^2 \to R^2$ is the rotation about the origin through $45°$; $B$ and $B'$ are the bases in Exercise 1 of Section 5.5.

**4.** $T: R^3 \to R^3$ is defined by

$$T\left(\begin{bmatrix} x_1 \\ x_2 \\ x_3 \end{bmatrix}\right) = \begin{bmatrix} x_1 + 2x_2 - x_3 \\ -x_2 \\ x_1 + 7x_3 \end{bmatrix}$$

$B$ is the standard basis for $R^3$ and $B' = \{v_1, v_2, v_3\}$, where

$$v_1 = \begin{bmatrix} 1 \\ 0 \\ 0 \end{bmatrix}, v_2 = \begin{bmatrix} 1 \\ 1 \\ 0 \end{bmatrix}, \text{ and } v_3 = \begin{bmatrix} 1 \\ 1 \\ 1 \end{bmatrix}.$$

5. $T: R^3 \to R^3$ is orthogonal projection on the $xy$-plane; $B$ and $B'$ are the bases in Exercise 4.

6. $T: R^2 \to R^2$ is defined by $T(x) = 5x$; $B$ and $B'$ are the bases in Exercise 2 of Section 5.5.

7. $T: P_1 \to P_1$ is defined by $T(a_0 + a_1 x) = a_0 + a_1(x + 1)$; $B$ and $B'$ are the bases in Exercise 5 of Section 5.5.

8. Prove that if $A$ and $B$ are similar matrices, then $\det(A) = \det(B)$.

9. Prove that similar matrices have the same rank and nullity.

10. Prove that if $A$ and $B$ are similar matrices, then $A^2$ and $B^2$ are also similar. More generally, prove that $A^k$ and $B^k$ are similar, where $k$ is any positive integer.

11. Let $C$ and $D$ be $m \times n$ matrices. Show that:
    (a) If $Cx = Dx$ for all $x$ in $R^n$, then $C = D$.
    (b) If $B = \{v_1, v_2, \ldots, v_n\}$ is a basis for a vector space $V$ and $C[x]_B = D[x]_B$ for all $x$ in $V$, then $C = D$.

# Chapter 6 | Eigenvalues, Eigenvectors, and Quadratic Forms

## 6.1 EIGENVALUES AND EIGENVECTORS

In many problems in science and mathematics, a linear operator $T: V \to V$ is given, and it is of importance to determine those scalars $\lambda$ for which the equation $T\mathbf{x} = \lambda\mathbf{x}$ has nonzero solutions. In this section we discuss this problem, and in later sections we shall investigate some of its applications.

**Definition.** If $A$ is a square matrix, then a scalar $\lambda$ is called an *eigenvalue* of $A$ if it satisfies the equation

$$\det(\lambda I - A) = 0$$

This equation is called the *characteristic equation* of $A$.

One of the meanings of the word "eigen" in German is "proper"; eigenvalues are also called *proper values, characteristic values,* or *latent roots* by some writers.

### Example 1

Find the eigenvalues of the matrix

$$A = \begin{bmatrix} 3 & 2 \\ -1 & 0 \end{bmatrix}$$

*Solution.* Since

$$\lambda I - A = \lambda \begin{bmatrix} 1 & 0 \\ 0 & 1 \end{bmatrix} - \begin{bmatrix} 3 & 2 \\ -1 & 0 \end{bmatrix} = \begin{bmatrix} \lambda - 3 & -2 \\ 1 & \lambda \end{bmatrix}$$

and

$$\det(\lambda I - A) = \det \begin{bmatrix} \lambda - 3 & -2 \\ 1 & \lambda \end{bmatrix} = \lambda^2 - 3\lambda + 2$$

the characteristic equation of $A$ is

$$\lambda^2 - 3\lambda + 2 = 0$$

The solutions of this equation are $\lambda = 1$ and $\lambda = 2$; these are the eigenvalues of $A$.

**Example 2**

Find the eigenvalues of the matrix

$$A = \begin{bmatrix} -2 & -1 \\ 5 & 2 \end{bmatrix}$$

*Solution.* Proceeding as in Example 1

$$\det(\lambda I - A) = \det \begin{bmatrix} \lambda + 2 & 1 \\ -5 & \lambda - 2 \end{bmatrix} = \lambda^2 + 1$$

The eigenvalues of $A$ must therefore satisfy the quadratic equation $\lambda^2 + 1 = 0$.

Since the only solutions to this equation are the imaginary numbers $\lambda = i$ and $\lambda = -i$, and since we are assuming that all our scalars are real numbers, $A$ has no eigenvalues.*

**Example 3**

Find the eigenvalues of

$$A = \begin{bmatrix} 2 & 1 & 0 \\ 3 & 2 & 0 \\ 0 & 0 & 4 \end{bmatrix}$$

*Solution.* As in the preceding examples

$$\det(\lambda I - A) = \det \begin{bmatrix} \lambda - 2 & -1 & 0 \\ -3 & \lambda - 2 & 0 \\ 0 & 0 & \lambda - 4 \end{bmatrix} = \lambda^3 - 8\lambda^2 + 17\lambda - 4$$

The eigenvalues of $A$ must therefore satisfy the cubic equation

$$\lambda^3 - 8\lambda^2 + 17\lambda - 4 = 0 \tag{6.1}$$

To solve this equation, we shall begin by searching for integer solutions. This task can be greatly simplified by exploiting the fact that all integer solutions (if there are any) to a polynomial equation with integer coefficients

$$c_0\lambda^n + c_1\lambda^{n-1} + \cdots + c_n = 0$$

---

*As we pointed out in Section 4.2, there are some applications that require complex scalars and complex vector spaces. In such cases, matrices are allowed to have complex eigenvalues. In this text, however, we consider only real eigenvalues.

must be divisors of the constant term, $c_n$. Thus the only possible integer solutions of (6.1) are the divisors of $-4$, that is, $\pm 1$, $\pm 2$, $\pm 4$. Successively substituting these values in (6.1) shows that $\lambda = 4$ is an integer solution. As a consequence, $\lambda - 4$ must be a factor of the left side of (6.1). Dividing $\lambda - 4$ into $\lambda^3 - 8\lambda^2 + 17\lambda - 4$ shows that (6.1) can be rewritten as

$$(\lambda - 4)(\lambda^2 - 4\lambda + 1) = 0$$

Thus the remaining solutions of (6.1) satisfy the quadratic equation

$$\lambda^2 - 4\lambda + 1 = 0$$

which can be solved by the quadratic formula. The eigenvalues of $A$ are, therefore,

$$\lambda = 4 \qquad \lambda = 2 + \sqrt{3} \qquad \text{and} \qquad \lambda = 2 - \sqrt{3}$$

In practical problems, the method illustrated in Example 3 cannot usually be used. The matrix $A$ is often so large that the determination of the characteristic equation is not practical. As a result, various approximation methods are used to obtain eigenvalues; some of these are discussed in Chapter 7.

If $A$ is a square matrix and $\lambda$ is a scalar, then

$$\det(\lambda I - A) = 0$$

if and only if the system

$$(\lambda I - A)\mathbf{x} = \mathbf{0} \tag{6.2}$$

has nontrivial solutions (Theorem 7 in Section 5.3). Since (6.2) can be rewritten as

$$A\mathbf{x} = \lambda\mathbf{x} \tag{6.3}$$

we have the following theorem.

**Theorem 1.** *If $A$ is an $n \times n$ matrix, then the following are equivalent.*

    (a) *$\lambda$ is an eigenvalue of $A$.*
    (b) *The system of equations $(\lambda I - A)\mathbf{x} = \mathbf{0}$ has nontrivial solutions.*
    (c) *There is a nonzero vector $\mathbf{x}$ in $R^n$ such that $A\mathbf{x} = \lambda\mathbf{x}$.*

If $\lambda$ is an eigenvalue of $A$, then the solution space for the system of equations $(\lambda I - A)\mathbf{x} = \mathbf{0}$ is called the **eigenspace** of $A$ corresponding to $\lambda$, and the *nonzero* vectors in the eigenspace are called the **eigenvectors** of $A$ corresponding to $\lambda$.

**Example 4**

Find bases for the eigenspaces of

$$A = \begin{bmatrix} 3 & -2 & 0 \\ -2 & 3 & 0 \\ 0 & 0 & 5 \end{bmatrix}$$

*Solution.* The characteristic equation of $A$ is $(\lambda - 1)(\lambda - 5)^2 = 0$, so that the eigenvalues of $A$ are $\lambda = 1$ and $\lambda = 5$.

By definition

$$\mathbf{x} = \begin{bmatrix} x_1 \\ x_2 \\ x_3 \end{bmatrix}$$

is an eigenvector of $A$ corresponding to $\lambda$ if and only if $\mathbf{x}$ is a nontrivial solution of $(\lambda I - A)\mathbf{x} = \mathbf{0}$, that is, of

$$\begin{bmatrix} \lambda - 3 & 2 & 0 \\ 2 & \lambda - 3 & 0 \\ 0 & 0 & \lambda - 5 \end{bmatrix} \begin{bmatrix} x_1 \\ x_2 \\ x_3 \end{bmatrix} = \begin{bmatrix} 0 \\ 0 \\ 0 \end{bmatrix} \tag{6.4}$$

If $\lambda = 5$, (6.4) becomes

$$\begin{bmatrix} 2 & 2 & 0 \\ 2 & 2 & 0 \\ 0 & 0 & 0 \end{bmatrix} \begin{bmatrix} x_1 \\ x_2 \\ x_3 \end{bmatrix} = \begin{bmatrix} 0 \\ 0 \\ 0 \end{bmatrix}$$

Solving this system yields

$$x_1 = -s \qquad x_2 = s \qquad x_3 = t$$

Thus the eigenvectors of $A$ corresponding to $\lambda = 5$ are the nonzero vectors of the form

$$\mathbf{x} = \begin{bmatrix} -s \\ s \\ t \end{bmatrix} = \begin{bmatrix} -s \\ s \\ 0 \end{bmatrix} + \begin{bmatrix} 0 \\ 0 \\ t \end{bmatrix} = s \begin{bmatrix} -1 \\ 1 \\ 0 \end{bmatrix} + t \begin{bmatrix} 0 \\ 0 \\ 1 \end{bmatrix}$$

Since

$$\begin{bmatrix} -1 \\ 1 \\ 0 \end{bmatrix} \qquad \text{and} \qquad \begin{bmatrix} 0 \\ 0 \\ 1 \end{bmatrix}$$

are linearly independent, they form a basis for the eigenspace corresponding to $\lambda = 5$.

If $\lambda = 1$, then (6.4) becomes

$$\begin{bmatrix} -2 & 2 & 0 \\ 2 & -2 & 0 \\ 0 & 0 & -4 \end{bmatrix} \begin{bmatrix} x_1 \\ x_2 \\ x_3 \end{bmatrix} = \begin{bmatrix} 0 \\ 0 \\ 0 \end{bmatrix}$$

Solving this system yields

$$x_1 = t \qquad x_2 = t \qquad x_3 = 0$$

Thus the eigenvectors corresponding to $\lambda = 1$ are the nonzero vectors of the form

$$\mathbf{x} = \begin{bmatrix} t \\ t \\ 0 \end{bmatrix} = t \begin{bmatrix} 1 \\ 1 \\ 0 \end{bmatrix}$$

so that

$$\begin{bmatrix} 1 \\ 1 \\ 0 \end{bmatrix}$$

is a basis for the eigenspace corresponding to $\lambda = 1$.

Eigenvalues and eigenvectors have an interesting geometric interpretation. If $\mathbf{x}$ is an eigenvector of $A$ corresponding to $\lambda$ then $A\mathbf{x} = \lambda\mathbf{x}$. Thus multiplication by $A$ maps $\mathbf{x}$ into a scalar multiple of itself. Therefore, depending on the value of $\lambda$, multiplication by $A$ dilates $\mathbf{x}$, contracts $\mathbf{x}$, or reverses the direction of $\mathbf{x}$ (Figure 6.1).

(a)            (b)            (c)

**Figure 6.1**   (*a*) Dilation $\lambda > 1$. (*b*) Contraction $0 < \lambda < 1$. (*c*) Reversal of direction $\lambda < 0$.

If $A$ is a $2 \times 2$ or $3 \times 3$ matrix, and $l$ is a line in $R^2$ or $R^3$ determined by an eigenvector $\mathbf{x}$ (Figure 6.2), then multiplication by $A$ maps each vector in $l$ back into $l$. To see why, suppose $\mathbf{x}$ corresponds to the eigenvalue $\lambda$. If $\mathbf{v}$ is any vector in $l$, then $\mathbf{v} = k\mathbf{x}$ for some scalar $k$; thus $A\mathbf{v} = A(k\mathbf{x}) = kA\mathbf{x} = (k\lambda)\mathbf{x}$. Thus, $A\mathbf{v}$ is a scalar multiple of $\mathbf{x}$; consequently $A\mathbf{v}$ also lies in $l$.

Eigenvectors and eigenvalues can be defined for linear operators as well as matrices. A scalar $\lambda$ is called an ***eigenvalue*** of a linear operator $T: V \rightarrow V$ if there is a nonzero vector $\mathbf{x}$ in $V$ such that $T\mathbf{x} = \lambda\mathbf{x}$. The vector $\mathbf{x}$ is called an ***eigenvector*** of $T$ corresponding to $\lambda$. Equivalently, the eigenvectors of $T$ corresponding to $\lambda$ are the nonzero vectors in the kernel of $\lambda I - T$ (Exercise 19). This kernel is called the ***eigenspace*** of $T$ corresponding to $\lambda$.

**Figure 6.2**

It can be shown that if $V$ is a finite dimensional space and $A$ is the matrix of $T$ with respect to *any* basis $B$, then:

1. The eigenvalues of $T$ are the eigenvalues of the matrix $A$.
2. A vector $\mathbf{x}$ is an eigenvector of $T$ corresponding to $\lambda$ if and only if its coordinate matrix $[\mathbf{x}]_B$ is an eigenvector of $A$ corresponding to $\lambda$.

We leave the proofs for the exercises.

**Example 5**

Find the eigenvalues and bases for the eigenspaces of the linear operator $T: P_2 \to P_2$ defined by

$$T(a + bx + cx^2) = (3a - 2b) + (-2a + 3b)x + (5c)x^2$$

*Solution.* The matrix of $T$ with respect to the standard basis $B = \{1, x, x^2\}$ is

$$A = \begin{bmatrix} 3 & -2 & 0 \\ -2 & 3 & 0 \\ 0 & 0 & 5 \end{bmatrix}$$

The eigenvalues of $T$ are the eigenvalues of $A$; namely $\lambda = 1$ and $\lambda = 5$ (Example 4). Also from Example 4 the eigenspace of $A$ corresponding to $\lambda = 5$ has the basis $\{\mathbf{u}_1, \mathbf{u}_2\}$ and that corresponding to $\lambda = 1$ has the basis $\{\mathbf{u}_3\}$, where

$$\mathbf{u}_1 = \begin{bmatrix} -1 \\ 1 \\ 0 \end{bmatrix} \qquad \mathbf{u}_2 = \begin{bmatrix} 0 \\ 0 \\ 1 \end{bmatrix} \qquad \mathbf{u}_3 = \begin{bmatrix} 1 \\ 1 \\ 0 \end{bmatrix}$$

These matrices are the coordinate matrices with respect to $B$ of

$$\mathbf{p}_1 = -1 + x \qquad \mathbf{p}_2 = x^2 \qquad \mathbf{p}_3 = 1 + x$$

Thus $\{-1 + x, x^2\}$ is a basis for the eigenspace of $T$ corresponding to $\lambda = 5$, and $\{1 + x\}$ is a basis for the eigenspace corresponding to $\lambda = 1$.

**EXERCISE SET 6.1**

1. Find the characteristic equations of the following matrices:

   (a) $\begin{bmatrix} 3 & 0 \\ 8 & -1 \end{bmatrix}$   (b) $\begin{bmatrix} 10 & -9 \\ 4 & -2 \end{bmatrix}$   (c) $\begin{bmatrix} 0 & 3 \\ 4 & 0 \end{bmatrix}$

   (d) $\begin{bmatrix} -2 & -7 \\ 1 & 2 \end{bmatrix}$   (e) $\begin{bmatrix} 0 & 0 \\ 0 & 0 \end{bmatrix}$   (f) $\begin{bmatrix} 1 & 0 \\ 0 & 1 \end{bmatrix}$

2. Find the eigenvalues of the matrices in Exercise 1.

3. Find bases for the eigenspaces of the matrices in Exercise 1.

4. In each part of Exercise 1, let $T: R^2 \to R^2$ be multiplication by the given matrix. In each case sketch the lines in $R^2$ that are mapped into themselves under $T$.

5. Find the characteristic equations of the following matrices:

(a) $\begin{bmatrix} 4 & 0 & 1 \\ -2 & 1 & 0 \\ -2 & 0 & 1 \end{bmatrix}$
(b) $\begin{bmatrix} 3 & 0 & -5 \\ \frac{1}{5} & -1 & 0 \\ 1 & 1 & -2 \end{bmatrix}$
(c) $\begin{bmatrix} -2 & 0 & 1 \\ -6 & -2 & 0 \\ 19 & 5 & -4 \end{bmatrix}$

(d) $\begin{bmatrix} -1 & 0 & 1 \\ -1 & 3 & 0 \\ -4 & 13 & -1 \end{bmatrix}$
(e) $\begin{bmatrix} 5 & 0 & 1 \\ 1 & 1 & 0 \\ -7 & 1 & 0 \end{bmatrix}$
(f) $\begin{bmatrix} 5 & 6 & 2 \\ 0 & -1 & -8 \\ 1 & 0 & -2 \end{bmatrix}$

6. Find the eigenvalues of the matrices in Exercise 5.

7. Find bases for the eigenspaces of the matrices in Exercise 5.

8. Find the characteristic equations of the following matrices:

(a) $\begin{bmatrix} 0 & 0 & 2 & 0 \\ 1 & 0 & 1 & 0 \\ 0 & 1 & -2 & 0 \\ 0 & 0 & 0 & 1 \end{bmatrix}$
(b) $\begin{bmatrix} 10 & -9 & 0 & 0 \\ 4 & -2 & 0 & 0 \\ 0 & 0 & -2 & -7 \\ 0 & 0 & 1 & 2 \end{bmatrix}$

9. Find the eigenvalues of the matrices in Exercise 8.

10. Find bases for the eigenspaces of the matrices in Exercise 8.

11. Let $T: P_2 \to P_2$ be defined by

$$T(a_0 + a_1 x + a_2 x^2) = (5a_0 + 6a_1 + 2a_2) - (a_1 + 8a_2)x + (a_0 - 2a_2)x^2.$$

(a) Find the eigenvalues of $T$.
(b) Find bases for the eigenspaces of $T$.

12. Let $T: M_{22} \to M_{22}$ be defined by

$$T\left(\begin{bmatrix} a & b \\ c & d \end{bmatrix}\right) = \begin{bmatrix} 2c & a + c \\ b - 2c & d \end{bmatrix}$$

(a) Find the eigenvalues of $T$.
(b) Find bases for the eigenspaces of $T$.

13. Prove that $\lambda = 0$ is an eigenvalue of a matrix $A$ if and only if $A$ is not invertible.

14. If $A$ is a square matrix, then $p(\lambda) = \det(\lambda I - A)$ is called the **characteristic polynomial of $A$**. Prove that the constant term in the characteristic polynomial of an $n \times n$ matrix $A$ is $(-1)^n \det(A)$. (**Hint.** The constant term is the value of the characteristic polynomial when $\lambda = 0$.)

15. The **trace** of a square matrix $A$ is the sum of the elements on the main diagonal. Show that the characteristic equation of a $2 \times 2$ matrix $A$ is $\lambda^2 - \mathrm{tr}(A)\lambda + \det(A) = 0$, where $\mathrm{tr}(A)$ is the trace of $A$.

16. Prove that the eigenvalues of a triangular matrix are the entries on the main diagonal.

**17.** Show that if $\lambda$ is an eigenvalue of $A$, then $\lambda^2$ is an eigenvalue of $A^2$; more generally, show that $\lambda^n$ is an eigenvalue of $A^n$ if $n$ is a positive integer.

**18.** Use the results of Exercises 16 and 17 to find the eigenvalues of $A^9$, where

$$A = \begin{bmatrix} 1 & 3 & 7 & 11 \\ 0 & -1 & 3 & 8 \\ 0 & 0 & -2 & 4 \\ 0 & 0 & 0 & 2 \end{bmatrix}$$

**19.** Let $\lambda$ be an eigenvalue of a linear operator $T: V \to V$. Prove that the eigenvectors of $T$ corresponding to $\lambda$ are the nonzero vectors in the kernel of $\lambda I - T$.

## 6.2 DIAGONALIZATION

In this section and the next, we shall be concerned with the following problems.

**Problem 1.** Given a linear operator $T: V \to V$ on a finite dimensional vector space, does there exist a basis for $V$ with respect to which the matrix of $T$ is diagonal?

**Problem 2.** Given a linear operator $T: V \to V$ on a finite dimensional inner product space, does there exist an orthonormal basis for $V$ with respect to which the matrix of $T$ is diagonal?

If $A$ is the matrix of $T: V \to V$ with respect to some basis, then Problem 1 is equivalent to asking if there is a change of basis such that the new matrix for $T$ is diagonal. By Theorem 11 in Section 5.6, the new matrix for $T$ will be $P^{-1}AP$ where $P$ is the appropriate transition matrix. If $V$ is an inner product space and the bases are orthonormal, then by Theorem 10 in Section 5.5, $P$ will be orthogonal. Thus we are led to the following matrix formulations of the above problems.

**Problem 1 (matrix form).** Given a square matrix $A$, does there exist an invertible matrix $P$ such that $P^{-1}AP$ is diagonal?

**Problem 2 (matrix form).** Given a square matrix $A$, does there exist an orthogonal matrix $P$ such that $P^{-1}AP$ $(=P^tAP)$ is diagonal?

These problems suggest the following definitions.

**Definition.** A square matrix $A$ is called ***diagonalizable*** if there is an invertible matrix $P$ such that $P^{-1}AP$ is diagonal; the matrix $P$ is said to ***diagonalize*** $A$.

If there is an orthogonal matrix $P$ such that $P^{-1}AP$ $(=P^tAP)$ is diagonal, then $A$ is called ***orthogonally diagonalizable,*** and $P$ is said to ***orthogonally diagonalize*** $A$.

The following theorem is the basic tool in the study of diagonalizability.

**Theorem 2.** *If A is an n × n matrix, then the following are equivalent.*

(a) *A is diagonalizable.*
(b) *A has n linearly independent eigenvectors.*

*Proof (a)* $\Rightarrow$ *(b)*. Since $A$ is assumed diagonalizable, there is an invertible matrix

$$P = \begin{bmatrix} p_{11} & p_{12} & \cdots & p_{1n} \\ p_{21} & p_{22} & \cdots & p_{2n} \\ \vdots & \vdots & & \vdots \\ p_{n1} & p_{n2} & \cdots & p_{nn} \end{bmatrix}$$

such that $P^{-1}AP$ is diagonal, say $P^{-1}AP = D$, where

$$D = \begin{bmatrix} \lambda_1 & 0 & \cdots & 0 \\ 0 & \lambda_2 & \cdots & 0 \\ \vdots & \vdots & & \vdots \\ 0 & 0 & \cdots & \lambda_n \end{bmatrix}$$

Therefore, $AP = PD$; that is

$$AP = \begin{bmatrix} p_{11} & p_{12} & \cdots & p_{1n} \\ p_{21} & p_{22} & \cdots & p_{2n} \\ \vdots & \vdots & & \vdots \\ p_{n1} & p_{n2} & \cdots & p_{nn} \end{bmatrix} \begin{bmatrix} \lambda_1 & 0 & \cdots & 0 \\ 0 & \lambda_2 & \cdots & 0 \\ \vdots & \vdots & & \vdots \\ 0 & 0 & \cdots & \lambda_n \end{bmatrix} = \begin{bmatrix} \lambda_1 p_{11} & \lambda_2 p_{12} & \cdots & \lambda_n p_{1n} \\ \lambda_1 p_{21} & \lambda_2 p_{22} & \cdots & \lambda_n p_{2n} \\ \vdots & \vdots & & \vdots \\ \lambda_1 p_{n1} & \lambda_2 p_{n2} & \cdots & \lambda_n p_{nn} \end{bmatrix}$$

$$(6.5)$$

Let $\mathbf{p}_1, \mathbf{p}_2, \ldots, \mathbf{p}_n$ denote the column vectors of $P$; then from (6.5) and Example 17 in Section 1.4, it follows that

$$A\mathbf{p}_1 = \lambda_1 \mathbf{p}_1, A\mathbf{p}_2 = \lambda_2 \mathbf{p}_2, \ldots, A\mathbf{p}_n = \lambda_n \mathbf{p}_n \qquad (6.6)$$

Since $P$ is invertible, its column vectors are all nonzero; thus by (6.6), $\lambda_1, \lambda_2, \ldots, \lambda_n$ are eigenvalues of $A$, and $\mathbf{p}_1, \mathbf{p}_2, \ldots, \mathbf{p}_n$ are corresponding eigenvectors. Since $P$ is invertible, it follows from Theorem 7 in Section 5.3 that $\mathbf{p}_1, \mathbf{p}_2, \ldots, \mathbf{p}_n$ are linearly independent. Thus $A$ has $n$ linearly independent eigenvectors.

*(b)* $\Rightarrow$ *(a)* Assume that $A$ has $n$ linearly independent eigenvectors, $\mathbf{p}_1, \mathbf{p}_2, \ldots, \mathbf{p}_n$, with corresponding eigenvalues $\lambda_1, \lambda_2, \ldots, \lambda_n$, and let

$$P = \begin{bmatrix} p_{11} & p_{12} & \cdots & p_{1n} \\ p_{21} & p_{22} & \cdots & p_{2n} \\ \vdots & \vdots & & \vdots \\ p_{n1} & p_{n2} & \cdots & p_{nn} \end{bmatrix}$$

be the matrix whose column vectors are $\mathbf{p}_1, \mathbf{p}_2, \ldots, \mathbf{p}_n$. By Example 17 in Section 1.4, the columns of the product $AP$ are

$$Ap_1, Ap_2, \ldots, Ap_n$$

But

$$A\mathbf{p}_1 = \lambda_1 \mathbf{p}_1, A\mathbf{p}_2 = \lambda_2 \mathbf{p}_2, \ldots, A\mathbf{p}_n = \lambda \mathbf{p}_n$$

so that

$$AP = \begin{bmatrix} \lambda_1 p_{11} & \lambda_2 p_{12} & \cdots & \lambda_n p_{1n} \\ \lambda_1 p_{21} & \lambda_2 p_{22} & \cdots & \lambda_n p_{2n} \\ \vdots & \vdots & & \vdots \\ \lambda_1 p_{n1} & \lambda_2 p_{n2} & \cdots & \lambda_n p_{nn} \end{bmatrix} = \begin{bmatrix} p_{11} & p_{12} & \cdots & p_{1n} \\ p_{21} & p_{22} & \cdots & p_{2n} \\ \vdots & \vdots & & \vdots \\ p_{n1} & p_{n2} & \cdots & p_{nn} \end{bmatrix} \begin{bmatrix} \lambda_1 & 0 & \cdots & 0 \\ 0 & \lambda_2 & \cdots & 0 \\ \vdots & \vdots & & \vdots \\ 0 & 0 & \cdots & \lambda_n \end{bmatrix}$$

$$= PD \tag{6.7}$$

where $D$ is the diagonal matrix having the eigenvalues $\lambda_1, \lambda_2, \ldots, \lambda_n$ on the main diagonal. Since the column vectors of $P$ are linearly independent, $P$ is invertible; thus (6.7) can be rewritten as $P^{-1}AP = D$; that is, $A$ is diagonalizable. ∎

**REMARK.** From this proof we obtain the following procedure for diagonalizing a diagonalizable $n \times n$ matrix $A$.

1. Find $n$ linearly independent eigenvectors of $A$, $\mathbf{p}_1, \mathbf{p}_2, \ldots, \mathbf{p}_n$.
2. Form the matrix $P$ having $\mathbf{p}_1, \mathbf{p}_2, \ldots, \mathbf{p}_n$ as its column vectors.
3. The matrix $P^{-1}AP$ will then be diagonal with $\lambda_1, \lambda_2, \ldots, \lambda_n$ as its successive diagonal entries, where $\lambda_i$ is the eigenvalue corresponding to $\mathbf{p}_i$, $i = 1, 2, \ldots, n$.

**Example 6**

Find a matrix $P$ that diagonalizes

$$A = \begin{bmatrix} 3 & -2 & 0 \\ -2 & 3 & 0 \\ 0 & 0 & 5 \end{bmatrix}$$

*Solution.* From Example 4 the eigenvalues of $A$ are $\lambda = 1$ and $\lambda = 5$. Further, it was shown that the vectors

$$\mathbf{p}_1 = \begin{bmatrix} -1 \\ 1 \\ 0 \end{bmatrix} \quad \text{and} \quad \mathbf{p}_2 = \begin{bmatrix} 0 \\ 0 \\ 1 \end{bmatrix}$$

form a basis for the eigenspace corresponding to $\lambda = 5$ and that

$$\mathbf{p}_3 = \begin{bmatrix} 1 \\ 1 \\ 0 \end{bmatrix}$$

is a basis for the eigenspace corresponding to $\lambda = 1$. It is easy to check that $\{\mathbf{p}_1, \mathbf{p}_2, \mathbf{p}_3\}$ is linearly independent, so that

$$P = \begin{bmatrix} -1 & 0 & 1 \\ 1 & 0 & 1 \\ 0 & 1 & 0 \end{bmatrix}$$

diagonalizes $A$. As a check, the reader should verify that

$$P^{-1}AP = \begin{bmatrix} -1/2 & 1/2 & 0 \\ 0 & 0 & 1 \\ 1/2 & 1/2 & 0 \end{bmatrix} \begin{bmatrix} 3 & -2 & 0 \\ -2 & 3 & 0 \\ 0 & 0 & 5 \end{bmatrix} \begin{bmatrix} -1 & 0 & 1 \\ 1 & 0 & 1 \\ 0 & 1 & 0 \end{bmatrix} = \begin{bmatrix} 5 & 0 & 0 \\ 0 & 5 & 0 \\ 0 & 0 & 1 \end{bmatrix}$$

There is no preferred order for the columns of $P$. Since the $i$th diagonal entry of $P^{-1}AP$ is an eigenvalue for the $i$th column vector of $P$, changing the order of the columns of $P$ just changes the order of the eigenvalues on the diagonal of $P^{-1}AP$. Thus, had we written

$$P = \begin{bmatrix} -1 & 1 & 0 \\ 1 & 1 & 0 \\ 0 & 0 & 1 \end{bmatrix}$$

in the last example, we would have obtained

$$P^{-1}AP = \begin{bmatrix} 5 & 0 & 0 \\ 0 & 1 & 0 \\ 0 & 0 & 5 \end{bmatrix}$$

## Example 7

The characteristic equation of

$$A = \begin{bmatrix} -3 & 2 \\ -2 & 1 \end{bmatrix}$$

is

$$\det(\lambda I - A) = \det \begin{bmatrix} \lambda + 3 & -2 \\ 2 & \lambda - 1 \end{bmatrix} = (\lambda + 1)^2 = 0$$

Thus $\lambda = -1$ is the only eigenvalue of $A$; the eigenvectors corresponding to $\lambda = -1$ are the solutions of $(-I - A)\mathbf{x} = \mathbf{0}$; that is, of

$$2x_1 - 2x_2 = 0$$
$$2x_1 - 2x_2 = 0$$

The solutions of this system are $x_1 = t$, $x_2 = t$; hence the eigenspace consists of all vectors of the form

$$\begin{bmatrix} t \\ t \end{bmatrix} = t \begin{bmatrix} 1 \\ 1 \end{bmatrix}$$

Since this space is 1-dimensional, $A$ does not have two linearly independent eigenvectors, and is therefore not diagonalizable.

The next result will be helpful in our study of diagonalizability. We defer its proof to the end of the section.

**Theorem 3.** *Eigenvectors corresponding to distinct eigenvalues of a matrix are linearly independent.*

As a consequence of this theorem, we obtain the following useful result.

**Theorem 4.** *If an $n \times n$ matrix $A$ has $n$ distinct eigenvalues, then $A$ is diagonalizable.*

*Proof.* If $\mathbf{v}_1, \mathbf{v}_2, \ldots, \mathbf{v}_n$ are eigenvectors corresponding to the distinct eigenvalues $\lambda_1, \lambda_2, \ldots, \lambda_n$, then by Theorem 3, $\mathbf{v}_1, \mathbf{v}_2, \ldots, \mathbf{v}_n$ are linearly independent. Thus $A$ is diagonalizable, by Theorem 2. ■

**Example 8**

We saw in Example 3 that

$$A = \begin{bmatrix} 2 & 1 & 0 \\ 3 & 2 & 0 \\ 0 & 0 & 4 \end{bmatrix}$$

has 3 distinct eigenvalues, $\lambda = 4, \lambda = 2 + \sqrt{3}, \lambda = 2 - \sqrt{3}$. Therefore, $A$ is diagonalizable. Further

$$P^{-1}AP = \begin{bmatrix} 4 & 0 & 0 \\ 0 & 2 + \sqrt{3} & 0 \\ 0 & 0 & 2 - \sqrt{3} \end{bmatrix}$$

for some invertible matrix $P$. The matrix $P$ can be found using the method shown in Example 6.

**Optional**

We conclude this section with a proof of Theorem 3.

*Proof.* Let $\mathbf{v}_1, \mathbf{v}_2, \ldots, \mathbf{v}_k$ be eigenvectors of $A$ corresponding to distinct eigenvalues $\lambda_1, \lambda_2, \ldots, \lambda_k$. We shall assume that $\mathbf{v}_1, \mathbf{v}_2, \ldots, \mathbf{v}_k$ are linearly dependent and obtain a contradiction. We can then conclude that $\mathbf{v}_1, \mathbf{v}_2, \ldots, \mathbf{v}_k$ are linearly independent.

Since an eigenvector is, by definition, nonzero, $\{\mathbf{v}_1\}$ is linearly independent. Let $r$ be the largest integer such that $\{\mathbf{v}_1, \mathbf{v}_2, \ldots, \mathbf{v}_r\}$ is linearly independent. Since we are assuming that $\{\mathbf{v}_1, \mathbf{v}_2, \ldots, \mathbf{v}_k\}$ is linearly dependent, $r$ satisfies $1 \leq r < k$. Moreover, by definition of $r$, $\{\mathbf{v}_1, \mathbf{v}_2, \ldots, \mathbf{v}_{r+1}\}$ is linearly dependent. Thus there are scalars $c_1, c_2, \ldots, c_{r+1}$, not all zero, such that

$$c_1\mathbf{v}_1 + c_2\mathbf{v}_2 + \cdots + c_{r+1}\mathbf{v}_{r+1} = \mathbf{0} \tag{6.8}$$

Multiplying both sides of (6.8) by $A$ and using

$$A\mathbf{v}_1 = \lambda_1\mathbf{v}_1, A\mathbf{v}_2 = \lambda_2\mathbf{v}_2, \ldots, A\mathbf{v}_{r+1} = \lambda_{r+1}\mathbf{v}_{r+1}$$

we obtain

$$c_1\lambda_1\mathbf{v}_1 + c_2\lambda_2\mathbf{v}_2 + \cdots + c_{r+1}\lambda_{r+1}\mathbf{v}_{r+1} = \mathbf{0} \tag{6.9}$$

Multiplying both sides of (6.8) by $\lambda_{r+1}$ and subtracting the resulting equation from (6.9) yields

$$c_1(\lambda_1 - \lambda_{r+1})\mathbf{v}_1 + c_2(\lambda_2 - \lambda_{r+1})\mathbf{v}_2 + \cdots + c_r(\lambda_r - \lambda_{r+1})\mathbf{v}_r = \mathbf{0}$$

Since $\{\mathbf{v}_1, \mathbf{v}_2, \ldots, \mathbf{v}_r\}$ is linearly independent, this equation implies that

$$c_1(\lambda_1 - \lambda_{r+1}) = c_2(\lambda_2 - \lambda_{r+1}) = \cdots = c_r(\lambda_r - \lambda_{r+1}) = 0$$

and since $\lambda_1, \lambda_2, \ldots, \lambda_{r+1}$ are distinct, it follows that

$$c_1 = c_2 = \cdots = c_r = 0 \tag{6.10}$$

Substituting these values in (6.8) yields

$$c_{r+1}\mathbf{v}_{r+1} = \mathbf{0}$$

Since the eigenvector $\mathbf{v}_{r+1}$ is nonzero.

$$c_{r+1} = 0 \tag{6.11}$$

Equations (6.10) and (6.11) contradict the fact that $c_1, c_2, \ldots, c_{r+1}$ are not all zero; this completes the proof. ▌

## EXERCISE SET 6.2

Show that the matrices in Exercises 1–4 are not diagonalizable.

**1.** $\begin{bmatrix} 2 & 0 \\ 1 & 2 \end{bmatrix}$   **2.** $\begin{bmatrix} 2 & -3 \\ 1 & -1 \end{bmatrix}$   **3.** $\begin{bmatrix} 3 & 0 & 0 \\ 0 & 2 & 0 \\ 0 & 1 & 2 \end{bmatrix}$   **4.** $\begin{bmatrix} -1 & 0 & 1 \\ -1 & 3 & 0 \\ -4 & 13 & -1 \end{bmatrix}$

In Exercises 5–8 find a matrix $P$ that diagonalizes $A$, and determine $P^{-1}AP$.

**5.** $A = \begin{bmatrix} -14 & 12 \\ -20 & 17 \end{bmatrix}$   **6.** $A = \begin{bmatrix} 1 & 0 \\ 6 & -1 \end{bmatrix}$

**7.** $A = \begin{bmatrix} 1 & 0 & 0 \\ 0 & 1 & 1 \\ 0 & 1 & 1 \end{bmatrix}$   **8.** $A = \begin{bmatrix} 2 & 0 & -2 \\ 0 & 3 & 0 \\ 0 & 0 & 3 \end{bmatrix}$

In Exercises 9–14 determine if $A$ is diagonalizable. If so, find a matrix $P$ that diagonalizes $A$, and determine $P^{-1}AP$.

**9.** $A = \begin{bmatrix} 19 & -9 & -6 \\ 25 & -11 & -9 \\ 17 & -9 & -4 \end{bmatrix}$

**10.** $A = \begin{bmatrix} -1 & 4 & -2 \\ -3 & 4 & 0 \\ -3 & 1 & 3 \end{bmatrix}$

**11.** $A = \begin{bmatrix} 5 & 0 & 0 \\ 1 & 5 & 0 \\ 0 & 1 & 5 \end{bmatrix}$

**12.** $A = \begin{bmatrix} 0 & 0 & 0 \\ 0 & 0 & 0 \\ 3 & 0 & 1 \end{bmatrix}$

**13.** $A = \begin{bmatrix} -2 & 0 & 0 & 0 \\ 0 & -2 & 0 & 0 \\ 0 & 0 & 3 & 0 \\ 0 & 0 & 1 & 3 \end{bmatrix}$

**14.** $A = \begin{bmatrix} -2 & 0 & 0 & 0 \\ 0 & -2 & 5 & -5 \\ 0 & 0 & 3 & 0 \\ 0 & 0 & 0 & 3 \end{bmatrix}$

**15.** Let $T: P_1 \to P_1$ be the linear operator defined by

$$T(a_0 + a_1 x) = a_0 + (6a_0 - a_1)x$$

Find a basis for $P_1$ with respect to which the matrix for $T$ is diagonal.

**16.** Let $A$ be an $n \times n$ matrix and $P$ an invertible $n \times n$ matrix. Show:
(a) $(P^{-1}AP)^2 = P^{-1}A^2P$.
(b) $(P^{-1}AP)^k = P^{-1}A^kP$ ($k$ a positive integer).

**17.** Use Exercise 16 to help compute $A^{10}$, where

$$A = \begin{bmatrix} 1 & 0 \\ -1 & 2 \end{bmatrix}$$

(**Hint.** Find a matrix $P$ that diagonalizes $A$ and compute $(P^{-1}AP)^{10}$.)

**18.** Let

$$A = \begin{bmatrix} a & b \\ c & d \end{bmatrix}$$

Show:
(a) $A$ is diagonalizable if $(a - d)^2 + 4bc > 0$.
(b) $A$ is not diagonalizable if $(a - d)^2 + 4bc < 0$.

## 6.3 ORTHOGONAL DIAGONALIZATION; SYMMETRIC MATRICES

In this section we investigate the problem of orthogonally diagonalizing a matrix. Our study will lead us to the consideration of an important class of matrices called symmetric matrices.

Through this section *orthogonal* means orthogonal with respect to the Euclidean inner product on $R^n$.

**Theorem 5.** *If $A$ is an $n \times n$ matrix, then the following are equivalent.*

*(a) $A$ is orthogonally diagonalizable.*

*(b) $A$ has an orthonormal set of $n$ eigenvectors.*

*Proof* (a) $\Rightarrow$ (b). Since $A$ is orthogonally diagonalizable, there is an orthogonal matrix $P$ such that $P^{-1}AP$ is diagonal. As shown in the proof of Theorem 2, the $n$ column vectors of $P$ are eigenvectors of $A$. Since $P$ is orthogonal, these column vectors are orthonormal (see the remark preceding Example 32 in Section 5.5) so that $A$ has $n$ orthonormal eigenvectors.

(b) $\Rightarrow$ (a). Assume that $A$ has an orthonormal set of $n$ eigenvectors $\{\mathbf{p}_1, \mathbf{p}_2, \ldots, \mathbf{p}_n\}$. As shown in the proof of Theorem 2, the matrix $P$ with these eigenvectors as columns diagonalizes $A$. Since these eigenvectors are orthonormal, $P$ is orthogonal and thus orthogonally diagonalizes $A$. ▮

We shall show below that an orthogonally diagonalizable matrix $A$ must satisfy the condition

$$A = A^t$$

A matrix with this property is said to be **symmetric.**

**Example 9**

The matrix

$$A = \begin{bmatrix} 1 & -4 & 5 \\ -4 & 3 & 0 \\ 5 & 0 & 7 \end{bmatrix}$$

is symmetric.

The proof of Theorem 5 shows that an orthogonally diagonalizable $n \times n$ matrix $A$ is orthogonally diagonalized by any $n \times n$ matrix $P$ whose columns form an orthonormal set of eigenvectors of $A$. Let $D$ be the diagonal matrix

$$D = P^{-1}AP$$

Thus

$$A = PDP^{-1}$$

or since $P$ is orthogonal

$$A = PDP^t$$

Therefore

$$A^t = (PDP^t)^t = PD^tP^t = PDP^t = A$$

This shows that an orthogonally diagonalizable matrix is symmetric. The converse is also true; however, we omit the proof, since it would take us beyond the scope of this text. The following theorem summarizes our discussion.

**Theorem 6.** *Every symmetric matrix is orthogonally diagonalizable. Conversely, every orthogonally diagonalizable matrix is symmetric.*

**Example 10**

Find an orthogonal matrix $P$ that diagonalizes

$$A = \begin{bmatrix} 4 & 2 & 2 \\ 2 & 4 & 2 \\ 2 & 2 & 4 \end{bmatrix}$$

*Solution.* We must construct a matrix $P$ whose column vectors form an orthonormal set of eigenvectors of $A$. This can be done as follows.

1. Find a basis for each eigenspace of $A$.
2. Apply the Gram-Schmidt process to each of these bases to obtain an orthonormal basis for each eigenspace.
3. Form the matrix $P$ whose columns are the basis vectors constructed in step 2; this matrix orthogonally diagonalizes $A$.

The characteristic equation of $A$ is

$$\det(\lambda I - A) = \det \begin{bmatrix} \lambda - 4 & -2 & -2 \\ -2 & \lambda - 4 & -2 \\ -2 & -2 & \lambda - 4 \end{bmatrix} = (\lambda - 2)^2(\lambda - 8) = 0$$

Thus the eigenvectors of $A$ are $\lambda = 2$ and $\lambda = 8$. By the method used in Example 4, it can be shown that

$$\mathbf{u}_1 = \begin{bmatrix} -1 \\ 1 \\ 0 \end{bmatrix} \quad \text{and} \quad \mathbf{u}_2 = \begin{bmatrix} -1 \\ 0 \\ 1 \end{bmatrix}$$

form a basis for the eigenspace corresponding to $\lambda = 2$. Applying the Gram-Schmidt process to $\{\mathbf{u}_1, \mathbf{u}_2\}$ yields the orthonormal eigenvectors

$$\mathbf{v}_1 = \begin{bmatrix} -\dfrac{1}{\sqrt{2}} \\ \dfrac{1}{\sqrt{2}} \\ 0 \end{bmatrix} \quad \text{and} \quad \mathbf{v}_2 = \begin{bmatrix} -\dfrac{1}{\sqrt{6}} \\ -\dfrac{1}{\sqrt{6}} \\ \dfrac{2}{\sqrt{6}} \end{bmatrix}$$

The eigenspace corresponding to $\lambda = 8$ has

$$\mathbf{u}_3 = \begin{bmatrix} 1 \\ 1 \\ 1 \end{bmatrix}$$

as a basis. Applying the Gram-Schmidt process to $\{\mathbf{u}_3\}$ yields

$$\mathbf{v}_3 = \begin{bmatrix} \dfrac{1}{\sqrt{3}} \\[2mm] \dfrac{1}{\sqrt{3}} \\[2mm] \dfrac{1}{\sqrt{3}} \end{bmatrix}$$

By construction $\langle \mathbf{v}_1, \mathbf{v}_2 \rangle = 0$; further (seemingly by luck), $\langle \mathbf{v}_1, \mathbf{v}_3 \rangle = \langle \mathbf{v}_2, \mathbf{v}_3 \rangle = 0$, so that $\{\mathbf{v}_1, \mathbf{v}_2, \mathbf{v}_3\}$ is an orthonormal set of eigenvectors. Thus

$$P = \begin{bmatrix} -\dfrac{1}{\sqrt{2}} & -\dfrac{1}{\sqrt{6}} & \dfrac{1}{\sqrt{3}} \\[2mm] \dfrac{1}{\sqrt{2}} & -\dfrac{1}{\sqrt{6}} & \dfrac{1}{\sqrt{3}} \\[2mm] 0 & \dfrac{2}{\sqrt{6}} & \dfrac{1}{\sqrt{3}} \end{bmatrix}$$

orthogonally diagonalizes $A$.

In this example it was crucial that the eigenvector $\mathbf{v}_3$ corresponding to $\lambda = 8$ turned out to be orthogonal to the eigenvectors $\mathbf{v}_1$ and $\mathbf{v}_2$ corresponding to $\lambda = 2$; the next theorem shows that this orthogonality was not accidental, but a consequence of the symmetry of $A$. The proof is optional and appears at the end of this section.

**Theorem 7.** *If $A$ is a symmetric matrix, then eigenvectors from different eigenspaces are orthogonal.*

We conclude this section by stating two important properties of symmetric matrices. We omit the proofs.

**Theorem 8.**

(a) *The characteristic equation of a symmetric matrix $A$ has only real roots.*

(b) *If an eigenvalue $\lambda$ of a symmetric matrix $A$ is repeated $k$ times as a root of the characteristic equation, then the eigenspace corresponding to $\lambda$ is $k$-dimensional.*

### Example 11

The characteristic equation of the symmetric matrix

$$A = \begin{bmatrix} 3 & 1 & 0 & 0 & 0 \\ 1 & 3 & 0 & 0 & 0 \\ 0 & 0 & 2 & 1 & 1 \\ 0 & 0 & 1 & 2 & 1 \\ 0 & 0 & 1 & 1 & 2 \end{bmatrix}$$

is

$$(\lambda - 4)^2(\lambda - 1)^2(\lambda - 2) = 0$$

so that the eigenvalues are $\lambda = 4$, $\lambda = 1$, and $\lambda = 2$, where $\lambda = 4$ and $\lambda = 1$ are repeated twice and $\lambda = 2$ occurs once. Thus the eigenspaces corresponding to $\lambda = 4$ and $\lambda = 1$ are 2-dimensional and the eigenspace corresponding to $\lambda = 1$ is 1-dimensional.

### Optional

**Proof of Theorem 7.** Let $\lambda_1$ and $\lambda_2$ be two different eigenvalues of the $n \times n$ symmetric matrix $A$, and let

$$\mathbf{v}_1 = \begin{bmatrix} v_1 \\ v_2 \\ \vdots \\ v_n \end{bmatrix} \quad \text{and} \quad \mathbf{v}_2 = \begin{bmatrix} v_1' \\ v_2' \\ \vdots \\ v_n' \end{bmatrix}$$

be the corresponding eigenvectors. We want to show that

$$\langle \mathbf{v}_1, \mathbf{v}_2 \rangle = v_1 v_1' + v_2 v_2' + \cdots + v_n v_n' = 0$$

Since $\mathbf{v}_1{}^t\mathbf{v}_2$ is a $1 \times 1$ matrix having $\langle \mathbf{v}_1, \mathbf{v}_2 \rangle$ as its only entry, we can complete the proof by showing that $\mathbf{v}_1{}^t\mathbf{v}_2 = 0$.

Since $\mathbf{v}_1{}^tA\mathbf{v}_2$ is a $1 \times 1$ matrix and every $1 \times 1$ matrix is obviously symmetric,

$$\begin{aligned} \mathbf{v}_1{}^tA\mathbf{v}_2 &= (\mathbf{v}_1{}^tA\mathbf{v}_2)^t \\ &= \mathbf{v}_2{}^tA^t\mathbf{v}_1 \quad \text{(by a property of the transpose;} \\ &\qquad\qquad\qquad \text{see Section 2.3)} \\ &= \mathbf{v}_2{}^tA\mathbf{v}_1 \quad \text{(since } A \text{ is symmetric)} \end{aligned}$$

Also

$$\mathbf{v}_1{}^tA\mathbf{v}_2 = \mathbf{v}_1{}^t\lambda_2\mathbf{v}_2 = \lambda_2\mathbf{v}_1{}^t\mathbf{v}_2$$

and

$$\begin{aligned} \mathbf{v}_2{}^tA\mathbf{v}_1 &= \mathbf{v}_2{}^t\lambda_1\mathbf{v}_1 = \lambda_1\mathbf{v}_2{}^t\mathbf{v}_1 \\ &= \lambda_1(\mathbf{v}_2{}^t\mathbf{v}_1)^t = \lambda_1\mathbf{v}_1{}^t\mathbf{v}_2 \end{aligned}$$

Thus

$$\lambda_1\mathbf{v}_1{}^t\mathbf{v}_2 = \lambda_2\mathbf{v}_1{}^t\mathbf{v}_2$$

or

$$(\lambda_1 - \lambda_2)\mathbf{v}_1{}^t\mathbf{v}_2 = 0$$

Since $\lambda_1 \neq \lambda_2$, it follows that $\mathbf{v}_1{}^t\mathbf{v}_2 = 0$. ▮

## EXERCISE SET 6.3

1. Use part (b) of Theorem 8 to find the dimensions of the eigenspaces of the following symmetric matrices.

(a) $\begin{bmatrix} 1 & 1 \\ 1 & 1 \end{bmatrix}$

(b) $\begin{bmatrix} \dfrac{7}{25} & 0 & -\dfrac{24}{25} \\ 0 & -1 & 0 \\ -\dfrac{24}{25} & 0 & \dfrac{7}{25} \end{bmatrix}$

(c) $\begin{bmatrix} 1 & 1 & 1 \\ 1 & 1 & 1 \\ 1 & 1 & 1 \end{bmatrix}$

(d) $\begin{bmatrix} 6 & 0 & 0 \\ 0 & 3 & 3 \\ 0 & 3 & 3 \end{bmatrix}$

(e) $\begin{bmatrix} 4 & 4 & 0 & 0 \\ 4 & 4 & 0 & 0 \\ 0 & 0 & 0 & 0 \\ 0 & 0 & 0 & 0 \end{bmatrix}$

(f) $\begin{bmatrix} \dfrac{10}{3} & -\dfrac{4}{3} & 0 & -\dfrac{4}{3} \\ -\dfrac{4}{3} & -\dfrac{5}{3} & 0 & \dfrac{1}{3} \\ 0 & 0 & -2 & 0 \\ -\dfrac{4}{3} & \dfrac{1}{3} & 0 & -\dfrac{5}{3} \end{bmatrix}$

In Exercises 2–9, find a matrix $P$ that orthogonally diagonalizes $A$, and determine $P^{-1}AP$.

2. $A = \begin{bmatrix} 3 & 1 \\ 1 & 3 \end{bmatrix}$

3. $A = \begin{bmatrix} 5 & 3\sqrt{3} \\ 3\sqrt{3} & -1 \end{bmatrix}$

4. $A = \begin{bmatrix} -7 & 24 \\ 24 & 7 \end{bmatrix}$

5. $A = \begin{bmatrix} -2 & 0 & -36 \\ 0 & -3 & 0 \\ -36 & 0 & -23 \end{bmatrix}$

6. $A = \begin{bmatrix} \dfrac{1}{9} & \dfrac{4}{3\sqrt{2}} & \dfrac{4}{9\sqrt{2}} \\ \dfrac{4}{3\sqrt{2}} & 0 & -\dfrac{1}{3} \\ \dfrac{4}{9\sqrt{2}} & -\dfrac{1}{3} & \dfrac{8}{9} \end{bmatrix}$

7. $A = \begin{bmatrix} \dfrac{8}{3} & -\dfrac{2}{\sqrt{6}} & -\dfrac{2}{\sqrt{18}} \\ -\dfrac{2}{\sqrt{6}} & 3 & \dfrac{1}{\sqrt{3}} \\ -\dfrac{2}{\sqrt{18}} & \dfrac{1}{\sqrt{3}} & \dfrac{7}{3} \end{bmatrix}$

**8.** $A = \begin{bmatrix} \frac{2}{3} & 0 & -\frac{1}{3} & \frac{1}{3} \\ 0 & 0 & 0 & 0 \\ -\frac{1}{3} & 0 & \frac{2}{3} & \frac{2}{3} \\ \frac{1}{3} & 0 & \frac{2}{3} & \frac{2}{3} \end{bmatrix}$   **9.** $A = \begin{bmatrix} \frac{30}{18} & -\frac{12}{18} & 0 & -\frac{12}{18} \\ -\frac{12}{18} & -\frac{15}{18} & 0 & \frac{3}{18} \\ 0 & 0 & -1 & 0 \\ -\frac{12}{18} & \frac{3}{18} & 0 & -\frac{15}{18} \end{bmatrix}$

**10.** Find a matrix that orthogonally diagonalizes

$$\begin{bmatrix} a & b \\ b & a \end{bmatrix}$$

where $b \neq 0$.

**11.** Two $n \times n$ matrices, $A$ and $B$, are called **orthogonally similar** if there is an orthogonal matrix $P$ such that $B = P^{-1}AP$. Show that if $A$ is symmetric and $A$ and $B$ are orthogonally similar, then $B$ is symmetric.

**12.** Show that if $A$ and $B$ are orthogonally similar and $B$ and $C$ are orthogonally similar, then $A$ and $C$ are orthogonally similar (Exercise 11).

**13.** Show that two symmetric matrices are orthogonally similar if and only if they have the same characteristic polynomial (Exercise 11).

## 6.4   QUADRATIC FORMS; CONIC SECTIONS

In this section we apply our results on symmetric matrices to the study of quadratic equations and quadratic forms. Quadratic forms arise in a variety of important problems concerned with such diverse areas as vibrations, relativity, geometry, and statistics.

An equation of the form

$$ax^2 + 2bxy + cy^2 + dx + ey + f = 0 \tag{6.12}$$

where $a, b, \ldots, f$ are real numbers and at least one of the numbers $a, b, c$ is not zero, is called a **quadratic equation in x and y;** the expression

$$ax^2 + 2bxy + cy^2$$

is called the **associated quadratic form.**

**Example 12**

In the quadratic equation

$$3x^2 + 5xy - 7y^2 + 2x + 7 = 0$$

the constants in (6.12) are

$$a = 3 \qquad b = \frac{5}{2} \qquad c = -7 \qquad d = 2 \qquad e = 0 \qquad f = 7$$

## Example 13

| Quadratic equation | Associated quadratic form |
|---|---|
| $3x^2 + 5xy - 7y^2 + 2x + 7 = 0$ | $3x^2 + 5xy - 7y^2$ |
| $4x^2 - 5y^2 + 8y + 9 = 0$ | $4x^2 - 5y^2$ |
| $xy + y = 0$ | $xy$ |

   Graphs of quadratic equations in $x$ and $y$ are called **conics** or **conic sections.** The most important conics are ellipses, circles, hyperbolas, and parabolas; these are called the **nondegenerate** conics. The remaining conics are called **degenerate** and include single points and pairs of lines (see Exercise 13).
   A nondegenerate conic is said to be in **standard position** relative to the coordinate axes if its equation can be rewritten in one of the forms given in Figure 6.3.

## Example 14

The equation

$$\frac{x^2}{4} + \frac{y^2}{9} = 1 \text{ is of the form } \frac{x^2}{k^2} + \frac{y^2}{l^2} = 1 \text{ with } k = 2, l = 3$$

Its graph is thus an ellipse in standard position intersecting the $x$-axis at $(-2, 0)$ and $(2, 0)$ and intersecting the $y$-axis at $(0, -3)$ and $(0, 3)$.
   The equation $x^2 - 8y^2 = -16$ can be rewritten as $y^2/2 - x^2/16 = 1$, which is of the form $y^2/k^2 - x^2/l^2 = 1$ with $k = \sqrt{2}, l = 4$. Its graph is thus a hyperbola in standard position intersecting the $y$-axis at $(0, -\sqrt{2})$ and $(0, \sqrt{2})$.
   The equation $5x^2 + 2y = 0$ can be rewritten as $x^2 = -(2/5)y$, which is of the form $x^2 = ky$ with $k = -2/5$. Since $k < 0$, its graph is a parabola in standard position opening downward.

   Observe that no conic in standard position has an $xy$-term (called a **cross-product term**) in its equation; the presence of an $xy$-term in the equation of a nondegenerate conic indicates that the conic is rotated out of standard position (Fig. 6.4a). Also, no conic in standard position has both an $x^2$ and $x$ term or both a $y^2$ and $y$ term. The occurrence of either of these pairs in the equation of a nondegenerate conic usually indicates that the conic is translated out of standard position (Fig. 6.4b).
   One technique for identifying the graph of a nondegenerate conic that is not in standard position consists of rotating and translating the $xy$-coordinate axes to obtain an $x'y'$-coordinate system relative to which the conic is in standard position. Once

**Equation**                                                    **Graph**

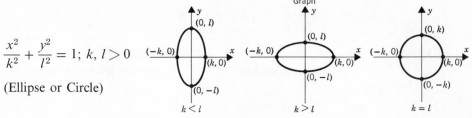

$$\frac{x^2}{k^2} + \frac{y^2}{l^2} = 1; \; k, \, l > 0$$

(Ellipse or Circle)

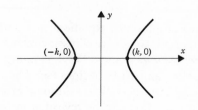

$$\frac{x^2}{k^2} - \frac{y^2}{l^2} = 1; \; k, \, l > 0$$

(Hyperbola)

$$\frac{y^2}{k^2} - \frac{x^2}{l^2} = 1; \; k, \, l > 0$$

(Hyperbola)

$$y^2 = kx$$

(Parabola)

$$x^2 = ky$$

(Parabola)

**Figure 6.3**

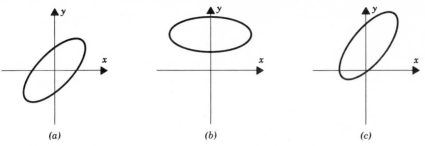

**Figure 6.4** (*a*) Rotated. (*b*) Translated. (*c*) Rotated and translated.

this is done, the equation of the conic in the $x'y'$-system will have one of the forms given in Figure 6.3 and can then easily be identified.

**Example 15**

Since the quadratic equation

$$2x^2 + y^2 - 12x - 4y + 18 = 0$$

contains $x^2$, $x$, $y^2$, and $y$ terms but no cross-product term, its graph is a conic that is translated out of standard position, but not rotated. This conic can be brought into standard position by properly translating the coordinate axes. To do this, first collect $x$ and $y$ terms. This yields

$$(2x^2 - 12x) + (y^2 - 4y) + 18 = 0$$

or

$$2(x^2 - 6x) + (y^2 - 4y) = -18$$

By completing the squares* on the two expressions in parentheses, we obtain

$$2(x^2 - 6x + 9) + (y^2 - 4y + 4) = -18 + 18 + 4$$

or

$$2(x - 3)^2 + (y - 2)^2 = 4 \tag{6.13}$$

If we translate the coordinate axes by means of the translation equations

$$x' = x - 3 \qquad y' = y - 2$$

(Example 3 in Section 3.1), then (6.13) becomes

$$2x'^2 + y'^2 = 4$$

---

*To complete the square on an expression of the form $x^2 + px$, add and subtract the constant $(p/2)^2$ to obtain

$$x^2 + px = x^2 + px + \left(\frac{p}{2}\right)^2 - \left(\frac{p}{2}\right)^2 = \left(x + \frac{p}{2}\right)^2 - \left(\frac{p}{2}\right)^2$$

or

$$\frac{x'^2}{2} + \frac{y'^2}{4} = 1$$

which is the equation of an ellipse in standard position in the $x'y'$-system. This ellipse is sketched in Fig. 6.5.

We now consider the identification of conics that are rotated out of standard position. For the remainder of this text we follow a standard convention of omitting the brackets on all $1 \times 1$ matrices. Thus the symbol 8 can denote either the number 8 or the $1 \times 1$ matrix whose entry is the number 8. It will always be possible to tell from the context which is meant. With this convention, (6.12) can be written in the matrix form

$$[x \ \ y]\begin{bmatrix} a & b \\ b & c \end{bmatrix}\begin{bmatrix} x \\ y \end{bmatrix} + [d \ \ e]\begin{bmatrix} x \\ y \end{bmatrix} + f = 0$$

or

$$\mathbf{x}^t A \mathbf{x} + K\mathbf{x} + f = 0 \qquad\qquad (6.14)$$

where

$$\mathbf{x} = \begin{bmatrix} x \\ y \end{bmatrix} \qquad A = \begin{bmatrix} a & b \\ b & c \end{bmatrix} \qquad K = [d \ \ e]$$

In this notation the quadratic form associated with (6.14) is

$$\mathbf{x}^t A \mathbf{x}$$

The symmetric matrix $A$ is called the ***matrix of the quadratic form*** $\mathbf{x}^t A\mathbf{x}$.

**Example 16**

The matrices of the quadratic forms

$$3x^2 + 5xy - 7y^2 \qquad \text{and} \qquad 8x^2 - 4y^2$$

**Figure 6.5**

are

$$\begin{bmatrix} 3 & \dfrac{5}{2} \\ \dfrac{5}{2} & 7 \end{bmatrix} \quad \text{and} \quad \begin{bmatrix} 8 & 0 \\ 0 & -4 \end{bmatrix}$$

Consider a conic $C$ with the equation

$$\mathbf{x}^t A \mathbf{x} + K \mathbf{x} + f = 0 \tag{6.15}$$

We now show it is possible to rotate the $xy$-coordinate axes so that the equation of the conic in the $x'y'$-coordinate system has no cross product term.

**Step (1).** Find a matrix

$$P = \begin{bmatrix} p_{11} & p_{12} \\ p_{21} & p_{22} \end{bmatrix}$$

which orthogonally diagonalizes $A$.

**Step (2).** Interchange the columns of $P$, if necessary, to make $\det(P) = 1$. This assures that the orthogonal coordinate transformation

$$\mathbf{x} = P\mathbf{x}', \text{ that is, } \begin{bmatrix} x \\ y \end{bmatrix} = P \begin{bmatrix} x' \\ y' \end{bmatrix} \tag{6.16}$$

is a rotation (Section 5.5).

**Step (3).** To obtain the equation for $C$ in the $x'y'$-system, substitute (6.16) into (6.15). This yields

$$(P\mathbf{x}')^t A(P\mathbf{x}') + K(P\mathbf{x}') + f = 0$$

or

$$\mathbf{x}'^t (P^t A P)\mathbf{x}' + (KP)\mathbf{x}' + f = 0 \tag{6.17}$$

Since $P$ orthogonally diagonalizes $A$

$$P^t A P = \begin{bmatrix} \lambda_1 & 0 \\ 0 & \lambda_2 \end{bmatrix}$$

where $\lambda_1$ and $\lambda_2$ are eigenvalues of $A$. Thus (6.17) can be rewritten as

$$[x' \ y'] \begin{bmatrix} \lambda_1 & 0 \\ 0 & \lambda_2 \end{bmatrix} \begin{bmatrix} x' \\ y' \end{bmatrix} + [d \ e] \begin{bmatrix} p_{11} & p_{12} \\ p_{21} & p_{22} \end{bmatrix} \begin{bmatrix} x' \\ y' \end{bmatrix} + f = 0$$

or

$$\lambda_1 x'^2 + \lambda_2 y'^2 + d'x' + e'y' + f = 0$$

(where $d' = dp_{11} + ep_{21}$ and $e' = dp_{12} + ep_{22}$). This equation has no cross product term.

The following theorem summarizes this discussion.

**Theorem 9.** (*Principal Axes Theorem for $R^2$*). *Let*

$$ax^2 + 2bxy + cy^2 + dx + ey + f = 0$$

*be the equation of a conic C, and let*

$$x^t Ax = ax^2 + 2bxy + cy^2$$

*be the associated quadratic form. Then the coordinate axes can be rotated so the equation for C in the new x'y'-coordinate system has the form*

$$\lambda_1 x'^2 + \lambda_2 y'^2 + d'x' + e'y' + f = 0$$

*where $\lambda_1$ and $\lambda_2$ are the eigenvalues of A. The rotation can be accomplished by the substitution*

$$x = Px'$$

*where P orthogonally diagonalizes A and $\det(P) = 1$.*

**Example 17**

Describe the conic $C$ whose equation is $5x^2 - 4xy + 8y^2 - 36 = 0$.

*Solution.* The matrix form of this equation is

$$x^t Ax - 36 = 0 \tag{6.18}$$

where

$$A = \begin{bmatrix} 5 & -2 \\ -2 & 8 \end{bmatrix}$$

The characteristic equation of $A$ is

$$\det(\lambda I - A) = \det \begin{bmatrix} \lambda - 5 & 2 \\ 2 & \lambda - 8 \end{bmatrix} = (\lambda - 9)(\lambda - 4) = 0$$

Thus the eigenvalues of $A$ are $\lambda = 4$ and $\lambda = 9$.

The eigenvectors corresponding to $\lambda = 4$ are the nonzero solutions of

$$\begin{bmatrix} -1 & 2 \\ 2 & -4 \end{bmatrix} \begin{bmatrix} x \\ y \end{bmatrix} = \begin{bmatrix} 0 \\ 0 \end{bmatrix}$$

Solving this system yields

$$\begin{bmatrix} x \\ y \end{bmatrix} = \begin{bmatrix} 2t \\ t \end{bmatrix} = t \begin{bmatrix} 2 \\ 1 \end{bmatrix}$$

Thus

$$\begin{bmatrix} 2 \\ 1 \end{bmatrix}$$

is a basis for the eigenspace corresponding to $\lambda = 4$. Normalizing this vector to obtain an orthonormal basis for this eigenspace, we get

$$\mathbf{v}_1 = \begin{bmatrix} \dfrac{2}{\sqrt{5}} \\ \dfrac{1}{\sqrt{5}} \end{bmatrix}$$

Similarly

$$\mathbf{v}_2 = \begin{bmatrix} -\dfrac{1}{\sqrt{5}} \\ \dfrac{2}{\sqrt{5}} \end{bmatrix}$$

is an orthonormal basis for the eigenspace corresponding to $\lambda = 9$.
Therefore

$$P = \begin{bmatrix} \dfrac{2}{\sqrt{5}} & -\dfrac{1}{\sqrt{5}} \\ \dfrac{1}{\sqrt{5}} & \dfrac{2}{\sqrt{5}} \end{bmatrix}$$

orthogonally diagonalizes $A$. Moreover, $\det(P) = 1$ so that the orthogonal coordinate transformation

$$\mathbf{x} = P\mathbf{x}' \tag{6.19}$$

is a rotation. Substituting (6.19) into (6.18) yields

$$(P\mathbf{x}')^t A(P\mathbf{x}') - 36 = 0$$

or

$$(\mathbf{x}')^t (P^t A P)\mathbf{x}' - 36 = 0$$

Since

$$P^t A P = \begin{bmatrix} 4 & 0 \\ 0 & 9 \end{bmatrix}$$

this equation can be written as

$$[x' \quad y'] \begin{bmatrix} 4 & 0 \\ 0 & 9 \end{bmatrix} \begin{bmatrix} x' \\ y' \end{bmatrix} - 36 = 0$$

or

$$4x'^2 + 9y'^2 - 36 = 0$$

This equation can be rewritten as

$$\frac{x'^2}{9} + \frac{y'^2}{4} = 1$$

which is the equation of the ellipse sketched in Figure 6.6.

**Figure 6.6**

### Example 18

Describe the conic $C$ whose equation is

$$5x^2 - 4xy + 8y^2 + \frac{20}{\sqrt{5}}x - \frac{80}{\sqrt{5}}y + 4 = 0$$

*Solution.* The matrix form of this equation is

$$\mathbf{x}^t A \mathbf{x} + K \mathbf{x} + 4 = 0 \tag{6.20}$$

where

$$A = \begin{bmatrix} 5 & -2 \\ -2 & 8 \end{bmatrix} \quad \text{and} \quad K = \begin{bmatrix} \dfrac{20}{\sqrt{5}} & \dfrac{-80}{\sqrt{5}} \end{bmatrix}$$

As shown in Example 17

$$P = \begin{bmatrix} \dfrac{2}{\sqrt{5}} & -\dfrac{1}{\sqrt{5}} \\ \dfrac{1}{\sqrt{5}} & \dfrac{2}{\sqrt{5}} \end{bmatrix}$$

orthogonally diagonalizes $A$. Substituting $\mathbf{x} = P\mathbf{x}'$ into (6.20) gives

$$(P\mathbf{x}')^t A(P\mathbf{x}') + K(P\mathbf{x}') + 4 = 0$$

or

$$(\mathbf{x}')^t(P^t A P)\mathbf{x}' + (KP)\mathbf{x}' + 4 = 0 \tag{6.21}$$

Since

$$P^t AP = \begin{bmatrix} 4 & 0 \\ 0 & 9 \end{bmatrix} \quad \text{and} \quad KP = \begin{bmatrix} \dfrac{20}{\sqrt{5}} & \dfrac{-80}{\sqrt{5}} \end{bmatrix} \begin{bmatrix} \dfrac{2}{\sqrt{5}} & -\dfrac{1}{\sqrt{5}} \\ \dfrac{1}{\sqrt{5}} & \dfrac{2}{\sqrt{5}} \end{bmatrix} = [-8, -36]$$

(6.21) can be written as

$$4x'^2 + 9y'^2 - 8x' - 36y' + 4 = 0 \tag{6.22}$$

To bring the conic into standard position, the $x'y'$-axes must be translated. Proceeding as in Example 15, we rewrite (6.22) as

$$4(x'^2 - 2x') + 9(y'^2 - 4y') = -4$$

Completing the squares yields

$$4(x'^2 - 2x' + 1) + 9(y'^2 - 4y' + 4) = -4 + 4 + 36$$

or

$$4(x' - 1)^2 + 9(y' - 2)^2 = 36 \tag{6.23}$$

If we translate the coordinate axes by means of the translation equations

$$x'' = x' - 1 \qquad y'' = y' - 2$$

then (6.23) becomes

$$4x''^2 + 9y''^2 = 36$$

or

$$\frac{x''^2}{9} + \frac{y''^2}{4} = 1$$

which is the equation of the ellipse sketched in Figure 6.7.

## EXERCISE SET 6.4

1. Find the quadratic forms associated with the following quadratic equations.
   (a) $2x^2 - 3xy + 4y^2 - 7x + 2y + 7 = 0$
   (b) $x^2 - xy + 5x + 8y - 3 = 0$
   (c) $5xy = 8$
   (d) $4x^2 - 2y^2 = 7$
   (e) $y^2 + 7x - 8y - 5 = 0$

2. Find the matrices of the quadratic forms in Exercise 1.

3. Express each of the quadratic equations in Exercise 1 in the matrix form $\mathbf{x}^t A\mathbf{x} + K\mathbf{x} + f = 0$.

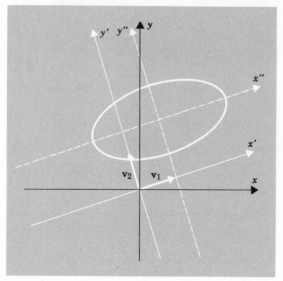

**Figure 6.7**

4. Name the following conics.
   (a) $2x^2 + 5y^2 = 20$      (b) $4x^2 + 9y^2 = 1$
   (c) $x^2 - y^2 - 8 = 0$      (d) $4y^2 - 5x^2 = 20$
   (e) $x^2 + y^2 - 25 = 0$      (f) $7y^2 - 2x = 0$
   (g) $-x^2 = 2y$      (h) $3x - 11y^2 = 0$
   (i) $y - x^2 = 0$      (j) $x^2 - 3 = -y^2$

5. In each part a translation will put the conic in standard position. Name the conic and give its equation in the translated coordinate system.
   (a) $9x^2 + 4y^2 - 36x - 24y + 36 = 0$
   (b) $x^2 - 16y^2 + 8x + 128y = 256$
   (c) $y^2 - 8x - 14y + 49 = 0$
   (d) $x^2 + y^2 + 6x - 10y + 18 = 0$
   (e) $2x^2 - 3y^2 + 6x + 20y = -41$
   (f) $x^2 + 10x + 7y = -32$

6. The following nondegenerate conics are rotated out of standard position. In each part rotate the coordinate axes to remove the $xy$-term. Name the conic and give its equation in the rotated coordinate system.
   (a) $2x^2 - 4xy - y^2 + 8 = 0$
   (b) $x^2 + 2xy + y^2 + 8x + y = 0$
   (c) $5x^2 + 4xy + 5y^2 = 9$
   (d) $11x^2 + 24xy + 4y^2 - 15 = 0$

In Exercises 7–12 translate and rotate the coordinate axes, if necessary, to put the conic in standard position. Name the conic and give its equation in the final coordinate system.

7. $9x^2 - 4xy + 6y^2 - 10x - 20y = 5$

8. $3x^2 - 8xy - 12y^2 - 30x - 64y = 0$

9. $2x^2 - 4xy - y^2 - 4x - 8y = -14$

10. $21x^2 + 6xy + 13y^2 - 114x + 34y + 73 = 0$

11. $x^2 - 6xy - 7y^2 + 10x + 2y + 9 = 0$

12. $4x^2 - 20xy + 25y^2 - 15x - 6y = 0$

13. The graph of a quadratic equation in $x$ and $y$ can, in certain cases, be a point, a line, or a pair of lines. These are called *degenerate* conics. It is also possible that the equation is not satisfied by any real values of $x$ and $y$. In such cases the equation has no graph; it is said to represent an *imaginary conic.* Each of the following represents a degenerate or imaginary conic. Where possible, sketch the graph.

(a) $x^2 - y^2 = 0$

(b) $x^2 + 3y^2 + 7 = 0$

(c) $8x^2 + 7y^2 = 0$

(d) $x^2 - 2xy + y^2 = 0$

(e) $9x^2 + 12xy + 4y^2 - 52 = 0$

(f) $x^2 + y^2 - 2x - 4y = -5$

## 6.5 QUADRIC SURFACES

In this section the techniques of the previous section are extended to quadratic equations in three variables.

An equation of the form

$$ax^2 + by^2 + cz^2 + 2dxy + 2exz + 2fyz + gx + hy + iz + j = 0 \quad (6.24)$$

where $a, b, \ldots, f$ are not all zero, is called a *quadratic equation in x, y, and z;* the expression

$$ax^2 + by^2 + cz^2 + 2dxy + 2exz + 2fyz$$

is called the *associated quadratic form.*

Equation (6.24) can be written in the matrix form

$$[x \ \ y \ \ z] \begin{bmatrix} a & d & e \\ d & b & f \\ e & f & c \end{bmatrix} \begin{bmatrix} x \\ y \\ z \end{bmatrix} + [g \ \ h \ \ i] \begin{bmatrix} x \\ y \\ z \end{bmatrix} + j = 0$$

or

$$x^t A x + K x + j = 0$$

where

$$x = \begin{bmatrix} x \\ y \\ z \end{bmatrix} \qquad A = \begin{bmatrix} a & d & e \\ d & b & f \\ e & f & c \end{bmatrix} \qquad K = [g \ \ h \ \ i]$$

The symmetric matrix $A$ is called the **matrix of the quadratic form**

$$\mathbf{x}^t A \mathbf{x} = ax^2 + by^2 + cz^2 + 2dxy + 2exz + 2fyz$$

**Example 19**

The quadratic form associated with the quadratic equation

$$3x^2 + 2y^2 - z^2 + 4xy + 3xz - 8yz + 7x + 2y + 3z - 7 = 0$$

is

$$3x^2 + 2y^2 - z^2 + 4xy + 3xz - 8yz$$

The matrix of this quadratic form is

$$\begin{bmatrix} 3 & 2 & \dfrac{3}{2} \\ 2 & 2 & -4 \\ \dfrac{3}{2} & -4 & -1 \end{bmatrix}$$

Graphs of quadratic equations in $x$, $y$, and $z$ are called **quadrics** or **quadric surfaces.** We now give some examples of quadrics and their equations.

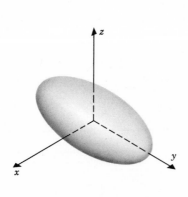

$$\frac{x^2}{l^2} + \frac{y^2}{m^2} + \frac{z^2}{n^2} = 1$$

**Ellipsoid**

$$\frac{x^2}{l^2} + \frac{y^2}{m^2} - \frac{z^2}{n^2} = 1$$

**Hyperboloid of One sheet**

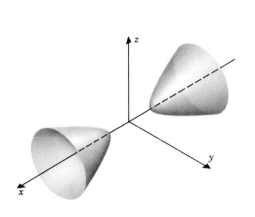

$$\frac{x^2}{l^2} - \frac{y^2}{m^2} - \frac{z^2}{n^2} = 1$$

**Hyperboloid of Two Sheets**

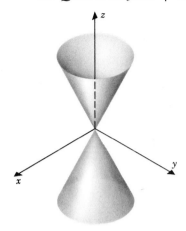

$$\frac{x^2}{l^2} + \frac{y^2}{m^2} - \frac{z^2}{n^2} = 0$$

**Elliptic Cone**

$$\frac{x^2}{l^2} + \frac{y^2}{m^2} - z = 0$$

**Elliptic Paraboloid**

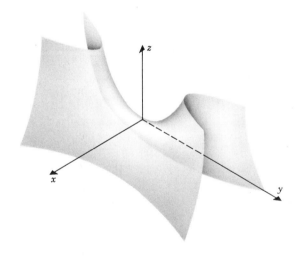

$$\frac{x^2}{l^2} - \frac{y^2}{m^2} + z = 0$$

**Hyperbolic Paraboloid**

**Figure 6.8**

A quadric whose equation has one of the forms that are given in Figure 6.8 is said to be in **standard position** relative to the coordinate axes. The presence of one or more of the cross-product terms $xy$, $xz$, and $yz$ in the equation of a nondegenerate quadric indicates that the quadric is rotated out of standard position; the presence of both $x^2$ and $x$ terms, $y^2$ and $y$ terms, or $z^2$ and $z$ terms usually indicates the quadric is translated out of standard position.

**Example 20**

Describe the quadric surface whose equation is

$$4x^2 + 36y^2 - 9z^2 - 16x - 216y + 304 = 0$$

*Solution.* Rearranging terms gives

$$4(x^2 - 4x) + 36(y^2 - 6y) - 9z^2 = -304$$

Completing the squares yields

$$4(x^2 - 4x + 4) + 36(y^2 - 6y + 9) - 9z^2 = -304 + 16 + 324$$

or

$$4(x - 2)^2 + 36(y - 3)^2 - 9z^2 = 36$$

or

$$\frac{(x - 2)^2}{9} + (y - 3)^2 - \frac{z^2}{4} = 1$$

Translating the axes by means of the translation equations

$$x' = x - 2 \qquad y' = y - 3 \qquad z' = z$$

yields

$$\frac{x'^2}{9} + y'^2 - \frac{z'^2}{4} = 1$$

which is the equation of a hyperboloid of one sheet.

The next result shows that it is always possible to eliminate the cross-product terms from the equation of a quadric by rotating the coordinate axes.

**Theorem 10.** (*Principal Axes Theorem for $R^3$*). *Let*

$$ax^2 + by^2 + cz^2 + 2dxy + 2exz + 2fyz + gx + hy + iz + j = 0 \qquad (6.25)$$

*be the equation of a quadric Q, and let*

$$x^t Ax = ax^2 + by^2 + cz^2 + 2dxy + 2exz + 2fyz$$

*be the associated quadratic form. Then the coordinate axes can be rotated so that the equation of Q in the x'y'z'-coordinate system has the form*

$$\lambda_1 x'^2 + \lambda_2 y'^2 + \lambda_3 z'^2 + g'x' + h'y' + i'z' + j = 0 \qquad (6.26)$$

*where* $\lambda_1$, $\lambda_2$, *and* $\lambda_3$ *are the eigenvalues of A. The rotation can be accomplished by the substitution*

$$\mathbf{x} = P\mathbf{x}'$$

*where P orthogonally diagonalizes A and det* $(P) = 1$.

This theorem suggests the following procedure for removing the cross-product terms from a quadratic equation in $x$, $y$, and $z$.

**Step (1).** Find a matrix $P$ that orthogonally diagonalizes $A$.

**Step (2).** Interchange two columns of $P$, if necessary, to make $\det(P) = 1$. This assures that the orthogonal coordinate transformation

$$\begin{bmatrix} x \\ y \\ z \end{bmatrix} = P \begin{bmatrix} x' \\ y' \\ z' \end{bmatrix} \tag{6.27}$$

is a rotation.

**Step (3).** Substitute (6.27) into (6.26).

The proof that the new equation has the form given in (6.26) is similar to the proof given in the previous section; it is left as an exercise.

**Example 21.**

Describe the quadric surface whose equation is

$$4x^2 + 4y^2 + 4z^2 + 4xy + 4xz + 4yz - 3 = 0$$

*Solution.* The matrix form of the above quadratic equation is

$$\mathbf{x}^t A\mathbf{x} - 3 = 0 \tag{6.28}$$

where

$$A = \begin{bmatrix} 4 & 2 & 2 \\ 2 & 4 & 2 \\ 2 & 2 & 4 \end{bmatrix}$$

As shown in Example 10, the eigenvalues of $A$ are $\lambda = 2$ and $\lambda = 8$, and $A$ is orthogonally diagonalized by the matrix

$$P = \begin{bmatrix} -\dfrac{1}{\sqrt{2}} & -\dfrac{1}{\sqrt{6}} & \dfrac{1}{\sqrt{3}} \\[2mm] \dfrac{1}{\sqrt{2}} & -\dfrac{1}{\sqrt{6}} & \dfrac{1}{\sqrt{3}} \\[2mm] 0 & \dfrac{2}{\sqrt{6}} & \dfrac{1}{\sqrt{3}} \end{bmatrix}$$

where the first two column vectors in $P$ are eigenvectors corresponding to $\lambda = 2$ and the third column vector is an eigenvector corresponding to $\lambda = 8$.

Since $\det(P) = 1$, the orthogonal coordinate transformation

$$\mathbf{x} = P\mathbf{x}', \text{ that is, } \begin{bmatrix} x \\ y \\ z \end{bmatrix} = P \begin{bmatrix} x' \\ y' \\ z' \end{bmatrix} \tag{6.29}$$

is a rotation.

Substituting (6.29) into (6.28) yields

$$(P\mathbf{x}')^t A (P\mathbf{x}') - 3 = 0$$

or equivalently

$$(\mathbf{x}')^t (P^t A P)\mathbf{x}' - 3 = 0 \tag{6.30}$$

Since

$$P^t A P = \begin{bmatrix} 2 & 0 & 0 \\ 0 & 2 & 0 \\ 0 & 0 & 8 \end{bmatrix}$$

(6.30) becomes

$$[x' \quad y' \quad z'] \begin{bmatrix} 2 & 0 & 0 \\ 0 & 2 & 0 \\ 0 & 0 & 8 \end{bmatrix} \begin{bmatrix} x' \\ y' \\ z' \end{bmatrix} - 3 = 0$$

or

$$2x'^2 + 2y'^2 + 8z'^2 = 3$$

This can be rewritten as

$$\frac{x'^2}{3/2} + \frac{y'^2}{3/2} + \frac{z'^2}{3/8} = 1$$

which is the equation of an ellipsoid.

## EXERCISE SET 6.5

1. Find the quadratic forms associated with the following quadratic equations.
   (a) $x^2 + 2y^2 - z^2 + 4xy - 5yz + 7x + 2z = 3$
   (b) $3x^2 + 7z^2 + 2xy - 3xz + 4yz - 3x = 4$
   (c) $xy + xz + yz = 1$
   (d) $x^2 + y^2 - z^2 = 7$
   (e) $3z^2 + 3xz - 14y + 9 = 0$
   (f) $2z^2 + 2xz + y^2 + 2x - y + 3z = 0$

2. Find the matrices of the quadratic forms in Exercise 1.

3. Express each of the quadratic equations in Exercise 1 in the matrix form $\mathbf{x}^t A \mathbf{x} + K\mathbf{x} + f = 0$.

**4.** Name the following quadrics.

(a) $36x^2 + 9y^2 + 4z^2 - 36 = 0$

(b) $2x^2 + 6y^2 - 3z^2 = 18$

(c) $6x^2 - 3y^2 - 2z^2 - 6 = 0$

(d) $9x^2 + 4y^2 - z^2 = 0$

(e) $16x^2 + y^2 = 16z$

(f) $7x^2 - 3y^2 + z = 0$

(g) $x^2 + y^2 + z^2 = 25$

**5.** In each part determine the translation equations that will put the quadric in standard position. Name the quadric.

(a) $9x^2 + 36y^2 + 4z^2 - 18x - 144y - 24z + 153 = 0$

(b) $6x^2 + 3y^2 - 2z^2 + 12x - 18y - 8z = -7$

(c) $3x^2 - 3y^2 - z^2 + 42x + 144 = 0$

(d) $4x^2 + 9y^2 - z^2 - 54y - 50z = 544$

(e) $x^2 + 16y^2 + 2x - 32y - 16z - 15 = 0$

(f) $7x^2 - 3y^2 + 126x + 72y + z + 135 = 0$

(g) $x^2 + y^2 + z^2 - 2x + 4y - 6z = 11$

**6.** In each part find a rotation $\mathbf{x} = P\mathbf{x}'$ that removes the cross-product terms. Name the quadric and give its equation in the $x'y'z'$-system.

(a) $2x^2 + 3y^2 + 23z^2 + 7xz + 150 = 0$

(b) $4x^2 + 4y^2 + 4z^2 + 4xy + 4xz + 4yz - 5 = 0$

(c) $144x^2 + 100y^2 + 81z^2 - 216xz - 540x - 720z = 0$

(d) $2xy + z = 0$

In Exercises 7–10 translate and rotate the coordinate axes to put the quadric in standard position. Name the quadric and give its equation in the final coordinate system.

**7.** $2xy + 2xz + 2yz - 6x - 6y - 4z = -9$

**8.** $7x^2 + 7y^2 + 10z^2 - 2xy - 4xz + 4yz - 12x + 12y + 60z = 24$

**9.** $2xy - 6x + 10y + z - 31 = 0$

**10.** $2x^2 + 2y^2 + 5z^2 - 4xy - 2xz + 2yz + 10x - 26y - 2z = 0$

**11.** Prove Theorem 10.

# Introduction to Numerical Methods of Linear Algebra

## 7.1 GAUSSIAN ELIMINATION WITH PIVOTAL CONDENSATION

In this section we discuss some practical aspects of solving systems of $n$ linear equations in $n$ unknowns. In practice, systems of linear equations are often solved on digital computers. Since computers are limited in the number of decimal places they can carry, they round off or truncate most numerical quantities. For example, a computer designed to store eight decimal places might record 2/3 either as .66666667 (rounded off) or .66666666 (truncated). In either case, an error is introduced that we shall call **rounding error.**

The main practical considerations in solving systems of linear equations on digital computers are:

1. Minimizing inaccuracies due to rounding errors.
2. Minimizing the computer time (and thus cost) needed to obtain the solution.

Except when the coefficient matrix has a specialized structure (for example, a large number of zeros), Gaussian elimination is usually the best method for solving the system. In this section we introduce a variation of Gaussian elimination designed to minimize the effect of rounding error.

Most computer arithmetic is performed using **normalized floating point numbers.** This means the numbers are expressed in the form*

$$\pm M \times 10^k \tag{7.1}$$

where $k$ is an integer and $M$ is a fraction that satisfies

$$.1 \leq M < 1$$

The fraction $M$ is called the **mantissa.**

---

*Most computers convert decimal numbers (base 10) to binary numbers (base 2). For simplicity, however, we shall think in terms of decimals.

**Example 1**

The following numbers are expressed in normalized floating point form.

$$73 = .73 \times 10^2$$
$$-.000152 = -.152 \times 10^{-3}$$
$$1{,}579 = .1579 \times 10^4$$
$$-1/4 = -.25 \times 10^0$$

The number of decimal places in the mantissa and the allowable size of the exponent $k$ in (7.1) depend on the computer being used. For example, the IBM 360 stores the equivalent of seven decimal digits in the mantissa and allows $10^k$ to range from $10^{-75}$ to $10^{75}$. A computer that uses $n$ decimal places in the mantissa is said to round off numbers to *n significant digits.*

**Example 2**

The following numbers are rounded off to three significant digits.

| Number | Normalized Floating Point Form | Rounded Value |
|---|---|---|
| 7/3 | $.233 \times 10^1$ | 2.33 |
| 1,758 | $.176 \times 10^4$ | 1,760 |
| .0000092143 | $.921 \times 10^{-5}$ | .00000921 |
| $-.12$ | $-.120 \times 10^0$ | $-.12$ |
| 13.850 | $.138 \times 10^2$ | 13.8 |
| $-.08495$ | $-.850 \times 10^{-1}$ | $-.085$ |

(If, as in the last two cases, the portion of decimal to be discarded in the rounding process is exactly half a unit, we shall adopt the convention of rounding so that the last retained digit is even. In practice, the treatment of this situation varies from computer to computer.)

We shall now introduce a variation of Gaussian elimination called *pivotal condensation* or *Gaussian elimination with pivoting;* it is designed to minimize the cumulative effect of rounding error in solving $n$ linear equations in $n$ unknowns. We assume that the system has a unique solution. As we describe each step, we shall illustrate the idea using the augmented matrix for the system

$$3x_1 + 2x_2 - x_3 = 1$$
$$6x_1 + 6x_2 + 2x_3 = 12$$
$$3x_1 - 2x_2 + x_3 = 11$$

**Step 1.** Locate the leftmost column that does not consist entirely of zeros, and find the entry in that column that has the largest absolute value. This is called the *pivot entry.*

$$\begin{bmatrix} 3 & 2 & -1 & 1 \\ 6 & 6 & 2 & 12 \\ 3 & -2 & 1 & 11 \end{bmatrix}$$

Pivot entry ← (6 is circled)

Leftmost nonzero column ↑

**Step 2.** Perform a row interchange, if necessary, to bring the pivot entry to the top of the column

$$\begin{bmatrix} 6 & 6 & 2 & 12 \\ 3 & 2 & -1 & 1 \\ 3 & -2 & 1 & 11 \end{bmatrix} \qquad \left( \begin{array}{l} \text{The first and second rows} \\ \text{of the previous matrix} \\ \text{were interchanged} \end{array} \right)$$

**Step 3.** If the pivot entry is $a$, multiply the top row by $1/a$.

$$\begin{bmatrix} 1 & 1 & \frac{1}{3} & 2 \\ 3 & 2 & -1 & 1 \\ 3 & -2 & 1 & 11 \end{bmatrix} \qquad \left( \begin{array}{l} \text{The first row of the} \\ \text{previous matrix was} \\ \text{multiplied by } 1/6 \end{array} \right)$$

**Step 4.** Add suitable multiples of the top row to the rows below so that in the column located in Step 1, all the entries below the top become zeros.

$$\begin{bmatrix} 1 & 1 & \frac{1}{3} & 2 \\ 0 & -1 & -2 & -5 \\ 0 & -5 & 0 & 5 \end{bmatrix} \qquad \left( \begin{array}{l} -3 \text{ times the first row of the} \\ \text{previous matrix was added to} \\ \text{the second and third rows} \end{array} \right)$$

**Step 5.** Cover the top row in the matrix and begin again with Step 1, applied to the submatrix that remains. Continue in this way until the *entire* matrix is in row-echelon form.

$$\begin{bmatrix} 1 & 1 & \frac{1}{3} & 2 \\ 0 & -1 & -2 & -5 \\ 0 & -5 & 0 & 5 \end{bmatrix}$$

Pivot entry ←  Left most nonzero column in the *sub*matrix ↑

$$\begin{bmatrix} 1 & 1 & \frac{1}{3} & 2 \\ 0 & -5 & 0 & 5 \\ 0 & -1 & -2 & -5 \end{bmatrix} \qquad \left( \begin{array}{l} \text{The first and second} \\ \text{rows of the *sub*matrix} \\ \text{were interchanged} \end{array} \right)$$

$$\begin{bmatrix} 1 & 1 & \frac{1}{3} & 2 \\ 0 & 1 & 0 & -1 \\ 0 & -1 & -2 & -5 \end{bmatrix}$$

$\left(\begin{array}{l}\text{The first row of}\\ \text{the }sub\text{matrix was}\\ \text{multiplied by }-1/5\end{array}\right)$

$$\begin{bmatrix} 1 & 1 & \frac{1}{3} & 2 \\ 0 & 1 & 0 & -1 \\ 0 & 0 & -2 & -6 \end{bmatrix}$$

$\left(\begin{array}{l}\text{The first row of the}\\ sub\text{matrix was added}\\ \text{to the second row}\end{array}\right)$

$$\begin{bmatrix} 1 & 1 & \frac{1}{3} & 2 \\ 0 & 1 & 0 & -1 \\ 0 & 0 & -2 & -6 \end{bmatrix}$$

$\left(\begin{array}{l}\text{The first row of the}\\ sub\text{matrix was covered and}\\ \text{we returned again to Step 1}\end{array}\right)$

Pivot — entry           — Leftmost nonzero column in the new submatrix

$$\begin{bmatrix} 1 & 1 & \frac{1}{3} & 2 \\ 0 & 1 & 0 & -1 \\ 0 & 0 & 1 & 3 \end{bmatrix}$$

$\left(\begin{array}{l}\text{The first row of the}\\ new\text{ submatrix was}\\ \text{multiplied by }-1/2\end{array}\right)$

The entire matrix is now in row-echelon form.

**Step 6.** Solve the corresponding system of equations by back substitution. The corresponding system of equations is

$$\begin{aligned} x_1 + x_2 + \frac{1}{3}x_3 &= 2 \\ x_2 &= -1 \\ x_3 &= 3 \end{aligned}$$

Solving by back substitution yields

$$x_3 = 3 \qquad x_2 = -1 \qquad x_1 = 2$$

Since the above computations are exact, this example does not illustrate the effectiveness of pivotal condensation in the reduction of rounding error; the next example does.

## Example 3

Solve the following system by Gaussian elimination with pivoting. After each calculation round off the result to three significant digits.

$$
\begin{aligned}
.00044x_1 + .0003x_2 - .0001x_3 &= .00046 \\
4x_1 + x_2 + x_3 &= 1.5 \\
3x_1 - 9.2x_2 - .5x_3 &= -8.2
\end{aligned}
\qquad (7.2)
$$

*Solution* (with pivoting). The augmented matrix is

$$
\begin{bmatrix}
.00044 & .0003 & -.0001 & .00046 \\
4 & 1 & 1 & 1.5 \\
3 & -9.2 & -.5 & -8.2
\end{bmatrix}
$$

To bring the pivot entry to the top of the first column, we interchange the first and second rows; this yields

$$
\begin{bmatrix}
4 & 1 & 1 & 1.5 \\
.00044 & .0003 & -.0001 & .00046 \\
3 & -9.2 & -.5 & -8.2
\end{bmatrix}
$$

Dividing each entry in the first row by 4 yields

$$
\begin{bmatrix}
1 & .25 & .25 & .375 \\
.00044 & .0003 & -.0001 & .00046 \\
3 & -9.2 & -.5 & -8.2
\end{bmatrix}
$$

Adding $-.00044$ times the first row to the second and $-3$ times the first row to the third yields (after rounding to three significant digits),

$$
\begin{bmatrix}
1 & .25 & .25 & .375 \\
0 & .000190 & -.00021 & .000295 \\
0 & -9.95 & -1.25 & -9.32
\end{bmatrix}
$$

Interchanging the second and third rows yields

$$
\begin{bmatrix}
1 & .25 & .25 & .375 \\
0 & -9.95 & -1.25 & -9.32 \\
0 & -.000190 & -.00021 & .000295
\end{bmatrix}
$$

Dividing each entry in the second row by $-9.95$ yields

$$
\begin{bmatrix}
1 & .25 & .25 & .375 \\
0 & 1 & .126 & .937 \\
0 & -.000190 & -.00021 & .000295
\end{bmatrix}
$$

Adding .000190 times the second row to the third yields

$$\begin{bmatrix} 1 & .25 & .25 & .375 \\ 0 & 1 & .126 & .937 \\ 0 & 0 & -.000234 & .000117 \end{bmatrix}$$

Dividing each entry in the third row by $-.000234$ yields the row-echelon form

$$\begin{bmatrix} 1 & .25 & .25 & .375 \\ 0 & 1 & .126 & .937 \\ 0 & 0 & 1 & -.5 \end{bmatrix}$$

The corresponding system of equations is

$$x_1 + .25x_2 + .25x_3 = .375$$
$$x_2 + .126x_3 = .937$$
$$x_3 = -.5$$

Solving by back substitution yields (to three significant digits)

$$x_1 = .250 \qquad x_2 = 1.00 \qquad x_3 = -.500 \tag{7.3}$$

If (7.2) is solved by Gaussian elimination *without* pivoting and each calculation is rounded to three significant digits, one obtains

$$x_1 = .245 \qquad x_2 = 1.01 \qquad x_3 = -.492 \tag{7.4}$$

Comparing (7.3) and (7.4) to the exact solution

$$x_1 = \frac{1}{4} \qquad x_2 = 1 \qquad x_3 = -\frac{1}{2}$$

shows that the use of pivoting yields more accurate results.

In spite of the fact that pivotal condensation can reduce the cumulative effect of rounding error, there are certain systems of equations, called **ill conditioned** systems, which are so extremely sensitive that even the slightest errors in the coefficients can result in major inaccuracies in the solution. For example, consider the system

$$x_1 + x_2 = -3$$
$$x_1 + 1.016x_2 = 5 \tag{7.5}$$

If we assume that this system is to be solved on a computer that rounds off to three significant digits, the computer will store this system as

$$x_1 + x_2 = -3$$
$$x_1 + 1.02x_2 = 5 \tag{7.6}$$

The exact solution to (7.5) is $x_1 = -503$, $x_2 = 500$, and the exact solution to (7.6) is $x_1 = -403$, $x_2 = 400$. Thus a rounding error of only .004 in one coefficient of (7.5) results in a gross error in the solution.

There is little that can be done computationally to avoid large errors in the solutions of ill-conditioned linear systems. However, in physical problems, where ill-conditioned systems arise, it is sometimes possible to reformulate the problem giving rise to the system to avoid ill-conditioning. Some of the texts referenced at the end of this chapter explain how to recognize ill-conditioned systems.

## EXERCISE SET 7.1

1. Express the following in normalized floating point form.
   (a) 14/5    (b) 3,452    (c) .000003879
   (d) −.135    (e) 17.921    (f) −.0863

2. Round off the numbers in Exercise 1 to three significant digits.

3. Round off the numbers in Exercise 1 to two significant digits.

In Exercises 4–7 use Gaussian elimination with pivoting to solve the system exactly. Check your work by using Gaussian elimination without pivoting to solve the system.

4. $3x_1 + x_2 = -2$
   $-5x_1 + x_2 = 22$

5. $x_1 + x_2 + x_3 = 6$
   $2x_1 - x_2 + 4x_3 = 12$
   $-3x_1 + 2x_2 - x_3 = -4$

6. $2x_1 + 3x_2 - x_3 = 5$
   $4x_1 + 4x_2 - 3x_3 = 3$
   $2x_1 - 3x_2 + x_3 = -1$

7. $5x_1 + 6x_2 - x_3 + 2x_4 = -3$
   $2x_1 - x_2 + x_3 + x_4 = 0$
   $-8x_1 + x_2 + 2x_3 - x_4 = 3$
   $5x_1 + 2x_2 + 3x_3 - x_4 = 4$

In Exercises 8–9 solve the system by Gaussian elimination with pivoting. Round off all calculations to three significant digits.

8. $.21x_1 + .33x_2 = .54$
   $.70x_1 + .24x_2 = .94$

9. $.11x_1 - .13x_2 + .20x_3 = -.02$
   $.10x_1 + .36x_2 + .45x_3 = .25$
   $.50x_1 - .01x_2 + .30x_3 = -.70$

10. Solve

$$.0001x_1 + x_2 = 1$$
$$x_1 + x_2 = 2$$

by Gaussian elimination with and without pivoting. Round off all computations to three significant digits. Compare the results with the exact solution.

## 7.2 THE GAUSS-SEIDEL AND JACOBI METHODS

Gaussian elimination is usually the best technique for solving a system of linear equations. However, when the number of equations is large, say one hundred or more, and when the matrix has many zeros, other methods may be more effective; in this section we study two such methods.

Consider a system of $n$ linear equations in $n$ unknowns

$$
\begin{array}{cccccc}
a_{11}x_1 + a_{12}x_2 + & \cdots & + a_{1n}x_n = b_1 \\
a_{21}x_1 + a_{22}x_2 + & \cdots & + a_{2n}x_n = b_2 \\
\vdots \qquad \vdots & & \vdots \qquad \vdots \\
a_{n1}x_1 + a_{n2}x_2 + & \cdots & + a_{nn}x_n = b_n
\end{array}
\tag{7.7}
$$

We shall assume that the diagonal entries $a_{11}, a_{22}, \ldots, a_{nn}$ are nonzero and that the system has exactly one solution.

The first method we shall discuss is called **Jacobi iteration** or the **method of simultaneous displacements.** To begin, rewrite system (7.7) by solving the first equation for $x_1$ in terms of the remaining unknowns, solving the second equation for $x_2$ in terms of the remaining unknowns, solving the third equation for $x_3$ in terms of the remaining unknowns, etc. This yields

$$
\begin{aligned}
x_1 &= \frac{1}{a_{11}}(b_1 - a_{12}x_2 - a_{13}x_3 - \cdots - a_{1n}x_n) \\
x_2 &= \frac{1}{a_{22}}(b_2 - a_{21}x_1 - a_{23}x_3 - \cdots - a_{2n}x_n) \\
&\vdots \\
x_n &= \frac{1}{a_{nn}}(b_n - a_{n1}x_1 - a_{n2}x_2 - \cdots - a_{nn-1}x_{n-1})
\end{aligned}
\tag{7.8}
$$

For example, the system

$$
\begin{array}{rcrcrcl}
20x_1 + & x_2 - & x_3 &=& 17 \\
x_1 - & 10x_2 + & x_3 &=& 13 \\
-x_1 + & x_2 + & 10x_3 &=& 18
\end{array}
\tag{7.9}
$$

would be rewritten

$$
\begin{aligned}
x_1 &= \frac{17}{20} - \frac{1}{20}x_2 + \frac{1}{20}x_3 \\
x_2 &= -\frac{13}{10} + \frac{1}{10}x_1 + \frac{1}{10}x_3 \\
x_3 &= \frac{18}{10} + \frac{1}{10}x_1 - \frac{1}{10}x_2
\end{aligned}
$$

or

$$
\begin{array}{rclclcl}
x_1 &=& .850 &-& .05x_2 &+& .05x_3 \\
x_2 &=& -1.3 &+& .1x_1 &+& .1x_3 \\
x_3 &=& 1.8 &+& .1x_1 &-& .1x_2
\end{array}
\tag{7.10}
$$

If an approximation to the solution of (7.7) is known, and these approximate values are substituted into the right side of (7.8), it is often the case that the values of $x_1$, $x_2, \ldots, x_n$ that result on the left side, form an even better approximation to the solution. This observation is the key to the Jacobi method.

To solve system (7.7) by Jacobi iteration, make an initial approximation to the solution. When no better choice is available use $x_1 = 0$, $x_2 = 0$, $x_3 = 0, \ldots$.

Substitute this initial approximation into the right side of (7.8) and use the values of $x_1$, $x_2, \ldots$ that result on the left side as a new approximation to the solution.

For example, to solve (7.9) by the Jacobi method, we would substitute the initial approximation $x_1 = 0$, $x_2 = 0$, $x_3 = 0$ into the right side of (7.10), and calculate the new approximation

$$x_1 = .850 \qquad x_2 = -1.3 \qquad x_3 = 1.8 \qquad (7.11)$$

To improve the approximation, we would repeat the substitution process.

For example, in solving (7.9), we would substitute the approximation (7.11) into the right side of (7.10) to obtain the next approximation

$$x_1 = .850 - .05(-1.3) + .05(1.8) = 1.005$$
$$x_2 = -1.3 + .1(1.850) + .1(1.8) = -1.035$$
$$x_3 = 1.8 + .1(.850) - .1(-1.3) = 2.0105$$

In this way a succession of approximations can be generated which, under certain conditions, get closer and closer to the exact solution of the system. In Figure 7.1 we have summarized the results obtained in solving system (7.9) by Jacobi iteration. All computations were rounded off to five significant digits. At the end of six substitutions (called *iterations*), the exact solution $x_1 = 1$, $x_2 = -1$, $x_3 = 2$ is accurately known to five significant digits.

We now discuss a minor modification of the Jacobi method that often reduces the number of iterations needed to obtain a given degree of accuracy. The technique is called **Gauss-Seidel iteration** or the method of **successive displacements.**

In each iteration of the Jacobi method, the new approximation is obtained by substituting the previous approximation into the right side of (7.8) and solving for

| | Initial approximation | First approximation | Second approximation | Third approximation | Fourth approximation | Fifth approximation | Sixth approximation |
|---|---|---|---|---|---|---|---|
| $x_1$ | 0 | .850 | 1.005 | 1.0025 | 1.0001 | .99997 | 1.0000 |
| $x_2$ | 0 | −1.3 | −1.035 | −.9980 | −.99935 | −.99999 | −1.0000 |
| $x_3$ | 0 | 1.8 | 2.015 | 2.004 | 2.0000 | 1.9999 | 2.0000 |

**Figure 7.1**

new values of $x_1, x_2, \ldots$. These new $x$-values are not all computed simultaneously; first $x_1$ is obtained from the top equation, then $x_2$ is obtained from the second equation, then $x_3$, etc. Since the new $x$-values are generally closer to the exact solution, this suggests that better accuracy might be obtained by using the new $x$-values as soon as they are known. To illustrate, consider system (7.9). In the first iteration of the Jacobi method, the initial approximation $x_1 = 0, x_2 = 0, x_3 = 0$ was substituted into each equation on the right side of (7.10) to obtain the new approximation

$$x_1 = .850 \qquad x_2 = -1.3 \qquad x_3 = 1.8 \qquad\qquad (7.12)$$

In the first iteration of the Gauss-Seidel method, the new approximation would be computed as follows. Substitute the initial approximation $x_1 = 0, x_2 = 0, x_3 = 0$ into the right side of the first equation in (7.10). This yields the new estimate $x_1 = .850$.

Use this new value of $x_1$ immediately by substituting

$$x_1 = .850 \qquad x_2 = 0 \qquad x_3 = 0$$

into the right side of the second equation in (7.10). This yields the new estimate $x_2 = -1.215$.

Use this new value of $x_2$ immediately by substituting

$$x_1 = .850 \qquad x_2 = -1.215 \qquad x_3 = 0$$

into the right side of the third equation in (7.10). This yields the new estimate $x_3 = 2.0065$.

Thus at the end of the first iteration of the Gauss-Seidel method the new approximation is

$$x_1 = .850 \qquad x_2 = -1.215 \qquad x_3 = 2.0065 \qquad\qquad (7.13)$$

The computations for the second iteration would be carried out as follows.

Substituting (7.13) into the right side of the first equation in (7.10) and rounding off to five significant digits yields

$$x_1 = .850 - .05(-1.215) + .05(2.0065) = 1.0111$$

Substituting

$$x_1 = 1.0111 \qquad x_2 = -1.215 \qquad x_3 = 2.0065$$

into the right side of the second equation in (7.10) and rounding off to five significant digits yields

$$x_2 = -1.3 + .1(1.0111) + .1(2.0065) = -.99824$$

Substituting

$$x_1 = 1.0111 \qquad x_2 = -.99824 \qquad x_3 = 2.0065$$

into the right side of the third equation in (7.10) and rounding off to five significant digits yields

$$x_3 = 1.8 + .1(1.0111) - .1(-.99824) = 2.0009$$

Thus, at the end of the second iteration of the Gauss-Seidel method, the new approximation is

$$x_1 = 1.0111 \qquad x_2 = -.99824 \qquad x_3 = 2.0009$$

In Figure 7.2 we have summarized the results obtained using four iterations of the Gauss-Seidel method to solve (7.9). All numbers were rounded off to five significant digits.

Comparing the tables in Figures 7.1 and 7.2, we see that the Gauss-Seidel method produces the solution of (7.9) (accurate to five significant digits) in four iterations while six iterations are needed to attain the same accuracy with the Jacobi method. It should not be concluded from this example, however, that the Gauss-Seidel method is always better than the Jacobi. Although it may seem surprising, there are examples where the Jacobi method is better than the Gauss-Seidel method.

The Gauss-Seidel and Jacobi methods do not always work. In some cases, one or both of these methods can fail to produce a good approximation to the solution, regardless of the number of iterations performed. In such cases the approximations are said to **diverge;** otherwise they are said to **converge.**

We conclude this section by discussing a condition that will insure that the approximations generated by the two methods converge.

A square matrix

$$A = \begin{bmatrix} a_{11} & a_{12} & \cdots & a_{1n} \\ a_{21} & a_{22} & \cdots & a_{2n} \\ \vdots & \vdots & & \vdots \\ a_{n1} & a_{n2} & \cdots & a_{nn} \end{bmatrix}$$

is called **strictly diagonally dominant** if the absolute value of each diagonal entry is greater than the sum of the absolute values of the remaining entries in the same

|       | *Initial approximation* | *First approximation* | *Second approximation* | *Third approximation* | *Fourth approximation* |
|-------|-------------------------|-----------------------|------------------------|-----------------------|------------------------|
| $x_1$ | 0 | .850 | 1.0111 | .99995 | 1.0000 |
| $x_2$ | 0 | −1.215 | −.99824 | −.99992 | −1.0000 |
| $x_3$ | 0 | 2.0065 | 2.0009 | 2.0000 | 2.0000 |

**Figure 7.2**

row; that is,

$$|a_{11}| > |a_{12}| + |a_{13}| + \cdots + |a_{1n}|$$
$$|a_{22}| > |a_{21}| + |a_{23}| + \cdots + |a_{2n}|$$
$$\vdots \qquad \vdots \qquad \vdots \qquad \vdots$$
$$|a_{nn}| > |a_{n1}| + |a_{n2}| + \cdots + |a_{nn-1}|$$

**Example 4**

$$\begin{bmatrix} 7 & -2 & 3 \\ 4 & 1 & -6 \\ 5 & 12 & -4 \end{bmatrix}$$

is not diagonally dominant since in the second row, $|1|$ is not greater than $|4| + |-6|$, and in the third row $|-4|$ is not greater than $|5| + |12|$.

If the second and third rows are interchanged, however, the resulting matrix

$$\begin{bmatrix} 7 & -2 & 3 \\ 5 & 12 & -4 \\ 4 & 1 & -6 \end{bmatrix}$$

is strictly diagonally dominant since

$$|7| > |-2| + |3|$$
$$|12| > |5| + |-4|$$
$$|-6| > |4| + |1|$$

It can be shown that if $A$ is strictly diagonally dominant, then the Gauss-Seidel and Jacobi approximations to the solutions of $A\mathbf{x} = \mathbf{b}$ both converge.

## EXERCISE SET 7.2

In Exercises 1–4 solve the systems by Jacobi iteration. Start with $x_1 = 0$, $x_2 = 0$. Use four iterations and round off the computations to three significant digits. Compare your results to the exact solutions.

1. $2x_1 + x_2 = 7$  
   $x_1 - 2x_2 = 1$

2. $3x_1 - x_2 = 5$  
   $2x_1 + 3x_2 = -4$

3. $5x_1 - 2x_2 = -13$  
   $x_1 + 7x_2 = -10$

4. $.4x_1 + .1x_2 = .2$  
   $.3x_1 + .7x_2 = 1.4$

In Exercises 5–8 solve the systems by Gauss-Seidel iteration. Start with $x_1 = 0$, $x_2 = 0$. Use three iterations and round off the computations to three significant digits. Compare your results to the exact solutions.

5. Solve the system in Exercise 1.

6. Solve the system in Exercise 2.

7. Solve the system in Exercise 3.

**8.** Solve the system in Exercise 4.

In Exercises 9–10 solve the systems by Jacobi iteration. Start with $x_1 = 0, x_2 = 0, x_3 = 0$. Use three iterations and round off the computations to three significant digits. Compare your results to the exact solutions.

**9.** $10x_1 + x_2 + 2x_3 = 3$
 $x_1 + 10x_2 - x_3 = 3/2$
 $2x_1 + x_2 + 10x_3 = -9$

**10.** $20x_1 - x_2 + x_3 = 20$
 $2x_1 + 10x_2 - x_3 = 11$
 $x_1 + x_2 - 20x_3 = -18$

In Exercises 11–12 solve the systems by Gauss-Seidel iteration. Start with $x_1 = 0, x_2 = 0, x_3 = 0$. Use three iterations and round off the computations to three significant digits. Compare your results to the exact solutions.

**11.** Solve the system in Exercise 9.

**12.** Solve the system in Exercise 10.

**13.** Which of the following are strictly diagonally dominant?

(a) $\begin{bmatrix} 2 & 1 \\ -1 & 4 \end{bmatrix}$  (b) $\begin{bmatrix} 3 & -5 \\ 1 & 2 \end{bmatrix}$  (c) $\begin{bmatrix} 6 & 0 & 1 \\ 3 & 5 & 3 \\ 0 & 0 & 1 \end{bmatrix}$

(d) $\begin{bmatrix} 4 & 1 & 2 \\ 0 & 3 & 2 \\ 4 & 1 & -7 \end{bmatrix}$  (e) $\begin{bmatrix} 5 & 1 & 2 & 0 \\ 3 & -7 & 2 & 1 \\ 0 & 2 & 5 & 1 \\ 1 & 1 & 2 & -5 \end{bmatrix}$

**14.** Consider the system

$$x_1 + 3x_2 = 4$$
$$x_1 - x_2 = 0$$

(a) Show that the approximations obtained by Jacobi iteration diverge.
(b) Is the coefficient matrix

$$A = \begin{bmatrix} 1 & 3 \\ 1 & -1 \end{bmatrix}$$

strictly diagonally dominant?

**15.** Show that if one or more of the diagonal entries $a_{11}, a_{22}, \ldots, a_{nn}$ in (7.7) is zero, then it is possible to interchange equations and relabel the unknowns so the diagonal entries in the resulting system are all nonzero.

## 7.3 APPROXIMATING EIGENVALUES BY THE POWER METHOD

The eigenvalues of a matrix can be found by solving its characteristic equation. In practical problems this method is inefficient. Moreover, in many physical problems only the eigenvalue with the largest absolute value is needed. In this section we discuss a method for approximating this eigenvalue and a corresponding eigenvector.

In the next section we shall discuss the approximation of the remaining eigenvalues and eigenvectors.

**Definition.** An eigenvalue of a matrix $A$ is called the **_dominant eigenvalue_** of $A$ if its absolute value is larger than the absolute values of the remaining eigenvalues. An eigenvector corresponding to the dominant eigenvalue is called a **_dominant eigenvector_** of $A$.

**Example 5**

If a $4 \times 4$ matrix $A$ has eigenvalues

$$\lambda_1 = -4 \quad \lambda_2 = 3 \quad \lambda_3 = -2 \quad \lambda_4 = 2$$

then $\lambda_1 = -4$ is the dominant eigenvalue since

$$|-4| > |3| \quad |-4| > |-2| \quad \text{and} \quad |-4| > |2|$$

**Example 6**

A $3 \times 3$ matrix $A$ with eigenvalues

$$\lambda_1 = 7 \quad \lambda_2 = -7 \quad \lambda_3 = 2$$

has no dominant eigenvalue.

Let $A$ be a diagonalizable $n \times n$ matrix with a dominant eigenvalue. We shall show at the end of this section that if $\mathbf{x}_0$ is an arbitrary nonzero vector in $R^n$, then the vector

$$A^p \mathbf{x}_0 \tag{7.14}$$

is usually a good approximation to a dominant eigenvector of $A$ when the exponent $p$ is large. The following example illustrates this idea.

**Example 7**

As shown in Example 1 of Chapter 6, the matrix

$$A = \begin{bmatrix} 3 & 2 \\ -1 & 0 \end{bmatrix}$$

has eigenvalues $\lambda_1 = 2$ and $\lambda_2 = 1$.

The eigenspace corresponding to the dominant eigenvalue $\lambda_1 = 2$ is the solution space of the system

$$(2I - A)\mathbf{x} = \mathbf{0}$$

that is

$$\begin{bmatrix} -1 & -2 \\ 1 & 2 \end{bmatrix} \begin{bmatrix} x_1 \\ x_2 \end{bmatrix} = \begin{bmatrix} 0 \\ 0 \end{bmatrix}$$

Solving this system yields $x_1 = -2t$, $x_2 = t$. Thus the eigenvectors corresponding to $\lambda_1 = 2$ are the nonzero vectors of the form

$$\mathbf{x} = \begin{bmatrix} -2t \\ t \end{bmatrix} \tag{7.15}$$

We now illustrate a procedure for using (7.14) to estimate a dominant eigenvector of $A$. To start, we arbitrarily select

$$\mathbf{x}_0 = \begin{bmatrix} 1 \\ 1 \end{bmatrix}$$

Repeatedly multiplying $x_0$ by $A$ yields

$$A\mathbf{x}_0 = \begin{bmatrix} 3 & 2 \\ -1 & 0 \end{bmatrix} \begin{bmatrix} 1 \\ 1 \end{bmatrix} = \begin{bmatrix} 5 \\ -1 \end{bmatrix}$$

$$A^2\mathbf{x}_0 = A(A\mathbf{x}_0) = \begin{bmatrix} 3 & 2 \\ -1 & 0 \end{bmatrix} \begin{bmatrix} 5 \\ -1 \end{bmatrix} = \begin{bmatrix} 13 \\ -5 \end{bmatrix} = 5 \begin{bmatrix} 2.6 \\ -1 \end{bmatrix}$$

$$A^3\mathbf{x}_0 = A(A^2\mathbf{x}_0) = \begin{bmatrix} 3 & 2 \\ -1 & 0 \end{bmatrix} \begin{bmatrix} 13 \\ -5 \end{bmatrix} = \begin{bmatrix} 29 \\ -13 \end{bmatrix} \approx 13 \begin{bmatrix} 2.23 \\ -1 \end{bmatrix}$$

$$A^4\mathbf{x}_0 = A(A^3\mathbf{x}_0) = \begin{bmatrix} 3 & 2 \\ -1 & 0 \end{bmatrix} \begin{bmatrix} 29 \\ -13 \end{bmatrix} = \begin{bmatrix} 61 \\ -29 \end{bmatrix} \approx 29 \begin{bmatrix} 2.10 \\ -1 \end{bmatrix}$$

$$A^5\mathbf{x}_0 = A(A^4\mathbf{x}_0) = \begin{bmatrix} 3 & 2 \\ -1 & 0 \end{bmatrix} \begin{bmatrix} 61 \\ -29 \end{bmatrix} = \begin{bmatrix} 125 \\ -61 \end{bmatrix} \approx 61 \begin{bmatrix} 2.05 \\ -1 \end{bmatrix}$$

$$A^6\mathbf{x}_0 = A(A^5\mathbf{x}_0) = \begin{bmatrix} 3 & 2 \\ -1 & 0 \end{bmatrix} \begin{bmatrix} 125 \\ -61 \end{bmatrix} = \begin{bmatrix} 253 \\ -125 \end{bmatrix} \approx 125 \begin{bmatrix} 2.02 \\ -1 \end{bmatrix}$$

$$A^7\mathbf{x}_0 = A(A^6\mathbf{x}_0) = \begin{bmatrix} 3 & 2 \\ -1 & 0 \end{bmatrix} \begin{bmatrix} 253 \\ -125 \end{bmatrix} = \begin{bmatrix} 509 \\ -253 \end{bmatrix} \approx 253 \begin{bmatrix} 2.01 \\ -1 \end{bmatrix}$$

It is evident from these calculations that the products are getting closer and closer to scalar multiples of

$$\begin{bmatrix} 2 \\ -1 \end{bmatrix}$$

which is the dominant eigenvector of $A$ obtained by letting $t = -1$ in (7.15). Since a scalar multiple of a dominant eigenvector is also a dominant eigenvector, the above calculations are producing better and better approximations to a dominant eigenvector of $A$.

We now show how to approximate the dominant eigenvalue once an approximation to a dominant eigenvector is known. Let $\lambda$ be an eigenvalue of $A$ and $\mathbf{x}$

a corresponding eigenvector. If $\langle \; , \; \rangle$ denotes the Euclidean inner product, then

$$\frac{\langle \mathbf{x}, A\mathbf{x} \rangle}{\langle \mathbf{x}, \mathbf{x} \rangle} = \frac{\langle \mathbf{x}, \lambda\mathbf{x} \rangle}{\langle \mathbf{x}, \mathbf{x} \rangle} = \frac{\lambda\langle \mathbf{x}, \mathbf{x} \rangle}{\langle \mathbf{x}, \mathbf{x} \rangle} = \lambda$$

Thus, if $\tilde{\mathbf{x}}$ is an approximation to a dominant eigenvector, the dominant eigenvalue $\lambda_1$ can be approximated by

$$\lambda_1 \approx \frac{\langle \tilde{\mathbf{x}}, A\tilde{\mathbf{x}} \rangle}{\langle \tilde{\mathbf{x}}, \tilde{\mathbf{x}} \rangle} \tag{7.16}$$

The ratio in (7.16) is called the **Rayleigh quotient.***

**Example 8**

In Example 7 we obtained

$$\tilde{\mathbf{x}} = \begin{bmatrix} 509 \\ -253 \end{bmatrix}$$

as an approximation to a dominant eigenvector. Therefore

$$A\tilde{\mathbf{x}} = \begin{bmatrix} 3 & 2 \\ -1 & 0 \end{bmatrix} \begin{bmatrix} 509 \\ -253 \end{bmatrix} = \begin{bmatrix} 1021 \\ -509 \end{bmatrix}$$

Substituting in (7.16) we obtain

$$\lambda_1 \approx \frac{\langle \tilde{\mathbf{x}}, A\tilde{\mathbf{x}} \rangle}{\langle \tilde{\mathbf{x}}, \tilde{\mathbf{x}} \rangle} = \frac{(509)(1021) + (-253)(-509)}{(509)(509) + (-253)(-253)} \approx 2.007$$

which is a relatively good approximation to the dominant eigenvalue $\lambda_1 = 2$.

The technique illustrated in Examples 7 and 8 for approximating dominant eigenvalues and eigenvectors is often called the **power method** or **iteration method.**
As illustrated in Example 7 the power method often generates vectors that have inconveniently large components. To remedy this problem it is usual to "scale down" the approximate eigenvector at each step so its components lie between $+1$ and $-1$. This can be achieved by multiplying the approximate eigenvector by the reciprocal of the component having largest absolute value. To illustrate, in the first step of Example 7, the approximation to the dominant eigenvector is

$$\begin{bmatrix} 5 \\ -1 \end{bmatrix}$$

---

*John William Strutt Rayleigh* (1842–1919)—A British physicist, Rayleigh was awarded the Nobel prize in physics in 1904 for his part in the discovery of argon in 1894. His research ranged over almost the entire field of physics, including sound, wave theory, optics, color vision, electrodynamics, electromagnetism, scattering of light, viscosity, and photography.

The component with largest absolute value is 5; thus the scaled down eigenvector is

$$\frac{1}{5}\begin{bmatrix} 5 \\ -1 \end{bmatrix} = \begin{bmatrix} 1 \\ -.2 \end{bmatrix}$$

We now summarize the steps in the ***power method with scaling.***

**Step 0.** Pick an arbitrary nonzero vector $\mathbf{x}_0$.

**Step 1.** Compute $A\mathbf{x}_0$ and scale down to obtain the first approximation to a dominant eigenvector. Call it $\mathbf{x}_1$.

**Step 2.** Compute $A\mathbf{x}_1$ and scale down to obtain the second approximation, $\mathbf{x}_2$.

**Step 3.** Compute $A\mathbf{x}_2$ and scale down to obtain the third approximation, $\mathbf{x}_3$.

Continuing in this way, a succession, $\mathbf{x}_0$, $\mathbf{x}_1$, $\mathbf{x}_2$, . . . , of better and better approximations to a dominant eigenvector will be obtained.

**Example 9**

Use the power method with scaling to approximate a dominant eigenvector and the dominant eigenvalue of the matrix $A$ in Example 7.

*Solution.* We arbitrarily select

$$\mathbf{x}_0 = \begin{bmatrix} 1 \\ 1 \end{bmatrix}$$

as an initial approximation. Multiplying $\mathbf{x}_0$ by $A$ and scaling down yields

$$A\mathbf{x}_0 = \begin{bmatrix} 3 & 2 \\ -1 & 0 \end{bmatrix}\begin{bmatrix} 1 \\ 1 \end{bmatrix} = \begin{bmatrix} 5 \\ -1 \end{bmatrix} \qquad \mathbf{x}_1 = \frac{1}{5}\begin{bmatrix} 5 \\ -1 \end{bmatrix} = \begin{bmatrix} 1 \\ -.2 \end{bmatrix}$$

Multiplying $\mathbf{x}_1$ by $A$ and scaling down yields

$$A\mathbf{x}_1 = \begin{bmatrix} 3 & 2 \\ -1 & 0 \end{bmatrix}\begin{bmatrix} 1 \\ -.2 \end{bmatrix} = \begin{bmatrix} 2.6 \\ -1 \end{bmatrix} \qquad \mathbf{x}_2 = \frac{1}{2.6}\begin{bmatrix} 2.6 \\ -1 \end{bmatrix} = \begin{bmatrix} 1 \\ -.385 \end{bmatrix}$$

From the Rayleigh quotient the first estimate of the dominant eigenvalue is

$$\lambda_1 \approx \frac{\langle \mathbf{x}_1, A\mathbf{x}_1 \rangle}{\langle \mathbf{x}_1, \mathbf{x}_1 \rangle} = \frac{(1)(2.6) + (-.2)(-1)}{(1)(1) + (-.2)(-.2)} = 2.692$$

Multiplying $\mathbf{x}_2$ by $A$ and scaling down yields

$$A\mathbf{x}_2 = \begin{bmatrix} 3 & 2 \\ -1 & 0 \end{bmatrix}\begin{bmatrix} 1 \\ -.385 \end{bmatrix} = \begin{bmatrix} 2.23 \\ -1 \end{bmatrix} \qquad \mathbf{x}_3 = \frac{1}{2.23}\begin{bmatrix} 2.23 \\ -1 \end{bmatrix} \approx \begin{bmatrix} 1 \\ -.448 \end{bmatrix}$$

From the Rayleigh quotient the second estimate of the dominant eigenvalue is

$$\lambda_1 \approx \frac{\langle \mathbf{x}_2, A\mathbf{x}_2 \rangle}{\langle \mathbf{x}_2, \mathbf{x}_2 \rangle} = \frac{(1)(2.23) + (-.385)(-1)}{(1)(1) + (-.385)(-.385)} = 2.278$$

Multiplying $x_3$ by $A$ and scaling down gives

$$A\mathbf{x}_3 = \begin{bmatrix} 3 & 2 \\ -1 & 0 \end{bmatrix}\begin{bmatrix} 1 \\ -.448 \end{bmatrix} = \begin{bmatrix} 2.104 \\ -1 \end{bmatrix} \qquad \mathbf{x}_4 = \frac{1}{2.104}\begin{bmatrix} 2.104 \\ -1 \end{bmatrix} = \begin{bmatrix} 1 \\ -.475 \end{bmatrix}$$

The third estimate of the dominant eigenvalue is

$$\lambda_1 \approx \frac{\langle \mathbf{x}_3, A\mathbf{x}_3 \rangle}{\langle \mathbf{x}_3, \mathbf{x}_3 \rangle} = \frac{(1)(2.104) + (-.448)(-1)}{(1)(1) + (-.448)(-.448)} = 2.125$$

Continuing in this way, we generate a succession of approximations to a dominant eigenvector and the dominant eigenvalue.

The values calculated above, together with results of further estimates are tabulated in Figure 7.3.

There are no hard and fast rules for determining how many steps to use in the power method. We shall consider one possible procedure that is widely used.

If $\tilde{q}$ denotes an approximation to a quantity $q$, then the **relative error** in the approximation is defined to be

$$\left| \frac{q - \tilde{q}}{q} \right| \tag{7.17}$$

| Step $i$ | 0 | 1 | 2 | 3 | 4 | 5 | 6 | 7 |
|---|---|---|---|---|---|---|---|---|
| $x_i$ The scaled down approximation to a dominant eigenvalue | $\begin{bmatrix} 1 \\ 1 \end{bmatrix}$ | $\begin{bmatrix} 1 \\ -.2 \end{bmatrix}$ | $\begin{bmatrix} 1 \\ -.385 \end{bmatrix}$ | $\begin{bmatrix} 1 \\ -.448 \end{bmatrix}$ | $\begin{bmatrix} 1 \\ -.475 \end{bmatrix}$ | $\begin{bmatrix} 1 \\ -.488 \end{bmatrix}$ | $\begin{bmatrix} 1 \\ -.494 \end{bmatrix}$ | $\begin{bmatrix} 1 \\ -.497 \end{bmatrix}$ |
| $Ax_i$ | | $\begin{bmatrix} 5 \\ -1 \end{bmatrix}$ | $\begin{bmatrix} 2.6 \\ -1 \end{bmatrix}$ | $\begin{bmatrix} 2.23 \\ -1 \end{bmatrix}$ | $\begin{bmatrix} 2.104 \\ -1 \end{bmatrix}$ | $\begin{bmatrix} 2.050 \\ -1 \end{bmatrix}$ | $\begin{bmatrix} 2.024 \\ -1 \end{bmatrix}$ | $\begin{bmatrix} 2.012 \\ -1 \end{bmatrix}$ | — |
| Approximation to $\lambda_1$ | — | 2.692 | 2.278 | 2.125 | 2.060 | 2.029 | 2.014 | — |

Figure 7.3

and the **percentage error** in the approximation is defined to be

$$\left|\frac{q - \tilde{q}}{q}\right| \times 100\%$$

### Example 10

If the exact value of a certain eigenvalue is $\lambda = 5$, and if $\tilde{\lambda} = 5.1$ is an approximation to $\lambda$, then the relative error is

$$\left|\frac{\lambda - \tilde{\lambda}}{\lambda}\right| = \left|\frac{5 - 5.1}{5}\right| = |-.02| = .02$$

and the percentage error is

$$(.02) \times 100\% = 2\%$$

In the power method one would ideally like to decide in advance the relative error $E$ that can be tolerated in the eigenvalue, and then stop the computations once the relative error is less than $E$. Thus, if $\tilde{\lambda}(i)$ denotes the approximation to the dominant eigenvalue $\lambda_1$ at the $i$th step, the computations would be stopped once the condition

$$\left|\frac{\lambda_1 - \tilde{\lambda}(i)}{\lambda_1}\right| < E$$

is satisfied. Unfortunately it is not possible to carry out this idea since the exact value $\lambda_1$ is unknown. To remedy this, it is usual to estimate $\lambda_1$ by $\tilde{\lambda}(i)$ and stop the computations at the $i$th step if

$$\left|\frac{\tilde{\lambda}(i) - \tilde{\lambda}(i - 1)}{\tilde{\lambda}(i)}\right| < E \tag{7.18}$$

The quantity on the left side of (7.18) is called the **estimated relative error.** When multiplied by 100%, it is called the **estimated percentage error.**

### Example 11

In Example 9, how many steps should be used to assure that the estimated percentage error in the dominant eigenvalue is less than 2%?

*Solution.* Let $\tilde{\lambda}(i)$ denote the approximation to the dominant eigenvalue at the $i$th step. Thus from Figure 7.3

$$\tilde{\lambda}(1) = 2.692, \qquad \tilde{\lambda}(2) = 2.278, \qquad \tilde{\lambda}(3) = 2.125, \text{ etc.}$$

From (7.18) the estimated relative error after two steps is

$$\left|\frac{\tilde{\lambda}(2) - \tilde{\lambda}(1)}{\tilde{\lambda}(2)}\right| = \left|\frac{2.278 - 2.692}{2.278}\right| \approx |-.182| = .182$$

The estimated percentage error after two steps is therefore 18.2%. The estimated relative error after three steps is

$$\left| \frac{\tilde{\lambda}(3) - \tilde{\lambda}(2)}{\tilde{\lambda}(3)} \right| = \left| \frac{2.125 - 2.278}{2.125} \right| \approx |-.072| = .072$$

and the estimated percentage error is 7.2%. The remaining percentage errors are listed in the table in Figure 7.4. From this table we see that the estimated percentage is less than 2% at the end of the fifth step.

| $i$ | 2 | 3 | 4 | 5 | 6 |
|---|---|---|---|---|---|
| $\tilde{\lambda}(i)$ | 2.278 | 2.125 | 2.060 | 2.029 | 2.014 |
| Estimated relative error after $i$ steps | .182 | .072 | .032 | .015 | .007 |
| Estimated percentage error after $i$ steps | 18.2% | 7.2% | 3.2% | 1.5% | .7% |

**Figure 7.4**

## Optional

We conclude this section with a proof that the power method works when $A$ is a diagonalizable matrix with a dominant eigenvalue.

Let $A$ be a diagonalizable $n \times n$ matrix. By Theorem 2 in Section 6.2, $A$ has $n$ linearly independent eigenvectors $\mathbf{v}_1, \mathbf{v}_2, \ldots, \mathbf{v}_n$. Let $\lambda_1, \lambda_2, \ldots, \lambda_n$ be the corresponding eigenvalues, and assume

$$|\lambda_1| > |\lambda_2| \geq \cdots \geq |\lambda_n| \tag{7.19}$$

By Theorem 9 in Section 4.5 the eigenvectors $\mathbf{v}_1, \mathbf{v}_2, \ldots, \mathbf{v}_n$ form a basis for $R^n$; thus an arbitrary vector $\mathbf{x}_0$ in $R^n$ can be expressed in the form:

$$\mathbf{x}_0 = k_1 \mathbf{v}_1 + k_2 \mathbf{v}_2 + \cdots + k_n \mathbf{v}_n \tag{7.20}$$

Multiplying both sides on the left by $A$ gives

$$\begin{aligned} A\mathbf{x}_0 &= A(k_1 \mathbf{v}_1 + k_2 \mathbf{v}_2 + \cdots + k_n \mathbf{v}_n) \\ &= k_1(A\mathbf{v}_1) + k_2(A\mathbf{v}_2) + \cdots + k_n(A\mathbf{v}_n) \\ &= k_1 \lambda_1 \mathbf{v}_1 + k_2 \lambda_2 \mathbf{v}_2 + \cdots + k_n \lambda_n \mathbf{v}_n \end{aligned}$$

Multiplying by $A$ again gives

$$
\begin{aligned}
A^2\mathbf{x}_0 &= A(k_1\lambda_1\mathbf{v}_1 + k_2\lambda_2\mathbf{v}_2 + \cdots + k_n\lambda_n\mathbf{v}_n) \\
&= k_1\lambda_1(A\mathbf{v}_1) + k_2\lambda_2(A\mathbf{v}_2) + \cdots + k_n\lambda_n(A\mathbf{v}_n) \\
&= k_1\lambda_1{}^2\mathbf{v}_1 + k_2\lambda_2{}^2\mathbf{v}_2 + \cdots + k_n\lambda_n{}^2\mathbf{v}_n
\end{aligned}
$$

Continuing, we would obtain after $p$ multiplications by $A$

$$A^p\mathbf{x}_0 = k_1\lambda_1{}^p\mathbf{v}_1 + k_2\lambda_2{}^p\mathbf{v}_2 + \cdots + k_n\lambda_n{}^p\mathbf{v}_n \tag{7.21}$$

Since $\lambda_1 \neq 0$ (see 7.19), (7.21) can be rewritten as

$$A^p\mathbf{x}_0 = \lambda_1{}^p\left(k_1\mathbf{v}_1 + k_2\left(\frac{\lambda_2}{\lambda_1}\right)^p\mathbf{v}_2 + \cdots + k_n\left(\frac{\lambda_n}{\lambda_1}\right)^p\mathbf{v}_n\right) \tag{7.22}$$

It follows from (7.19) that

$$\frac{\lambda_2}{\lambda_1}, \ldots, \frac{\lambda_n}{\lambda_1}$$

are all less than one in absolute value; thus $(\lambda_2/\lambda_1)^p, \ldots, (\lambda_n/\lambda_1)^p$ get steadily closer to zero as $p$ increases, and from (7.22), the approximation

$$A^p\mathbf{x}_0 \approx \lambda_1{}^p k_1\mathbf{v}_1 \tag{7.23}$$

gets better and better.

If $k_1 \neq 0^*$, then $\lambda_1{}^p k_1\mathbf{v}_1$ is a nonzero scalar multiple of the dominant eigenvector $\mathbf{v}_1$; thus, $\lambda_1{}^p k_1\mathbf{v}_1$ is also a dominant eigenvector. Therefore, by (7.23), $A^p\mathbf{x}_0$ becomes a better and better estimate of a dominant eigenvector as $p$ is increased. ▪

## EXERCISE SET 7.3

**1.** Find the dominant eigenvalue (if it exists).

(a) $\begin{bmatrix} -1 & 4 \\ 1 & -1 \end{bmatrix}$  (b) $\begin{bmatrix} 0 & 1 \\ 4 & 0 \end{bmatrix}$

(c) $\begin{bmatrix} 4 & 2 & 1 \\ 0 & -5 & 3 \\ 0 & 0 & 6 \end{bmatrix}$  (d) $\begin{bmatrix} 1 & -12 & 0 \\ 1 & 0 & 0 \\ 0 & 0 & 3 \end{bmatrix}$

**2.** Let

$$A = \begin{bmatrix} 3 & 4 \\ 1 & 3 \end{bmatrix}$$

---

*One can not usually tell by inspection of the $x_0$ selected whether $k_1 \neq 0$. If, by accident, $k_1 = 0$, the power method still works in practical problems, since computer roundoff errors generally build up to make $k_1$ small but nonzero. This is one instance where errors help to obtain correct results!

(a) Use the power method with scaling to approximate a dominant eigenvector of $A$. Start with

$$\mathbf{x}_0 = \begin{bmatrix} 1 \\ 1 \end{bmatrix}$$

Round off all computations to three significant digits, and stop after three iterations (that is, three multiplications by $A$).

(b) Use the result of part (a) and the Rayleigh quotient to approximate the dominant eigenvalue of $A$.

(c) Find the exact values of the dominant eigenvector and eigenvalue.

(d) Find the percentage error in the approximation of the dominant eigenvalue.

In Exercises 3–4 follow the directions given in Exercise 2.

**3.** $A = \begin{bmatrix} 5 & 4 \\ 3 & 4 \end{bmatrix}$   **4.** $A = \begin{bmatrix} -3 & 2 \\ 2 & 0 \end{bmatrix}$

**5.** Let

$$A = \begin{bmatrix} 18 & 17 \\ 2 & 3 \end{bmatrix}$$

(a) Use the power method with scaling to approximate the dominant eigenvalue and a dominant eigenvector of $A$. Start with

$$\mathbf{x}_0 = \begin{bmatrix} 1 \\ 1 \end{bmatrix}$$

Round off all computations to three significant digits, and stop when the estimated percentage error in the dominant eigenvalue is less than 2%.

(b) Find the exact values of the dominant eigenvalue and eigenvector.

**6.** Repeat the directions of Exercise 5 with

$$A = \begin{bmatrix} -5 & 5 \\ 6 & -4 \end{bmatrix}$$

**7.** Let

$$A = \begin{bmatrix} 2 & 1 & 0 \\ 1 & 2 & 0 \\ 0 & 0 & 10 \end{bmatrix}$$

(a) Use the power method with scaling to approximate a dominant eigenvector of $A$. Start with

$$\mathbf{x}_0 = \begin{bmatrix} 1 \\ 1 \\ 1 \end{bmatrix}$$

Round off all computations to three significant digits, and stop after three iterations.

(b) Use the result of part (a) and the Rayleigh quotient to approximate the dominant eigenvalue of $A$.

(c) Find the exact values for the dominant eigenvalue and eigenvector.

(d) Find the percentage error in the approximation of the dominant eigenvalue.

## 7.4 APPROXIMATING NONDOMINANT EIGENVALUES BY DEFLATION

In this section we briefly outline a method for obtaining the nondominant eigenvectors and eigenvalues of a *symmetric* matrix.

We shall need the following theorem, which we state without proof.*

**Theorem 1.** *Let A be a symmetric $n \times n$ matrix with eigenvalues $\lambda_1, \lambda_2, \ldots, \lambda_n$. If $v_1$ is an eigenvector corresponding to $\lambda_1$, and $\|v_1\| = 1$, then:*

(a) *The matrix $B = A - \lambda_1 v_1 v_1^t$ has eigenvalues $0, \lambda_2, \ldots, \lambda_n$.*

(b) *If $v$ is an eigenvector of B corresponding to one of the eigenvalues $\lambda_2, \ldots, \lambda_n$, then $v$ is also an eigenvector of A corresponding to this eigenvalue.*

(We are assuming in Theorem 1 that $v_1$ is expressed as an $n \times 1$ matrix; thus $v_1 v_1^t$ is an $n \times n$ matrix.)

### Example 12

It was shown in Example 4 of Section 6.1 that

$$A = \begin{bmatrix} 3 & -2 & 0 \\ -2 & 3 & 0 \\ 0 & 0 & 5 \end{bmatrix}$$

has eigenvalues $\lambda_1 = 5$, $\lambda_2 = 5$, $\lambda_3 = 1$, and

$$v = \begin{bmatrix} -1 \\ 1 \\ 0 \end{bmatrix}$$

is an eigenvector corresponding to $\lambda_1 = 5$. Normalizing $v$ yields

$$v_1 = \frac{1}{\sqrt{2}} \begin{bmatrix} -1 \\ 1 \\ 0 \end{bmatrix} = \begin{bmatrix} -\dfrac{1}{\sqrt{2}} \\ \dfrac{1}{\sqrt{2}} \\ 0 \end{bmatrix}$$

which is an eigenvector of norm 1 corresponding to $\lambda_1 = 5$.

---

*Readers interested in the proof of this theorem are referred to the references given at the end of this section.

By Theorem 1 the matrix

$$
B = A - \lambda_1 \mathbf{v}_1 \mathbf{v}_1^t = \begin{bmatrix} 3 & -2 & 0 \\ -2 & 3 & 0 \\ 0 & 0 & 5 \end{bmatrix} - 5 \begin{bmatrix} -\dfrac{1}{\sqrt{2}} \\ \dfrac{1}{\sqrt{2}} \\ 0 \end{bmatrix} \begin{bmatrix} -\dfrac{1}{\sqrt{2}} & \dfrac{1}{\sqrt{2}} & 0 \end{bmatrix}
$$

$$
= \begin{bmatrix} 3 & -2 & 0 \\ -2 & 3 & 0 \\ 0 & 0 & 5 \end{bmatrix} - 5 \begin{bmatrix} \dfrac{1}{2} & -\dfrac{1}{2} & 0 \\ -\dfrac{1}{2} & \dfrac{1}{2} & 0 \\ 0 & 0 & 0 \end{bmatrix} = \begin{bmatrix} \dfrac{1}{2} & \dfrac{1}{2} & 0 \\ \dfrac{1}{2} & \dfrac{1}{2} & 0 \\ 0 & 0 & 5 \end{bmatrix}
$$

should have eigenvalues $\lambda = 0, 5$, and 1. As a check the characteristic equation of $B$ is

$$
\det(\lambda I - B) = \det \begin{bmatrix} \lambda - \dfrac{1}{2} & -\dfrac{1}{2} & 0 \\ -\dfrac{1}{2} & \lambda - \dfrac{1}{2} & 0 \\ 0 & 0 & \lambda - 5 \end{bmatrix} = \lambda(\lambda - 5)(\lambda - 1) = 0
$$

Hence the eigenvalues of $B$ are $\lambda = 0, \lambda = 5, \lambda = 1$ as predicted by Theorem 1. The eigenspace of $B$ corresponding to $\lambda = 5$ is the solution space of the system

$$(5I - B)\mathbf{x} = \mathbf{0}$$

that is

$$
\begin{bmatrix} \dfrac{9}{2} & -\dfrac{1}{2} & 0 \\ -\dfrac{1}{2} & \dfrac{9}{2} & 0 \\ 0 & 0 & 0 \end{bmatrix} \begin{bmatrix} x_1 \\ x_2 \\ x_3 \end{bmatrix} = \begin{bmatrix} 0 \\ 0 \\ 0 \end{bmatrix}
$$

Solving this system yields $x_1 = 0$, $x_2 = 0$, $x_3 = t$. Thus the eigenvectors of $B$ corresponding to $\lambda = 5$ are the nonzero vectors of the form

$$
\mathbf{x} = \begin{bmatrix} 0 \\ 0 \\ t \end{bmatrix}
$$

As predicted by part (b) of Theorem 1, these are also eigenvectors of $A$ corresponding to $\lambda = 5$, since

$$A \begin{bmatrix} 0 \\ 0 \\ t \end{bmatrix} = \begin{bmatrix} 3 & -2 & 0 \\ -2 & 3 & 0 \\ 0 & 0 & 5 \end{bmatrix} \begin{bmatrix} 0 \\ 0 \\ t \end{bmatrix} = \begin{bmatrix} 0 \\ 0 \\ 5t \end{bmatrix}$$

that is

$$A \begin{bmatrix} 0 \\ 0 \\ t \end{bmatrix} = 5 \begin{bmatrix} 0 \\ 0 \\ t \end{bmatrix}$$

Similarly the eigenvectors of $B$ corresponding to $\lambda = 1$ are also eigenvectors of $A$ corresponding to $\lambda = 1$.

Theorem 1, to a limited extent, makes it possible to determine the nondominant eigenvalues and eigenvectors of a *symmetric* $n \times n$ matrix $A$. To see how, assume the eigenvalues of $A$ can be ordered according to the size of their absolute values as follows.

$$|\lambda_1| > |\lambda_2| > |\lambda_3| \geq \cdots \geq |\lambda_n| \tag{7.24}$$

Suppose the dominant eigenvalue and a dominant eigenvector of $A$ have been obtained by the power method. By normalizing the dominant eigenvector, we can obtain a dominant eigenvector $\mathbf{v}_1$ having norm one. By Theorem 1 the eigenvalues of $B = A - \lambda_1 \mathbf{v}_1 \mathbf{v}_1^t$ will be $0, \lambda_2, \lambda_3, \ldots, \lambda_n$. From (7.24) these eigenvalues will be ordered according to their absolute values as follows.

$$|\lambda_2| > |\lambda_3| \geq \cdots \geq |\lambda_n| \geq 0$$

Thus $\lambda_2$ is the dominant eigenvalue of $B$. Now, by applying the power method to $B$, we can approximate the eigenvalue $\lambda_2$ and a corresponding eigenvector. This technique for approximating the eigenvalue with second largest absolute value is called *deflation.*

Unfortunately, there are practical limitations of the deflation method. Since $\lambda_1$ and $\mathbf{v}_1$ are only approximated in the power method, an error is introduced into $B$ when deflation is used. If the deflation process is applied again, the next matrix has additional errors introduced through the approximation of $\lambda_2$ and $\mathbf{v}_2$. As the process continues, this compounding of errors eventually destroys the accuracy of the results. In practice, one should generally avoid finding more than two or three eigenvalues by deflation.

When the ratio $|\lambda_2/\lambda_1|$ is close to one, the power method has a slow rate of convergence; that is, many steps are needed to obtain a reasonable degree of accuracy. Readers interested in studying techniques for "speeding up" this rate of convergence and learning more about the numerical methods of linear algebra may wish to consult the following texts:

*Analysis of Numerical Methods,* E. Isaacson and H. B. Keller, John Wiley and Sons, New York, 1966.

*Applied Linear Algebra,* B. Noble, Prentice Hall, Inc., 1969.
*Computational Methods of Linear Algebra,* V. N. Faddeeva, Dover, 1959.

Additional references appear in the bibliographies of these texts.

## EXERCISE SET 7.4

**1.** Let

$$A = \begin{bmatrix} 6 & 2 \\ 2 & 3 \end{bmatrix}$$

(a) Use the power method with scaling to approximate a dominant eigenvector. Start with

$$\mathbf{x}_0 = \begin{bmatrix} 1 \\ 1 \end{bmatrix}$$

Round off all computations to three significant digits, and stop after three iterations (i.e., three multiplications by $A$.)

(b) Use the result of part (a) and the Rayleigh quotient to approximate the dominant eigenvalue of $A$.

(c) Use deflation to approximate the remaining eigenvalue and a corresponding eigenvector; that is, apply the power method to

$$B = A - \tilde{\lambda}_1 \tilde{\mathbf{v}}_1 \tilde{\mathbf{v}}_1{}^t$$

where $\tilde{\mathbf{v}}_1$ and $\tilde{\lambda}_1$ are the approximations obtained in parts (a) and (b). Start with

$$\mathbf{x}_0 = \begin{bmatrix} 1 \\ 1 \end{bmatrix}$$

Round off all computations to three significant digits and stop after three iterations.

(d) Find the exact values of the eigenvalues and eigenvectors.

**2.** Follow the directions given in Exercise 1 with

$$A = \begin{bmatrix} 10 & 4 \\ 4 & 4 \end{bmatrix}$$

# Answers to Exercises

1. $b, d, f$
2. (a) $x = \frac{7}{6}t + 1/2$, $y = t$
   (b) $x_1 = -2s + \frac{7}{2}t + 4$, $x_2 = s$, $x_3 = t$
   (c) $x_1 = \frac{4}{3}r - \frac{7}{3}s + \frac{8}{3}t - 5/3$, $x_2 = r$, $x_3 = s$, $x_4 = t$
   (d) $v = \frac{1}{2}q - \frac{3}{2}r - \frac{1}{2}s + 2t$, $w = q$, $x = r$, $y = s$, $z = t$

3. (a) $\begin{bmatrix} 1 & -2 & 0 \\ 3 & 4 & -1 \\ 2 & -1 & 3 \end{bmatrix}$   (b) $\begin{bmatrix} 1 & 0 & 1 & 1 \\ -1 & 2 & -1 & 3 \end{bmatrix}$

   (c) $\begin{bmatrix} 1 & 0 & 1 & 0 & 0 & 1 \\ 0 & 2 & -1 & 0 & 1 & 2 \\ 0 & 0 & 2 & 1 & 0 & 3 \end{bmatrix}$   (d) $\begin{bmatrix} 1 & 0 & 1 \\ 0 & 1 & 2 \end{bmatrix}$

4. (a) $\begin{aligned} x_1 \quad - \quad x_3 &= 2 \\ 2x_1 + x_2 + x_3 &= 3 \\ -x_2 + 2x_3 &= 4 \end{aligned}$   (b) $\begin{aligned} x_1 \quad\quad &= 0 \\ x_2 &= 0 \\ x_1 - x_2 &= 1 \end{aligned}$

   (c) $\begin{aligned} x_1 + 2x_2 + 3x_3 + 4x_4 &= 5 \\ 5x_1 + 4x_2 + 3x_3 + 2x_4 &= 1 \end{aligned}$   (d) $\begin{aligned} x_1 \quad\quad\quad &= 1 \\ x_2 \quad\quad &= 2 \\ x_3 \quad &= 3 \\ x_4 &= 4 \end{aligned}$

5. $k = 6$ infinitely many solutions
   $k \neq 6$ no solutions

6. (a) The lines have no common point of intersection.
   (b) The lines intersect in exactly one point.
   (c) The three lines coincide.

**EXERCISE SET 1.2 (page 16).**

1. $b, d, f$     2. $b, c, f$
3. (a) $x_1 = 4, x_2 = 3, x_3 = 2$
   (b) $x_1 = 2 - 3t, x_2 = 4 + t, x_3 = 2 - t, x_4 = t$
   (c) $x_1 = -1 - 5s - 5t, x_2 = s, x_3 = 1 - 3t, x_4 = 2 - 4t, x_5 = t$
   (d) Inconsistent
4. (a) $x_1 = 4, x_2 = 3, x_3 = 2$
   (b) $x_1 = 2 - 3t, x_2 = 4 + t, x_3 = 2 - t, x_4 = t$
   (c) $x_1 = -1 - 5s - 5t, x_2 = s, x_3 = 1 - 3t, x_4 = 2 - 4t, x_5 = t$
   (d) Inconsistent
5. (a) $x_1 = 2, x_2 = 1, x_3 = 3$
   (b) $x_1 = -\frac{3}{7}t, x_2 = -\frac{4}{7}t, x_3 = t$
   (c) $x_1 = 1, x_2 = 2s, x_3 = s, x_4 = -3t, x_5 = t$
7. (a) Inconsistent     (b) $x_1 = -4, x_2 = 2, x_3 = 7$     (c) $x_1 = 3, x_2 = 0$
9. (a) $x_1 = 0, x_2 = -3t, x_3 = t$     (b) Inconsistent
11. (a) $x_1 = \frac{2}{3}a - \frac{1}{9}b, x_2 = -\frac{1}{3}a + \frac{2}{9}b$
    (b) $x_1 = a - \frac{1}{3}c, x_2 = a - \frac{1}{2}b, x_3 = -a + \frac{1}{2}b + \frac{1}{3}c$
12. $a = 4$ infinitely many, $a = -4$ none, $a \neq \pm 4$ exactly one.
13. $\alpha = \dfrac{\pi}{2}, \beta = \pi, \gamma = 0$.

**EXERCISE SET 1.3 (page 20).**

1. $a, c, d$
2. $x_1 = 0, x_2 = 0, x_3 = 0$
3. $x_1 = -\frac{1}{4}s, x_2 = -\frac{1}{4}s - t, x_3 = s, x_4 = t$
4. $x_1 = 0, x_2 = 0, x_3 = 0, x_4 = 0$
5. $x = t/8, y = 5t/16, z = t$
6. $\lambda = 4, \lambda = 2$

**EXERCISE SET 1.4 (page 27).**

1. (a) undefined     (b) $4 \times 2$     (c) undefined
   (d) undefined     (e) $5 \times 5$     (f) $5 \times 2$
3. $a = 5, b = -3, c = 4, d = 1$

**4.** (a) $\begin{bmatrix} 12 & -3 \\ -4 & 5 \\ 4 & 1 \end{bmatrix}$ (b) $\begin{bmatrix} 7 & 6 & 5 \\ -2 & 1 & 3 \\ 7 & 3 & 7 \end{bmatrix}$ (c) $\begin{bmatrix} -5 & 4 & -1 \\ 0 & -1 & -1 \\ -1 & 1 & 1 \end{bmatrix}$

(d) $\begin{bmatrix} 9 & 8 & 19 \\ -2 & 0 & 0 \\ 32 & 9 & 25 \end{bmatrix}$ (e) $\begin{bmatrix} 14 & 36 & 25 \\ 4 & -1 & 7 \\ 12 & 26 & 21 \end{bmatrix}$ (f) $\begin{bmatrix} -28 & 7 \\ 0 & -14 \end{bmatrix}$

**5.** (a) Undefined (b) $\begin{bmatrix} 42 & 108 & 75 \\ 12 & -3 & 21 \\ 36 & 78 & 63 \end{bmatrix}$

(c) $\begin{bmatrix} 3 & 45 & 9 \\ 11 & -11 & 17 \\ 7 & 17 & 13 \end{bmatrix}$ (d) $\begin{bmatrix} 3 & 45 & 9 \\ 11 & -11 & 17 \\ 7 & 17 & 13 \end{bmatrix}$

(e) Undefined (f) $\begin{bmatrix} 48 & 15 & 31 \\ 0 & 2 & 6 \\ 38 & 10 & 27 \end{bmatrix}$

**6.** (a) $[67 \quad 41 \quad 41]$ (b) $[63 \quad 67 \quad 57]$

(c) $\begin{bmatrix} 41 \\ 21 \\ 67 \end{bmatrix}$ (d) $\begin{bmatrix} 6 \\ 6 \\ 63 \end{bmatrix}$

(e) $[24 \quad 56 \quad 97]$ (f) $\begin{bmatrix} 76 \\ 98 \\ 97 \end{bmatrix}$

**7.** 182

## EXERCISE SET 1.5 (page 33).

**5.** $(A + B)^2 = A^2 + AB + BA + B^2$
**6.** No
**7.** (a) $C = \frac{1}{2}A - \frac{3}{2}B$ (b) $C = 2AB - 3I$
(c) $C = A^2 + 2I - B$ (d) $C = \frac{1}{4}BA - \frac{1}{2}A^2$
**8.** $0A$ and $A0$ may not have the same size.

**9.** $\begin{bmatrix} \pm 1 & 0 & 0 \\ 0 & \pm 1 & 0 \\ 0 & 0 & \pm 1 \end{bmatrix}$

**EXERCISE SET 1.6 (page 37).**

1. $A^{-1} = \begin{bmatrix} 2 & -1 \\ -5 & 3 \end{bmatrix}$  $B^{-1} = \begin{bmatrix} 1/5 & 3/20 \\ -1/5 & 1/10 \end{bmatrix}$  $C^{-1} = \begin{bmatrix} 1/2 & 0 \\ 0 & 1/3 \end{bmatrix}$

3. $\begin{bmatrix} -3 & 2 \\ 5/2 & -3/2 \end{bmatrix}$  4. $\begin{bmatrix} 1 & 2/7 \\ 4/7 & 1/7 \end{bmatrix}$

5. $A^3 = \begin{bmatrix} 1 & 0 \\ 26 & 27 \end{bmatrix}$  $A^{-3} = \begin{bmatrix} 1 & 0 \\ -26/27 & 1/27 \end{bmatrix}$  $A^2 - 2A + I = \begin{bmatrix} 0 & 0 \\ 4 & 4 \end{bmatrix}$

6. $A^{-1} = \begin{bmatrix} \dfrac{1}{2} & -\dfrac{1}{2} & \dfrac{1}{2} \\ \dfrac{1}{2} & \dfrac{1}{2} & -\dfrac{1}{2} \\ -\dfrac{1}{2} & \dfrac{1}{2} & \dfrac{1}{2} \end{bmatrix}$  8. $A^{-1} = \begin{bmatrix} 1/a_{11} & 0 & \cdots & 0 \\ 0 & 1/a_{22} & \cdots & 0 \\ \vdots & \vdots & & \vdots \\ 0 & 0 & \cdots & 1/a_{nn} \end{bmatrix}$

**EXERCISE SET 1.7 (page 45).**

1. $a, b, d, f, g$
2. (a) Add $-5$ times the first row to the second.
   (b) Interchange the first and third rows.
   (c) Multiply the second row by $1/8$.

3. (a) $E_1 = \begin{bmatrix} 0 & 0 & 1 \\ 0 & 1 & 0 \\ 1 & 0 & 0 \end{bmatrix}$  (b) $E_2 = \begin{bmatrix} 0 & 0 & 1 \\ 0 & 1 & 0 \\ 1 & 0 & 0 \end{bmatrix}$

   (c) $E_3 = \begin{bmatrix} 1 & 0 & 0 \\ 0 & 1 & 0 \\ 2 & 0 & 1 \end{bmatrix}$  (d) $E_4 = \begin{bmatrix} 1 & 0 & 0 \\ 0 & 1 & 0 \\ -2 & 0 & 1 \end{bmatrix}$

4. No, since $C$ cannot be obtained by performing a single row operation on $B$.

5. (a) $\begin{bmatrix} -5 & 2 \\ 3 & -1 \end{bmatrix}$  (b) $\begin{bmatrix} -5 & -3 \\ -3 & -2 \end{bmatrix}$  (c) Not invertible

**6. (a)** $\begin{bmatrix} \dfrac{3}{2} & -\dfrac{11}{10} & -\dfrac{6}{5} \\ -1 & 1 & 1 \\ -\dfrac{1}{2} & \dfrac{7}{10} & \dfrac{2}{5} \end{bmatrix}$  **(b)** Not invertible

**(c)** $\begin{bmatrix} \dfrac{1}{2} & -\dfrac{1}{2} & \dfrac{1}{2} \\ -\dfrac{1}{2} & \dfrac{1}{2} & \dfrac{1}{2} \\ \dfrac{1}{2} & \dfrac{1}{2} & -\dfrac{1}{2} \end{bmatrix}$  **(d)** $\begin{bmatrix} \dfrac{7}{2} & 0 & -3 \\ -1 & 1 & 0 \\ 0 & -1 & 1 \end{bmatrix}$

**(e)** $\begin{bmatrix} \dfrac{1}{2} & -\dfrac{1}{2} & \dfrac{1}{2} \\ 0 & 0 & 1 \\ \dfrac{1}{2} & \dfrac{1}{2} & -\dfrac{1}{2} \end{bmatrix}$  **(f)** $\begin{bmatrix} 1 & 0 & -2 \\ 3 & 1 & 2 \\ 1 & -1 & 0 \end{bmatrix}$

**7. (a)** $\begin{bmatrix} \dfrac{1}{2}\sqrt{2} & -\dfrac{1}{2}\sqrt{2} & 0 \\ \dfrac{1}{2}\sqrt{2} & \dfrac{1}{2}\sqrt{2} & 0 \\ 0 & 0 & 1 \end{bmatrix}$  **(b)** $\begin{bmatrix} 1 & 0 & 0 & 0 \\ -\dfrac{1}{2} & \dfrac{1}{2} & 0 & 0 \\ 0 & -\dfrac{1}{4} & \dfrac{1}{4} & 0 \\ 0 & 0 & -\dfrac{1}{8} & \dfrac{1}{8} \end{bmatrix}$

**(c)** Not invertible.

**8.** $A^{-1} = \begin{bmatrix} \cos\theta & -\sin\theta & 0 \\ \sin\theta & \cos\theta & 0 \\ 0 & 0 & 1 \end{bmatrix}$

**9. (a)** $E_1 = \begin{bmatrix} 1 & 0 \\ -3 & 1 \end{bmatrix}, E_2 = \begin{bmatrix} 1 & 0 \\ 0 & \dfrac{1}{4} \end{bmatrix}$  **(b)** $A^{-1} = \begin{bmatrix} 1 & 0 \\ 0 & \dfrac{1}{4} \end{bmatrix}\begin{bmatrix} 1 & 0 \\ -3 & 1 \end{bmatrix}$

**(c)** $A = \begin{bmatrix} 1 & 0 \\ 3 & 1 \end{bmatrix}\begin{bmatrix} 1 & 0 \\ 0 & 4 \end{bmatrix}$

**11.** $A = \begin{bmatrix} 1 & 0 & 0 \\ -2 & 1 & 0 \\ 0 & 0 & 1 \end{bmatrix}\begin{bmatrix} 1 & 0 & 0 \\ 0 & 1 & 0 \\ 0 & 1 & 1 \end{bmatrix}\begin{bmatrix} 1 & 3 & 3 & 8 \\ 0 & 1 & 7 & 8 \\ 0 & 0 & 0 & 0 \end{bmatrix}$

**13.** (a) $\begin{bmatrix} 1/k_1 & 0 & 0 & 0 \\ 0 & 1/k_2 & 0 & 0 \\ 0 & 0 & 1/k_3 & 0 \\ 0 & 0 & 0 & 1/k_4 \end{bmatrix}$  (b) $\begin{bmatrix} 0 & 0 & 0 & 1/k_4 \\ 0 & 0 & 1/k_3 & 0 \\ 0 & 1/k_2 & 0 & 0 \\ 1/k_1 & 0 & 0 & 0 \end{bmatrix}$

(c) $\begin{bmatrix} 1/k & 0 & 0 & 0 \\ -1/k^2 & 1/k & 0 & 0 \\ 1/k^3 & -1/k^2 & 1/k & 0 \\ -1/k^4 & 1/k^3 & -1/k^2 & 1/k \end{bmatrix}$

## EXERCISE SET 1.8 (page 52).

**1.** $x_1 = 41, x_2 = -17$

**2.** $x_1 = 46/27, x_2 = -13/27$

**3.** $x_1 = -7, x_2 = 4, x_3 = -1$

**4.** $x_1 = 1, x_2 = -11, x_3 = 16$

**5.** $x = 1, y = 5, z = -1$

**6.** $w = 1, x = -6, y = 10, z = -7$

**7.** (a) $x_1 = 16/3, x_2 = -4/3, x_3 = -11/3$

(b) $x_1 = -5/3, x_2 = 5/3, x_3 = 10/3$

(c) $x_1 = 3, x_2 = 0, x_3 = -4$

(d) $x_1 = 41/42, x_2 = -5/6, x_3 = 25/21$

**8.** (a) $b_2 = 3b_1, b_3 = -2b_1$

(b) $b_3 = b_2 - b_1, b_4 = 2b_1 - b_2$

**9.** (a) $X = \begin{bmatrix} 0 \\ 0 \\ 0 \end{bmatrix}$  (b) $X = \begin{bmatrix} 4t \\ 5t/2 \\ t \end{bmatrix}$

## EXERCISE SET 2.1 (page 61).

**1.** (a) 5  (b) 7  (c) 10  (d) 0  (e) 4  (f) 5

**2.** (a) Odd  (b) Odd  (c) Even  (d) Even  (e) Even  (f) Odd

**3.** 5.  **4.** 0.  **5.** 59.  **6.** $k^2 - 4k - 5$.  **7.** 0.  **8.** 425  **9.** 104

**10.** $-k^4 - k^3 + 18k^2 + 9k - 21$  **11.** (a) $\lambda = 3, \lambda = 2$  (b) $\lambda = 2, \lambda = 6$

**14.** 275

**15.** (a) 120  (b) $-120$

**EXERCISE SET 2.2 (page 66).**

**1.** (a) 6    (b) − 16    (c) 0    (d) 0
**2.** 21.    **3.** −5.`    **4.** −36.    **5.** 35
**6.** − 128.    **7.** −72.    **8.** 1/6.    **9.** 0
**10.** (a) 5    (b) 10    (c) 5    (d) 10

**EXERCISE SET 2.3 (page 73).**

**1.** (a) $\begin{bmatrix} 2 & -3 & 0 \\ 1 & 1 & 2 \end{bmatrix}$    (b) $\begin{bmatrix} 6 & -8 & 0 \\ 1 & 4 & 1 \\ 1 & 3 & 3 \end{bmatrix}$    (c) $\begin{bmatrix} 7 \\ 0 \\ 2 \end{bmatrix}$    (d) $\begin{bmatrix} a_{11} & a_{21} \\ a_{12} & a_{22} \\ a_{13} & a_{23} \end{bmatrix}$

**4.** (a) Invertible    (b) Not invertible
 (c) Not invertible    (d) Not invertible
**5.** (a) 135    (b) 8/5    (c) 1/40    (d) −5
**8.** (a) $k = \frac{1}{2}(5 + \sqrt{17}), k = \frac{1}{2}(5 - \sqrt{17})$    (b) $k = -1$

**EXERCISE SET 2.4 (page 83).**

**1.** (a) $M_{11} = 29, M_{12} = -11, M_{13} = -19, M_{21} = 21, M_{22} = 13, M_{23} = -19$
  $M_{31} = 27, M_{32} = -5, M_{33} = 19$
 (b) $C_{11} = 29, C_{12} = 11, C_{13} = -19, C_{21} = -21, C_{22} = 13$
  $C_{23} = 19, C_{31} = 27, C_{32} = 5, C_{33} = 19$
**2.** (a) $M_{13} = 36, C_{13} = 36$    (b) $M_{23} = 24, C_{23} = -24$
 (c) $M_{22} = -48, C_{22} = -48$    (d) $M_{21} = -108, C_{21} = 108$

**3.** 152

**4.** (a) $\begin{bmatrix} 29 & -21 & 27 \\ 11 & 13 & 5 \\ -19 & 19 & 19 \end{bmatrix}$    (b) $(1/152) \begin{bmatrix} 29 & -21 & 27 \\ 11 & 13 & 5 \\ -19 & 19 & 19 \end{bmatrix}$

**5.** 48.    **6.** −66.    **7.** 0.    **8.** $k^3 - 8k^2 - 10k + 95.$    **9.** − 120
**10.** 0

**11.** $A^{-1} = \begin{bmatrix} -4 & 3 & 0 & -1 \\ 2 & -1 & 0 & 0 \\ -7 & 0 & -1 & 8 \\ 6 & 0 & 1 & -7 \end{bmatrix}$

**12.** $x_1 = 1, x_2 = 2$
**13.** $x = 3/11, y = 2/11, z = -1/11$
**14.** $x = 26/21, y = 25/21, z = 5/7$
**15.** $x_1 = -30/11, x_2 = -38/11, x_3 = -40/11$
**16.** $x_1 = 3, x_2 = 5, x_3 = -1, x_4 = 8$
**17.** Cramer's rule does not apply.
**18.** $z = 2$

## EXERCISE SET 3.1 (page 93).

**3.** (a) $\overrightarrow{P_1P_2} = (-1, 3)$    (b) $\overrightarrow{P_1P_2} = (-7, 2)$
   (c) $\overrightarrow{P_1P_2} = (2, -12, -11)$    (d) $\overrightarrow{P_1P_2} = (-8, 7, 4)$
**4.** $\overrightarrow{PQ}$, where $Q = (9, 5, 1)$ is one possible answer.
**5.** $\overrightarrow{PQ}$, where $P = (0, 4, -8)$ is one possible answer.
**6.** (a) $(-2, 0, 4)$    (b) $(23, -15, 4)$
   (c) $(-1, -5, 2)$    (d) $(-39, 69, -12)$
   (e) $(30, -7, 5)$    (f) $(0, -10, 0)$
**7.** $\mathbf{x} = (-1/2, 5/6, 1)$
**8.** $c_1 = 1, c_2 = -2, c_3 = 3$
**10.** $c_1 = -t, c_2 = -t, c_3 = t$ (where $t$ is arbitrary)
**11.** (a) $(9/2, -1/2, -1/2)$    (b) $(23/4, -9/4, 1/4)$
**12.** (a) $x' = 5, y' = 8$    (b) $x = -1, y = 3$

## EXERCISE SET 3.2 (page 97).

**1.** (a) 5    (b) $5\sqrt{2}$    (c) 3    (d) $\sqrt{3}$    (e) $\sqrt{129}$    (f) 9
**2.** (a) $\sqrt{13}$    (b) $2\sqrt{26}$    (c) $\sqrt{209}$    (d) $\sqrt{93}$
**3.** (a) $2\sqrt{3}$    (b) $\sqrt{14} + \sqrt{2}$    (c) $4\sqrt{14}$    (d) $2\sqrt{37}$
   (e) $(1/\sqrt{6}, 1/\sqrt{6}, -2/\sqrt{6})$    (f) 1
**4.** $k = \pm\dfrac{3}{\sqrt{21}}$
**7.** $(1/\sqrt{3}, 1/\sqrt{3}, 1/\sqrt{3})$
**8.** A sphere of radius 1 centered at $(x_0, y_0, z_0)$

**EXERCISE SET 3.3 (page 104).**

1. (a) $-10$  (b) $-3$  (c) $0$  (d) $-20$
2. (a) $-1/\sqrt{5}$  (b) $-3/\sqrt{58}$  (c) $0$  (d) $-20/(3\sqrt{70})$
3. (a) obtuse  (b) acute  (c) obtuse  (d) orthogonal.
4. (a) $(12/13, -8/13)$  (b) $(0,0)$  (c) $(-80/13, 0, -16/13)$
   (d) $(32/89, 12/89, 16/89)$
5. (a) $(14/13, 21/13)$  (b) $(2,6)$  (c) $(-11/13, 1, 55/13)$
   (d) $(-32/89, -12/89, 73/89)$
7. $\pm(2/\sqrt{13}, 3/\sqrt{13})$
8. (a) $6$  (b) $36$  (c) $24\sqrt{5}$  (d) $24\sqrt{5}$
10. $\cos\theta_1 = 0$, $\cos\theta_2 = 3/\sqrt{10}$, $\cos\theta_3 = 1/\sqrt{10}$
13. $\theta = \cos^{-1}\dfrac{2}{\sqrt{6}}$
14. $\cos\beta = b/\sqrt{a^2 + b^2 + c^2}$, $\cos\gamma = c/\sqrt{a^2 + b^2 + c^2}$

**EXERCISE SET 3.4 (page 111).**

1. (a) $(-23, 7, -1)$  (b) $(-20, -67, -9)$  (c) $(-78, -52, -26)$
   (d) $(0, -56, -392)$  (e) $(24, 0, -16)$  (f) $(-12, -22, -8)$
2. (a) $(12, 30, -6)$  (b) $(-2, 0, 2)$
3. (a) $\frac{1}{2}\sqrt{374}$  (b) $9\sqrt{13}$
7. $x = (1/2 - \frac{1}{2}t, -1/2 + \frac{3}{2}t, t)$, where $t$ is arbitrary
9. $227$
10. (a) $\mathbf{u} = (0, \sqrt{2}, 1)$ in the $x'y'z'$-system
    $\mathbf{v} = (0, \sqrt{2}, 0)$ in the $x'y'z'$-system
    (b) $\mathbf{u} \times \mathbf{v} = (-1, 1, 0)$ in the $xyz$-system
    (c) $\mathbf{u} \times \mathbf{v} = (-\sqrt{2}, 0, 0)$ in the $x'y'z'$-system

**EXERCISE SET 3.5 (page 119).**

1. (a) $(x - 2) + 4(y - 6) + 2(z - 1) = 0$
   (b) $-(x + 1) + 7(y + 1) + 6(z - 2) = 0$
   (c) $z = 0$
   (d) $2x + 3y + 4z = 0$

**2.** (a) $x + 4y + 2z - 28 = 0$  (b) $-x + 7y + 6z - 6 = 0$
(c) $z = 0$  (d) $2x + 3y + 4z = 0$

**3.** (a) $(5, 0, 0)$ is a point in the plane and $\mathbf{n} = (2, -3, 7)$ is a normal vector so that $2(x - 5) - 3y + 7z = 0$ is a point normal form; other points and normals yield other correct answers.

(b) $x + 3z = 0$ is one possible answer.

**4.** (a) $2y - z - 1 = 0$  (b) $x + 9y - 5z - 16 = 0$

**5.** (a) $x = 2 + t, y = 4 + 2t, z = 6 + 5t$

(b) $x = -3 + 5t, y = 2 - 7t, z = -4 - 3t$

(c) $x = 1, y = 1, z = 5 + t$

(d) $x = t, y = t, z = t$

**6.** (a) $x - 2 = \dfrac{y - 4}{2} = \dfrac{z - 6}{5}$  (b) $\dfrac{x + 3}{5} = \dfrac{y - 2}{-7} = \dfrac{z + 4}{-3}$

**7.** (a) $x = 6 + t, y = -1 + 3t, z = 5 - 9t$ or $x = 7 + t, y = 2 + 3t$
$z = -4 - 9t$ are possible answers

(b) $x = -t, \ y = -t, \ z = -t$ or $x = -1 - t, \ y = -1 - t, \ z = -1 - t$ are possible answers

**8.** (a) $x = -11/7 + \frac{23}{7}t, \ y = -12/7 - \frac{1}{7}t, \ z = t$

(b) $x = \frac{5}{3}t, \ y = t, \ z = 0$

**9.** (a) $x - 2y - 17 = 0, y + 2z - 5 = 0$  (b) $3x - 5y = 0, 2y - z = 0$

**10.** $xy$-plane: $z = 0$, $xz$-plane: $y = 0$, $yz$-plane: $x = 0$

**12.** $(-222/7, -64/7, 78/7)$

**13.** $5x - 2y + z - 30 = 0$

**15.** $(-17, -1, 1)$  **16.** $x - 4y + 4z + 9 = 0$

**EXERCISE SET 4.1 (page 125).**

**1.** (a) $(-3, -4, -8, 4)$  (b) $(53, 34, 49, 20)$  (c) $(-1, 2, 7, -10)$
(d) $(-99, -84, -150, 30)$  (e) $(-63, -28, -21, -69)$  (f) $(2, 6, 15, -14)$

**2.** $\left(-\dfrac{7}{6}, -1, -\dfrac{3}{2}, -\dfrac{1}{3}\right)$  **3.** $c_1 = 1, c_2 = 1, c_3 = -1, c_4 = 1$

**5.** (a) 5  (b) $\sqrt{11}$  (c) $\sqrt{14}$  (d) $\sqrt{48}$

**6.** (a) $\sqrt{73}$  (b) $\sqrt{14} + 3\sqrt{7}$  (c) $4\sqrt{14}$  (d) $\sqrt{1801}$  (e) $\sqrt{6}$  (f) 1

**8.** $k = \pm 3/\sqrt{14}$

**9.** (a) $-1$  (b) $-1$  (c) 0  (d) 27

**10.** (a) $(2/\sqrt{5}, 1/\sqrt{5})$ and $(-2/\sqrt{5}, -1/\sqrt{5})$

**11.** (a) $\sqrt{10}$  (b) $3\sqrt{3}$  (c) $\sqrt{59}$  (d) 10

**EXERCISE SET 4.2 (page 130).**

1. Not a vector space. Axiom 8 fails.
2. Not a vector space. Axiom 10 fails.
3. Not a vector space. Axioms 9 and 10 fail.
4. The set is a vector space under the given operations.
5. The set is a vector space under the given operations.
6. Not a vector space. Axioms 5 and 6 fail.
7. The set is a vector space under the given operations.
8. Not a vector space. Axioms 7 and 8 fail.
9. The set is a vector space under the given operations.
10. Not a vector space. Axioms 1, 4, 5, and 6 fail.
11. The set is a vector space under the given operations.
12. The set is a vector space under the given operations.
13. The set is a vector space under the given operations.
14. The set is a vector space under the given operations.

**EXERCISE SET 4.3 (page 137).**

1. $a, c$     2. $b, c$     3. $a, b, d$     4. $b, d, e$     5. $a, b, d$
6. (a) $(5, 9, 5) = 3u - 4v + w$     (b) $(2, 0, 6) = 4u - 2w$
   (c) $(0, 0, 0) = 0u + 0v + 0w$     (d) $(2, 2, 3) = \frac{1}{2}u - \frac{1}{2}v + \frac{1}{2}w$
7. (a) $5 + 9x + 5x^2 = 3p_1 - 4p_2 + p_3$
   (b) $2 + 6x^2 = 4p_1 - 2p_3$
   (c) $0 = 0p_1 + 0p_2 + 0p_3$
   (d) $2 + 2x + 3x^2 = \frac{1}{2}p_1 - \frac{1}{2}p_2 + \frac{1}{2}p_3$
8. $a, c, d$
9. (a) The vectors span          (b) The vectors do not span
   (c) The vectors do not span    (d) The vectors span
10. $a, c$
11. The polynomials do not span $P_2$
12. $a, b, d$
13. $8x - 7y + z = 0$
14. $x = 2t, y = 7t, z = -t$, where $-\infty < t < +\infty$

**EXERCISE SET 4.4 (page 143).**

1. (a) $\mathbf{u}_2$ is a scalar multiple of $\mathbf{u}_1$.
   (b) The vectors are linearly dependent by Theorem 6.
   (c) $\mathbf{p}_2$ is a scalar multiple of $\mathbf{p}_1$.
   (d) $B$ is a scalar multiple of $A$.
2. (a) independent  (b) independent  (c) independent  (d) dependent
3. (a) independent  (b) independent  (c) independent  (d) independent
4. (a) independent  (b) independent  (c) independent  (d) dependent
5. (a) dependent  (b) independent  (c) independent  (d) dependent
   (e) dependent  (f) dependent
6. (a) They do not lie in a plane.
   (b) They do lie in a plane.
7. (a) They do not lie on the same line.
   (b) They do not lie on the same line.
   (c) They do lie on the same line.
8. $\lambda = -1/2, \lambda = 1$

**EXERCISE SET 4.5 (page 150).**

1. (a) A basis for $R^2$ has two vectors.
   (b) A basis for $R^3$ has three vectors.
   (c) A basis for $P_2$ has three vectors.
   (d) A basis for $M_{22}$ has four vectors.
2. $a, b$    3. $a, b$    4. $c, d$
7. Basis: $(0, 0, 0)$; dimension $= 0$
8. Basis: $(-1/4, -1/4, 1, 0), (0, -1, 0, 1)$; dimension $= 2$
9. Basis: $(0, 0, 0)$; dimension $= 0$
10. Basis: $(3, 1, 0), (-1, 0, 1)$; dimension $= 2$
11. Basis: $(0, 0, 0, 0)$; dimension $= 0$
12. Basis: $(0, 0, 0)$; dimension $= 0$
13. (a) $(2/3, 1, 0), (-5/3, 0, 1)$    (b) $(1, 1, 0), (0, 0, 1)$
    (c) $(2, -1, 4)$                   (d) $(1, 1, 0), (0, 1, 1)$
14. (a) 3-dimensional  (b) 2-dimensional  (c) 1-dimensional
15. 3-dimensional

**EXERCISE SET 4.6 (page 159).**

1. $(2, -1, 0, 1), (3, 5, 7, -1), (1, 4, 2, 7)$
2. $(1, -3)$.   3. $(1, 2, -1), (0, 0, 1)$
4. $(1, 12, 2, 0), (0, 1, 7/41, 2/41), (0, 0, 1, 12)$
5. $(1, -3, 2, 2, 1), (0, 1, 2, 0, -2/3), (0, 0, 1, 0, -1/3)$
6. (a) $(1, 1, -4, -3), (0, 1, -5, -2), (0, 0, 1, -1/2)$
   (b) $(1, -1, 2, 0), (0, 1, 0, 0), (0, 0, 1, -1/6)$
   (c) $(1, 1, 0, 0), (0, 1, 1, 1), (0, 0, 1, 1), (0, 0, 0, 1)$
7. (a) $1 - x + 2x^2, x, x^2$   (b) $1 + x, x + x^2, x^2$   (c) $1 + x - 3x^2$

8. (a) $\begin{bmatrix} 1 & 1 \\ -4 & -3 \end{bmatrix}, \begin{bmatrix} 0 & 1 \\ -5 & -2 \end{bmatrix}, \begin{bmatrix} 0 & 0 \\ 1 & -1/2 \end{bmatrix}$   (b) $\begin{bmatrix} 1 & -1 \\ 2 & 0 \end{bmatrix}, \begin{bmatrix} 0 & 1 \\ -2 & 1/3 \end{bmatrix}$

9. (a) $(\mathbf{v})_B = (3, -2, 1), [\mathbf{v}]_B = \begin{bmatrix} 3 \\ -2 \\ 1 \end{bmatrix}$

   (b) $(\mathbf{v})_B = (-2, 0, 1), [\mathbf{v}]_B = \begin{bmatrix} -2 \\ 0 \\ 1 \end{bmatrix}$

10. $(\mathbf{p})_B = (0, 2, -1), [\mathbf{p}]_B = \begin{bmatrix} 0 \\ 2 \\ -1 \end{bmatrix}$

11. $(A)_B = (-1, 1, -1, 3), [A]_B = \begin{bmatrix} -1 \\ 1 \\ -1 \\ 3 \end{bmatrix}$

**EXERCISE SET 4.7 (page 164).**

1. (a) $-12$   (b) 0   (c) 0   (d) 120
2. (a) $-5$   (b) 0   (c) 3   (d) 52
3. (a) 16   (b) 56
4. (a) $-6$   (b) 0
6. (a) Not an inner product, axiom 4 fails.
   (b) Not an inner product, axioms 2, 3 fail.
   (c) $\langle \mathbf{u}, \mathbf{v} \rangle$ is an inner product.
   (d) Not an inner product, axiom 4 fails.

**9.** $\langle x, y \rangle = 4x_4 y_4 + 3x_3 y_3 + 2x_2 y_2 + x_1 y_1 + x_3 y_4 + x_4 y_3$
$\qquad - x_1 y_2 - x_1 y_3 - x_1 y_4 - x_2 y_1 - x_3 y_1 - x_4 y_1$

**17.** (a) $-28/15$    (b) 0

**18.** (a) 0    (b) 1    (c) $-\dfrac{4}{\pi} \ln (1/\sqrt{2})$

**EXERCISE SET 4.8 (page 170).**

**1.** (a) $\sqrt{21}$    (b) $\sqrt{206}$    (c) $\sqrt{2}$    (d) 0

**2.** (a) $\sqrt{10}$    (b) $\sqrt{85}$    (c) 1    (d) 0

**3.** (a) $\sqrt{6}$    (b) 5

**4.** (a) $\sqrt{90}$    (b) 0

**5.** (a) $\sqrt{45}$    (b) 0

**6.** (a) $\sqrt{18}$    (b) 0

**7.** $\sqrt{18}$

**8.** (a) $\sqrt{98}$    (b) 0

**9.** (a) $-1/\sqrt{2}$    (b) $-3/\sqrt{73}$    (c) 0

   (d) $-\dfrac{20}{9\sqrt{10}}$    (e) $-1/\sqrt{2}$    (f) $2/\sqrt{55}$

**10.** (a) 0    (b) 0

**11.** (a) $\dfrac{19}{10\sqrt{7}}$    (b) 0

**12.** (a) $k = -3$    (b) $k = -2, k = -3$

**14.** $a, b, c$

**15.** $\pm \dfrac{1}{\sqrt{3249}} (-34, 44, -6, 11)$

**EXERCISE SET 4.9 (page 179).**

**1.** $b$    **2.** $b, d$    **3.** $a$    **4.** $a$

**6.** (a) $(2/\sqrt{2}, 2/\sqrt{6}, 5/\sqrt{21}, 1/\sqrt{7})$    (b) $(4/\sqrt{2}, 10/\sqrt{6}, 10/\sqrt{21}, -6/\sqrt{7})$

**7.** (a) $(1/\sqrt{10}, -3/\sqrt{10}), (3/\sqrt{10}, 1/\sqrt{10})$    (b) $(1, 0), (0, -1)$

**8.** (a) $(1/\sqrt{3}, 1/\sqrt{3}, 1/\sqrt{3}), (-1/\sqrt{2}, 1/\sqrt{2}, 0), (1/\sqrt{6}, 1/\sqrt{6}, -2/\sqrt{6})$

   (b) $(1, 0, 0), (0, 7/\sqrt{53}, -2/\sqrt{53}), (0, 30/\sqrt{11925}, 105/\sqrt{11925})$

**9.** $(0, 2/\sqrt{5}, 1/\sqrt{5}, 0), (5/\sqrt{30}, -1/\sqrt{30}, 2/\sqrt{30}, 0),$
$(1/\sqrt{10}, 1/\sqrt{10}, -2/\sqrt{10}, -2/\sqrt{10}), (1/\sqrt{15}, 1/\sqrt{15}, -2/\sqrt{15}, 3/\sqrt{15})$

**10.** $(0, 1/\sqrt{5}, 2/\sqrt{5}), (-\sqrt{5}/\sqrt{6}, -2/\sqrt{30}, 1/\sqrt{30})$

**11.** $(1/\sqrt{6}, 1/\sqrt{6}, 1/\sqrt{6}), (1/\sqrt{6}, 1/\sqrt{6}, -1/\sqrt{6}), (2/\sqrt{6}, -1/\sqrt{6}, 0)$

**12.** $w_1 = (-4/5, 2, 3/5), w_2 = (9/5, 0, 12/5)$

**13.** $w_1 = (39/42, 93/42, 120/42), w_2 = (3/42, -9/42, 6/42)$

**14.** $w_1 = (-5/4, -1/4, 5/4, 9/4), w_2 = (1/4, 9/4, 19/4, -9/4)$

## EXERCISE SET 5.1 (page 189).

**1.** Linear     **2.** Nonlinear     **3.** Linear     **4.** Linear

**5.** Nonlinear     **6.** Linear     **7.** Linear     **8.** Nonlinear

**9.** Linear     **10.** Linear     **11.** Nonlinear     **12.** Linear

**13.** Linear     **14.** Nonlinear     **15.** Linear     **16.** Nonlinear

**17.** Linear     **18.** Linear     **19.** Linear     **20.** Nonlinear

**21.** $F(x, y) = (-x, y)$

**23.** (a) $\begin{bmatrix} 1 & 3 & 4 \\ 1 & 0 & -7 \end{bmatrix}$    (b) $\begin{bmatrix} 42 \\ -55 \end{bmatrix}$    (c) $\begin{bmatrix} x + 3y + 4z \\ x - 7y \end{bmatrix}$

**24.** (a) $(2, 0, -1)$     (b) $T(x, y, z) = (x, 0, z)$

**25.** (a) $T(3, 8, 4) = (-2, 3, -1)$
(b) $T(x, y, z) = \frac{1}{3}(2x - y - z, -x + 2y - z, -x - y + 2z)$

**26.** (a) $T(-1, 2) = (-3/\sqrt{2}, 1/\sqrt{2}); \; T(x, y) = (x/\sqrt{2} - y/\sqrt{2}, x/\sqrt{2} + y/\sqrt{2})$
(b) $T(-1, 2) = (1, -2); \; T(x, y) = (-x, -y)$
(c) $T(-1, 2) = \left(-\frac{\sqrt{3}}{2} - 1, \sqrt{3} - \frac{1}{2}\right);$

$$T(x, y) = \left(\frac{\sqrt{3}}{2}x - \frac{1}{2}y, \frac{1}{2}x + \frac{\sqrt{3}}{2}y\right)$$

(d) $T(-1, 2) = \left(-\frac{1}{2} + \sqrt{3}, \frac{\sqrt{3}}{2} + 1\right);$

$$T(x, y) = \left(\frac{1}{2}x + \frac{\sqrt{3}}{2}y, -\frac{\sqrt{3}}{2}x + \frac{1}{2}y\right)$$

**30.** (a) $T$ is a rotation about the $z$-axis through the angle $\theta$.
(b) $T$ is a rotation about the $y$-axis through the angle $\theta$.
(c) $T$ is a rotation about the $x$-axis through the angle $\theta$.

**EXERCISE SET 5.2 (page 194).**

**1.** $a, c$    **2.** $a$    **3.** $a, b, c$    **4.** $a$

**5.** $b$        **6.** $a$    **7.** $\ker(T) = \{0\}$; $R(T) = V$

**8.** Rank $(T) = 1$, nullity $(T) = 1$

**9.** Rank $(T) = 3$, nullity $(T) = 0$

**10.** (a) Rank $(T) = n$, nullity $(T) = 0$

  (b) Rank $(T) = 0$, nullity $(T) = n$

  (c) Rank $(T) = n$, nullity $(T) = 0$

**11.** $T(x, y, z) = (30x - 10y - 3z, -9x + 3y + z)$, $T(1, 1, 1) = (17, -5)$

**12.** $T(2 - 2x + 3x^2) = 8 + 8x - 7x^2$

**15.** (a) $14x - 8y - 5z = 0$     (b) $x = -t, y = -t, z = t, -\infty < t < +\infty$

**17.** $\ker(D)$ consists of all constant polynomials

**18.** $\ker(J)$ consists of all polynomials of the form $kx$

**EXERCISE SET 5.3 (page 204).**

**1.** $\begin{bmatrix} 2 \\ 3 \\ 1 \end{bmatrix}, \begin{bmatrix} -1 \\ 5 \\ 4 \end{bmatrix}, \begin{bmatrix} 0 \\ 7 \\ 2 \end{bmatrix}, \begin{bmatrix} 1 \\ -1 \\ 7 \end{bmatrix}$

**2.** (a) Rank $(T) = 3$, nullity $(T) = 1$

  (b) Rank $(T) = 3$, nullity $(T) = 2$

  (c) Rank $(T) = 3$, nullity $(T) = 3$

  (d) Rank $(T) = 2$, nullity $(T) = 5$

**3.** Rank $= 3$, nullity $= 1$     **4.** Rank $= 2$, nullity $= 1$

**5.** Rank $= 3$, nullity $= 1$     **6.** Rank $= 3$, nullity $= 1$

**8.** (a) Nullity $(T) = 2$     (b) Nullity $(T) = 4$

  (c) Nullity $(T) = 3$     (d) Nullity $(T) = 1$

**9.** Nullity $(T) = 0$, rank $(T) = 6$

**10.** (a) dimension $=$ nullity $(T) = 3$

  (b) No. In order for $A\mathbf{x} = \mathbf{b}$ to be consistent for all $b$ in $R^5$, we must have $R(T) = R^5$. But $R(T) \neq R^5$ since rank $(T) = \dim R(T) = 4$.

**11.** (a) $\begin{bmatrix} 1 \\ 5 \\ 7 \end{bmatrix}, \begin{bmatrix} 0 \\ 1 \\ 1 \end{bmatrix}$  (b) $\begin{bmatrix} -14/11 \\ 19/11 \\ 1 \end{bmatrix}$

  (c) Rank $(T) = 2$, nullity $(T) = 1$

**12.** (a) $\begin{bmatrix} 1 \\ 2 \\ 0 \end{bmatrix}$  (b) $\begin{bmatrix} 1/2 \\ 0 \\ 1 \end{bmatrix}, \begin{bmatrix} 0 \\ 1 \\ 0 \end{bmatrix}$

   (c) Rank $(T) = 1$, nullity $(T) = 2$.

**13.** (a) $\begin{bmatrix} 1 \\ 4 \end{bmatrix}, \begin{bmatrix} 0 \\ 1 \end{bmatrix}$  (b) $\begin{bmatrix} -1 \\ -1 \\ 1 \\ 0 \end{bmatrix}, \begin{bmatrix} -4/7 \\ 2/7 \\ 0 \\ 1 \end{bmatrix}$

   (c) Rank $(T) = 2$, nullity $(T) = 2$.

**14.** (a) $\begin{bmatrix} 1 \\ 3 \\ -1 \\ 2 \end{bmatrix}, \begin{bmatrix} 0 \\ 1 \\ -2/7 \\ 5/14 \end{bmatrix}, \begin{bmatrix} 0 \\ 0 \\ 0 \\ 1 \end{bmatrix}$  (b) $\begin{bmatrix} -1 \\ -1 \\ 1 \\ 0 \\ 0 \end{bmatrix}, \begin{bmatrix} -1 \\ -2 \\ 0 \\ 0 \\ 1 \end{bmatrix}$

   (c) Rank $(T) = 3$, nullity $(T) = 2$

## EXERCISE SET 5.4 (page 214).

**1.** (a) $\begin{bmatrix} 2 & -1 \\ 1 & 1 \end{bmatrix}$  (b) $\begin{bmatrix} 1 & 0 \\ 0 & 1 \end{bmatrix}$  (c) $\begin{bmatrix} 1 & 2 & 1 \\ 1 & 5 & 0 \\ 0 & 0 & 1 \end{bmatrix}$  (d) $\begin{bmatrix} 4 & 0 & 0 \\ 0 & 7 & 0 \\ 0 & 0 & -8 \end{bmatrix}$

**2.** (a) $\begin{bmatrix} 0 & 1 \\ -1 & 0 \\ 1 & 3 \\ 1 & -1 \end{bmatrix}$  (b) $\begin{bmatrix} 7 & 2 & -1 & 1 \\ 0 & 1 & 1 & 0 \\ -1 & 0 & 0 & 0 \end{bmatrix}$

   (c) $\begin{bmatrix} 0 & 0 & 0 \\ 0 & 0 & 0 \\ 0 & 0 & 0 \\ 0 & 0 & 0 \\ 0 & 0 & 0 \end{bmatrix}$  (d) $\begin{bmatrix} 0 & 0 & 0 & 1 \\ 1 & 0 & 0 & 0 \\ 0 & 0 & 1 & 0 \\ 0 & 1 & 0 & 0 \\ 1 & 0 & -1 & 0 \end{bmatrix}$

**3.** $\begin{bmatrix} 1 & 1 & 0 \\ 0 & -2 & -3 \end{bmatrix}$  **4.** $\begin{bmatrix} 1 & 0 & 0 & 0 \\ 0 & 0 & 1 & 0 \\ 0 & 1 & 0 & 0 \\ 0 & 0 & 0 & 1 \end{bmatrix}$  **5.** $\begin{bmatrix} 0 & 0 & 0 \\ 0 & 1 & 0 \\ 0 & 0 & 1 \end{bmatrix}$

**6.** (a) $\begin{bmatrix} 0 & 0 \\ -1/2 & 1 \\ 8/3 & 4/3 \end{bmatrix}$  (b) $\begin{bmatrix} 14 \\ -8 \end{bmatrix}$

**7.** (a) $\begin{bmatrix} 1 & -3/2 & 1/2 \\ -1 & 1/2 & 1/2 \\ 0 & 1/2 & -1/2 \end{bmatrix}$ (b) $\begin{bmatrix} 2 \\ -2 \\ 2 \end{bmatrix}$

**8.** (a) $\begin{bmatrix} 0 & 0 & 0 \\ 0 & 0 & 0 \\ 1 & 1 & 4 \\ 0 & 2 & 5 \\ 1 & 3 & 1 \end{bmatrix}$ (b) $-3x^2 + 5x^3 - 2x^4$

**9.** (a) $[T(\mathbf{v}_1)]_B = \begin{bmatrix} 1 \\ -2 \end{bmatrix}$, $[T(\mathbf{v}_2)]_B = \begin{bmatrix} 3 \\ 5 \end{bmatrix}$

   (b) $T(\mathbf{v}_1) = \begin{bmatrix} 3 \\ -5 \end{bmatrix}$, $T(\mathbf{v}_2) = \begin{bmatrix} -2 \\ 29 \end{bmatrix}$ (c) $\begin{bmatrix} 19/7 \\ -83/7 \end{bmatrix}$

**10.** (a) $[T(\mathbf{v}_1)]_{B'} = \begin{bmatrix} 3 \\ 1 \\ -3 \end{bmatrix}$, $[T(\mathbf{v}_2)]_{B'} = \begin{bmatrix} -2 \\ 6 \\ 0 \end{bmatrix}$

   $[T(\mathbf{v}_3)]_{B'} = \begin{bmatrix} 1 \\ 2 \\ 7 \end{bmatrix}$, $[T(\mathbf{v}_4)]_{B'} = \begin{bmatrix} 0 \\ 1 \\ 1 \end{bmatrix}$

   (b) $T(\mathbf{v}_1) = \begin{bmatrix} 11 \\ 5 \\ 22 \end{bmatrix}$, $T(\mathbf{v}_2) = \begin{bmatrix} -42 \\ 32 \\ -10 \end{bmatrix}$, $T(\mathbf{v}_3) = \begin{bmatrix} -56 \\ 87 \\ 17 \end{bmatrix}$

   $T(\mathbf{v}_4) = \begin{bmatrix} -13 \\ 17 \\ 2 \end{bmatrix}$ (c) $\begin{bmatrix} -31 \\ 37 \\ 12 \end{bmatrix}$

**11.** (a) $[T(\mathbf{v}_1)]_B = \begin{bmatrix} 1 \\ 2 \\ 6 \end{bmatrix}$, $[T(\mathbf{v}_2)]_B = \begin{bmatrix} 3 \\ 0 \\ -2 \end{bmatrix}$, $[T(\mathbf{v}_3)]_B = \begin{bmatrix} -1 \\ 5 \\ 4 \end{bmatrix}$

   (b) $T(\mathbf{v}_1) = 16 + 51x + 19x^2$, $T(\mathbf{v}_2) = -6 - 5x + 5x^2$, $T(\mathbf{v}_3) = 7 + 40x + 15x^2$

   (c) $T(1 + x^2) = 22 + 56x + 14x^2$

**15.** (a) $\begin{bmatrix} 0 & 1 & 0 \\ 0 & 0 & 2 \\ 0 & 0 & 0 \end{bmatrix}$ (b) $\begin{bmatrix} 0 & -3/2 & 23/6 \\ 0 & 0 & -16/3 \\ 0 & 0 & 0 \end{bmatrix}$ (c) $-6 + 48x$

**16.** (a) $\begin{bmatrix} 0 & 0 & 0 \\ 0 & 0 & -1 \\ 0 & 1 & 0 \end{bmatrix}$ (b) $\begin{bmatrix} 0 & 0 & 0 \\ 0 & 1 & 0 \\ 0 & 0 & 2 \end{bmatrix}$ (c) $\begin{bmatrix} 2 & 1 & 0 \\ 0 & 2 & 2 \\ 0 & 0 & 2 \end{bmatrix}$

## EXERCISE SET 5.5 (page 225).

**1.** (a) $\begin{bmatrix} 4/11 & 3/11 \\ -1/11 & 2/11 \end{bmatrix}$ (b) $\begin{bmatrix} -3/11 \\ -13/11 \end{bmatrix}$ (d) $\begin{bmatrix} 2 & -3 \\ 1 & 4 \end{bmatrix}$

**2.** (a) $\begin{bmatrix} 0 & -5/2 \\ -2 & -13/2 \end{bmatrix}$ (b) $\begin{bmatrix} -4 \\ -7 \end{bmatrix}$ (d) $\begin{bmatrix} 13/10 & -1/2 \\ -2/5 & 0 \end{bmatrix}$

**3.** (a) $\begin{bmatrix} 3/4 & 3/4 & 1/12 \\ -3/4 & -17/12 & -17/12 \\ 0 & 2/3 & 2/3 \end{bmatrix}$ (b) $\begin{bmatrix} 19/12 \\ -43/12 \\ 4/3 \end{bmatrix}$

**4.** (a) $\begin{bmatrix} 3 & 2 & 5/2 \\ -2 & -3 & -1/2 \\ 5 & 1 & 6 \end{bmatrix}$ (b) $\begin{bmatrix} -7/2 \\ 23/2 \\ 6 \end{bmatrix}$

**5.** (a) $\begin{bmatrix} 3/4 & 7/2 \\ 3/2 & 1 \end{bmatrix}$ (b) $\begin{bmatrix} -11/4 \\ 1/2 \end{bmatrix}$ (d) $\begin{bmatrix} -2/9 & 7/9 \\ 1/3 & -1/6 \end{bmatrix}$

**6.** (b) $\begin{bmatrix} 1/2 & 0 \\ -1/6 & 1/3 \end{bmatrix}$ (c) $\begin{bmatrix} 1 \\ -2 \end{bmatrix}$ (e) $\begin{bmatrix} 2 & 0 \\ 1 & 3 \end{bmatrix}$

**7.** $a, b, d, e$

**8.** (a) $\begin{bmatrix} 1 & 0 \\ 0 & 1 \end{bmatrix}$ (b) $\begin{bmatrix} 1/\sqrt{2} & 1/\sqrt{2} \\ -1/\sqrt{2} & 1/\sqrt{2} \end{bmatrix}$ (d) $\begin{bmatrix} -1/\sqrt{2} & 0 & 1/\sqrt{2} \\ 1/\sqrt{6} & -2/\sqrt{6} & 1/\sqrt{6} \\ 1/\sqrt{3} & 1/\sqrt{3} & 1/\sqrt{3} \end{bmatrix}$

(e) $\begin{bmatrix} 1/2 & 1/2 & 1/2 & 1/2 \\ 1/2 & -5/6 & 1/6 & 1/6 \\ 1/2 & 1/6 & 1/6 & -5/6 \\ 1/2 & 1/6 & -5/6 & 1/6 \end{bmatrix}$

**10.** (a) $\begin{bmatrix} \cos\theta & \sin\theta \\ -\sin\theta & \cos\theta \end{bmatrix}$ (b) $\begin{bmatrix} \cos\theta & \sin\theta & 0 \\ -\sin\theta & \cos\theta & 0 \\ 0 & 0 & 1 \end{bmatrix}$

**11.** (a) $(-2, -1)$ (b) $(-4/5, -22/5)$ (c) $(-11/5, 52/5)$ (d) $(0, 0)$

**13.** $a, c$

**15.** (a) $(12/5, -9/5, -7)$ (b) $(2, 1, 6)$ (c) $(-42/5, 19/5, -3)$
(d) $(0, 0, 0)$

**17.** $b$

**EXERCISE SET 5.6 (page 231).**

**1.** $[T]_B = \begin{bmatrix} 1 & -2 \\ 0 & -1 \end{bmatrix}$  $[T]_{B'} = \begin{bmatrix} -3/11 & -56/11 \\ -2/11 & 3/11 \end{bmatrix}$

**2.** $[T]_B = \begin{bmatrix} 4/5 & 61/10 \\ 18/5 & -19/5 \end{bmatrix}$  $[T]_{B'} = \begin{bmatrix} -155/10 & 9/2 \\ -375/10 & 25/2 \end{bmatrix}$

**3.** $[T]_B = \begin{bmatrix} 1/\sqrt{2} & -1/\sqrt{2} \\ 1/\sqrt{2} & 1/\sqrt{2} \end{bmatrix}$  $[T]_{B'} = \begin{bmatrix} \dfrac{13}{11\sqrt{2}} & -\dfrac{25}{11\sqrt{2}} \\ \dfrac{1}{11\sqrt{2}} & \dfrac{15}{11\sqrt{2}} \end{bmatrix}$

**4.** $[T]_B = \begin{bmatrix} 1 & 2 & -1 \\ 0 & -1 & 0 \\ 1 & 0 & 7 \end{bmatrix}$  $[T]_{B'} = \begin{bmatrix} 1 & 4 & 3 \\ -1 & -2 & -9 \\ 1 & 1 & 8 \end{bmatrix}$

**5.** $[T]_B = \begin{bmatrix} 1 & 0 & 0 \\ 0 & 1 & 0 \\ 0 & 0 & 0 \end{bmatrix}$  $[T]_{B'} = \begin{bmatrix} 1 & 0 & 0 \\ 0 & 1 & 1 \\ 0 & 0 & 0 \end{bmatrix}$

**6.** $[T]_B = \begin{bmatrix} 5 & 0 \\ 0 & 5 \end{bmatrix}$  $[T]_{B'} = \begin{bmatrix} 5 & 0 \\ 0 & 5 \end{bmatrix}$

**7.** $[T]_B = \begin{bmatrix} 2/3 & -2/9 \\ 1/2 & 4/3 \end{bmatrix}$  $[T]_{B'} = \begin{bmatrix} 1 & 1 \\ 0 & 1 \end{bmatrix}$

**EXERCISE SET 6.1 (page 238).**

**1.** (a) $\lambda^2 - 2\lambda - 3 = 0$  (b) $\lambda^2 - 8\lambda + 16 = 0$
  (c) $\lambda^2 - 12 = 0$  (d) $\lambda^2 + 3 = 0$
  (e) $\lambda^2 = 0$  (f) $\lambda^2 - 2\lambda + 1 = 0$
**2.** (a) $\lambda = 3, \lambda = -1$  (b) $\lambda = 4$
  (c) $\lambda = \sqrt{12}, \lambda = -\sqrt{12}$  (d) No real eigenvalues
  (e) $\lambda = 0$  (f) $\lambda = 1$

**3.** (a) Basis for eigenspace corresponding to $\lambda = 3$: $\begin{bmatrix} 1/2 \\ 1 \end{bmatrix}$

  basis for eigenspace corresponding to $\lambda = -1$: $\begin{bmatrix} 0 \\ 1 \end{bmatrix}$

  (b) Basis for eigenspace corresponding to $\lambda = 4$: $\begin{bmatrix} 3/2 \\ 1 \end{bmatrix}$

(c) Basis for eigenspace corresponding to $\lambda = \sqrt{12}$: $\begin{bmatrix} 3/\sqrt{12} \\ 1 \end{bmatrix}$

   basis for eigenspace corresponding to $\lambda = -\sqrt{12}$: $\begin{bmatrix} -3/\sqrt{12} \\ 1 \end{bmatrix}$

(d) There are no eigenspaces.

(e) Basis for eigenspace corresponding to $\lambda = 0$: $\begin{bmatrix} 1 \\ 0 \end{bmatrix}, \begin{bmatrix} 0 \\ 1 \end{bmatrix}$

(f) Basis for eigenspace corresponding to $\lambda = 1$: $\begin{bmatrix} 1 \\ 0 \end{bmatrix}, \begin{bmatrix} 0 \\ 1 \end{bmatrix}$

5. (a) $\lambda^3 - 6\lambda^2 + 11\lambda - 6 = 0$   (b) $\lambda^3 - 2\lambda = 0$
   (c) $\lambda^3 + 8\lambda^2 + \lambda + 8 = 0$   (d) $\lambda^3 - \lambda^2 - \lambda - 2 = 0$
   (e) $\lambda^3 - 6\lambda^2 + 12\lambda - 8 = 0$   (f) $\lambda^3 - 2\lambda^2 - 15\lambda + 36 = 0$

6. (a) $\lambda = 1, \lambda = 2, \lambda = 3$   (b) $\lambda = 0, \lambda = \sqrt{2}, \lambda = -\sqrt{2}$
   (c) $\lambda = -8$   (d) $\lambda = 2$   (e) $\lambda = 2$
   (f) $\lambda = -4, \lambda = 3$

7. (a) $\lambda = 1$: basis $\begin{bmatrix} 0 \\ 1 \\ 0 \end{bmatrix}$, $\lambda = 2$: basis $\begin{bmatrix} -1/2 \\ 1 \\ 1 \end{bmatrix}$, $\lambda = 3$: basis $\begin{bmatrix} -1 \\ 1 \\ 1 \end{bmatrix}$

   (b) $\lambda = 0$: basis $\begin{bmatrix} 5/3 \\ 1/3 \\ 1 \end{bmatrix}$, $\lambda = \sqrt{2}$: basis $\begin{bmatrix} \frac{1}{7}(15 + 5\sqrt{2}) \\ \frac{1}{7}(-1 + 2\sqrt{2}) \\ 1 \end{bmatrix}$

   $\lambda = -\sqrt{2}$: basis $\begin{bmatrix} \frac{1}{7}(15 - 5\sqrt{2}) \\ \frac{1}{7}(-1 - 2\sqrt{2}) \\ 1 \end{bmatrix}$

   (c) $\lambda = -8$: basis $\begin{bmatrix} -1/6 \\ -1/6 \\ 1 \end{bmatrix}$   (d) $\lambda = 2$: basis $\begin{bmatrix} 1/3 \\ 1/3 \\ 1 \end{bmatrix}$

   (e) $\lambda = 2$: basis $\begin{bmatrix} -1/3 \\ -1/3 \\ 1 \end{bmatrix}$   (f) $\lambda = -4$: basis $\begin{bmatrix} -2 \\ 8/3 \\ 1 \end{bmatrix}$, $\lambda = 3$: $\begin{bmatrix} 5 \\ -2 \\ 1 \end{bmatrix}$

**8.** (a) $(\lambda - 1)^2(\lambda + 2)(\lambda + 1) = 0$     (b) $(\lambda - 4)^2(\lambda^2 + 3) = 0$

**9.** (a) $\lambda = 1, \lambda = -2, \lambda = -1$     (b) $\lambda = 4$

**10.** (a) $\lambda = 1$: basis $\begin{bmatrix} 0 \\ 0 \\ 0 \\ 1 \end{bmatrix}$, $\lambda = -2$: basis $\begin{bmatrix} -1 \\ 0 \\ 1 \\ 0 \end{bmatrix}$, $\lambda = -1$: basis $\begin{bmatrix} -2 \\ 1 \\ 1 \\ 0 \end{bmatrix}$

     (b) $\lambda = 4$: basis $\begin{bmatrix} 3/2 \\ 1 \\ 0 \\ 0 \end{bmatrix}$

**11.** (a) $\lambda = -4, \lambda = 3$

     (b) Basis for eigenspace corresponding to $\lambda = -4$: $-2 + \frac{8}{3}x + x^2$
     basis for eigenspace corresponding to $\lambda = 3$: $5 - 2x + x^2$

**12.** (a) $\lambda = 1, \lambda = -2, \lambda = -1$

     (b) Basis for eigenspace corresponding to $\lambda = 1$: $\begin{bmatrix} 0 & 0 \\ 0 & 1 \end{bmatrix}$; basis for eigenspace

     corresponding to $\lambda = -2$: $\begin{bmatrix} -1 & 0 \\ 1 & 0 \end{bmatrix}$; basis for eigenspace corresponding

     to $\lambda = -1$: $\begin{bmatrix} -2 & 1 \\ 1 & 0 \end{bmatrix}$

**18.** $1, -1, -2^9, 2^9$.

**EXERCISE SET 6.2 (page 245).**

**5.** $P = \begin{bmatrix} 4/5 & 3/4 \\ 1 & 1 \end{bmatrix}$     $P^{-1}AP = \begin{bmatrix} 1 & 0 \\ 0 & 2 \end{bmatrix}$

**6.** $P = \begin{bmatrix} 1/3 & 0 \\ 1 & 1 \end{bmatrix}$     $P^{-1}AP = \begin{bmatrix} 1 & 0 \\ 0 & -1 \end{bmatrix}$

**7.** $P = \begin{bmatrix} 0 & 1 & 0 \\ 1 & 0 & 1 \\ -1 & 0 & 1 \end{bmatrix}$     $P^{-1}AP = \begin{bmatrix} 0 & 0 & 0 \\ 0 & 1 & 0 \\ 0 & 0 & 2 \end{bmatrix}$

**8.** $P = \begin{bmatrix} -2 & 0 & 1 \\ 0 & 1 & 0 \\ 1 & 0 & 0 \end{bmatrix}$     $P^{-1}AP = \begin{bmatrix} 3 & 0 & 0 \\ 0 & 3 & 0 \\ 0 & 0 & 2 \end{bmatrix}$

**9.** Not diagonalizable

**10.** $P = \begin{bmatrix} 1 & 2 & 1 \\ 1 & 3 & 3 \\ 1 & 3 & 4 \end{bmatrix}$   $P^{-1}AP = \begin{bmatrix} 1 & 0 & 0 \\ 0 & 2 & 0 \\ 0 & 0 & 3 \end{bmatrix}$

**11.** Not diagonalizable

**12.** $P = \begin{bmatrix} -1/3 & 0 & 0 \\ 0 & 1 & 0 \\ 1 & 0 & 1 \end{bmatrix}$   $P^{-1}AP = \begin{bmatrix} 0 & 0 & 0 \\ 0 & 0 & 0 \\ 0 & 0 & 1 \end{bmatrix}$

**13.** Not diagonalizable

**14.** $P = \begin{bmatrix} 1 & 1 & 0 & 0 \\ 0 & 1 & 1 & 0 \\ 0 & 0 & 1 & 1 \\ 0 & 0 & 0 & 1 \end{bmatrix}$   $P^{-1}AP = \begin{bmatrix} -2 & 0 & 0 & 0 \\ 0 & -2 & 0 & 0 \\ 0 & 0 & 3 & 0 \\ 0 & 0 & 0 & 3 \end{bmatrix}$

**15.** $\mathbf{p}_1 = 1/3 + x, \mathbf{p}_2 = x$

**17.** $\begin{bmatrix} 1 & 0 \\ -1023 & 1024 \end{bmatrix}$

## EXERCISE SET 6.3 (page 251).

**1.** (a) $\lambda = 0$: 1-dimensional, $\lambda = 2$: 1-dimensional
(b) $\lambda = 1$: 1-dimensional, $\lambda = -1$: 2-dimensional
(c) $\lambda = 3$: 1-dimensional, $\lambda = 0$: 2-dimensional
(d) $\lambda = 0$: 1-dimensional, $\lambda = 6$: 2-dimensional
(e) $\lambda = 0$: 3-dimensional, $\lambda = 8$: 1-dimensional
(f) $\lambda = -2$: 3-dimensional, $\lambda = 4$: 1-dimensional

**2.** $P = \begin{bmatrix} 1/\sqrt{2} & -1/\sqrt{2} \\ 1/\sqrt{2} & 1/\sqrt{2} \end{bmatrix}$   $P^{-1}AP = \begin{bmatrix} 4 & 0 \\ 0 & 2 \end{bmatrix}$

**3.** $P = \begin{bmatrix} \sqrt{3}/2 & -1/2 \\ 1/2 & \sqrt{3}/2 \end{bmatrix}$   $P^{-1}AP = \begin{bmatrix} 8 & 0 \\ 0 & -4 \end{bmatrix}$

**4.** $P = \begin{bmatrix} 3/5 & -4/5 \\ 4/5 & 3/5 \end{bmatrix}$   $P^{-1}AP = \begin{bmatrix} 25 & 0 \\ 0 & -25 \end{bmatrix}$

**5.** $P = \begin{bmatrix} -4/5 & 0 & 3/5 \\ 0 & 1 & 0 \\ 3/5 & 0 & 4/5 \end{bmatrix}$   $P^{-1}AP = \begin{bmatrix} 25 & 0 & 0 \\ 0 & -3 & 0 \\ 0 & 0 & -50 \end{bmatrix}$

**6.** $P = \begin{bmatrix} \dfrac{1}{3} & \dfrac{2}{3} & \dfrac{2}{3} \\ 0 & \dfrac{1}{\sqrt{2}} & -\dfrac{1}{\sqrt{2}} \\ \dfrac{4}{3\sqrt{2}} & -\dfrac{1}{3\sqrt{2}} & -\dfrac{1}{3\sqrt{2}} \end{bmatrix}$   $P^{-1}AP = \begin{bmatrix} 1 & 0 & 0 \\ 0 & 1 & 0 \\ 0 & 0 & -1 \end{bmatrix}$

**7.** $P = \begin{bmatrix} 1/\sqrt{3} & 1/\sqrt{3} & 1/\sqrt{3} \\ 0 & 1/\sqrt{2} & -1/\sqrt{2} \\ 2/\sqrt{6} & -1/\sqrt{6} & -1/\sqrt{6} \end{bmatrix}$   $P^{-1}AP = \begin{bmatrix} 2 & 0 & 0 \\ 0 & 2 & 0 \\ 0 & 0 & 4 \end{bmatrix}$

**8.** $P = \begin{bmatrix} 1/\sqrt{2} & -1/\sqrt{6} & 2/\sqrt{21} & -1/\sqrt{7} \\ 0 & 0 & 3/\sqrt{21} & 2/\sqrt{7} \\ 0 & 2/\sqrt{6} & 2/\sqrt{21} & -1/\sqrt{7} \\ 1/\sqrt{2} & 1/\sqrt{6} & -2/\sqrt{21} & 1/\sqrt{7} \end{bmatrix}$   $P^{-1}AP = \begin{bmatrix} 1 & 0 & 0 & 0 \\ 0 & 1 & 0 & 0 \\ 0 & 0 & 0 & 0 \\ 0 & 0 & 0 & 0 \end{bmatrix}$

**9.** $P = \begin{bmatrix} \dfrac{1}{3} & 0 & 0 & \dfrac{4}{3\sqrt{2}} \\ \dfrac{2}{3} & \dfrac{1}{\sqrt{2}} & 0 & -\dfrac{1}{3\sqrt{2}} \\ 0 & 0 & 1 & 0 \\ \dfrac{2}{3} & -\dfrac{1}{\sqrt{2}} & 0 & -\dfrac{1}{3\sqrt{2}} \end{bmatrix}$,   $P^{-1}AP = \begin{bmatrix} -1 & 0 & 0 & 0 \\ 0 & -1 & 0 & 0 \\ 0 & 0 & -1 & 0 \\ 0 & 0 & 0 & 2 \end{bmatrix}$

## EXERCISE SET 6.4 (page 261).

**1.** (a) $2x^2 - 3xy + 4y^2$   (b) $x^2 - xy$   (c) $5xy$   (d) $4x^2 - 2y^2$   (e) $y^2$

**2.** (a) $\begin{bmatrix} 2 & -3/2 \\ -3/2 & 4 \end{bmatrix}$   (b) $\begin{bmatrix} 1 & -1/2 \\ -1/2 & 0 \end{bmatrix}$

(c) $\begin{bmatrix} 0 & 5/2 \\ 5/2 & 0 \end{bmatrix}$   (d) $\begin{bmatrix} 4 & 0 \\ 0 & -2 \end{bmatrix}$   (e) $\begin{bmatrix} 0 & 0 \\ 0 & 1 \end{bmatrix}$

**3.** (a) $[x \ \ y] \begin{bmatrix} 2 & -3/2 \\ -3/2 & 4 \end{bmatrix} \begin{bmatrix} x \\ y \end{bmatrix} + [-7 \ \ 2] \begin{bmatrix} x \\ y \end{bmatrix} + 7 = 0$

(b) $[x \ \ y] \begin{bmatrix} 1 & -1/2 \\ -1/2 & 0 \end{bmatrix} \begin{bmatrix} x \\ y \end{bmatrix} + [5 \ \ 8] \begin{bmatrix} x \\ y \end{bmatrix} - 3 = 0$

(c) $[x \ \ y] \begin{bmatrix} 0 & 5/2 \\ 5/2 & 0 \end{bmatrix} \begin{bmatrix} x \\ y \end{bmatrix} - 8 = 0$

(d) $[x \quad y] \begin{bmatrix} 4 & 0 \\ 0 & -2 \end{bmatrix} \begin{bmatrix} x \\ y \end{bmatrix} - 7 = 0$

(e) $[x \quad y] \begin{bmatrix} 0 & 0 \\ 0 & 1 \end{bmatrix} \begin{bmatrix} x \\ y \end{bmatrix} + [7 \quad -8] \begin{bmatrix} x \\ y \end{bmatrix} - 5 = 0$

4. (a) ellipse    (b) ellipse    (c) hyperbola    (d) hyperbola

   (e) circle    (f) parabola    (g) parabola    (h) parabola

   (i) parabola    (j) circle

5. (a) $9x'^2 + 4y'^2 = 36$, ellipse

   (b) $x'^2 - 16y'^2 = 16$, hyperbola

   (c) $y'^2 = 8x'$, parabola

   (d) $x'^2 + y'^2 = 16$, circle

   (e) $18y'^2 - 12x'^2 = 419$, hyperbola

   (f) $y' = -\dfrac{1}{7}x'^2$, parabola

6. (a) $2y'^2 - 3x'^2 = 8$, hyperbola

   (b) $2\sqrt{2}x'^2 - 7x' + 9y' = 0$, parabola

   (c) $7x'^2 + 3y'^2 = 9$, ellipse

   (d) $4x'^2 - y'^2 = 3$, hyperbola

7. $2x''^2 + y''^2 = 6$, ellipse

8. $13y''^2 - 4x''^2 = 81$, hyperbola

9. $2x''^2 - 3y''^2 = 24$, hyperbola

10. $6x''^2 + 11y''^2 = 66$, ellipse

11. $4y''^2 - x''^2 = 0$, hyperbola

12. $\sqrt{29}y'^2 - 3x' = 0$, parabola

13. (a) Two intersecting lines, $y = x$ and $y = -x$

   (b) No graph

   (c) The graph is the single point $(0, 0)$

   (d) The graph is the line $y = x$

   (e) The graph consists of two parallel lines $y' = 2$ and $y' = -2$

   (f) The graph is the single point $(1, 2)$

## EXERCISE SET 6.5 (page 268).

1. (a) $x^2 + 2y^2 - z^2 + 4xy - 5yz$

   (b) $3x^2 + 7z^2 + 2xy - 3xz + 4yz$

   (c) $xy + xz + yz$

(d) $x^2 + y^2 - z^2$

(e) $3z^2 + 3xz$

(f) $2z^2 + 2xz + y^2$

2. (a) $\begin{bmatrix} 1 & 2 & 0 \\ 2 & 2 & -5/2 \\ 0 & -5/2 & -1 \end{bmatrix}$ (b) $\begin{bmatrix} 3 & 1 & -3/2 \\ 1 & 0 & 2 \\ -3/2 & 2 & 7 \end{bmatrix}$

(c) $\begin{bmatrix} 0 & 1/2 & 1/2 \\ 1/2 & 0 & 1/2 \\ 1/2 & 1/2 & 0 \end{bmatrix}$ (d) $\begin{bmatrix} 1 & 0 & 0 \\ 0 & 1 & 0 \\ 0 & 0 & -1 \end{bmatrix}$

(e) $\begin{bmatrix} 0 & 0 & 3/2 \\ 0 & 0 & 0 \\ 3/2 & 0 & 3 \end{bmatrix}$ (f) $\begin{bmatrix} 0 & 0 & 1 \\ 0 & 1 & 0 \\ 1 & 0 & 2 \end{bmatrix}$

3. (a) $[x \quad y \quad z] \begin{bmatrix} 1 & 2 & 0 \\ 2 & 2 & -5/2 \\ 0 & -5/2 & -1 \end{bmatrix} \begin{bmatrix} x \\ y \\ z \end{bmatrix} + [7 \quad 0 \quad 2] \begin{bmatrix} x \\ y \\ z \end{bmatrix} - 3 = 0$

(b) $[x \quad y \quad z] \begin{bmatrix} 3 & 1 & -3/2 \\ 1 & 0 & 2 \\ -3/2 & 2 & 7 \end{bmatrix} \begin{bmatrix} x \\ y \\ z \end{bmatrix} + [-3 \quad 0 \quad 0] \begin{bmatrix} x \\ y \\ z \end{bmatrix} - 4 = 0$

(c) $[x \quad y \quad z] \begin{bmatrix} 0 & 1/2 & 1/2 \\ 1/2 & 0 & 1/2 \\ 1/2 & 1/2 & 0 \end{bmatrix} \begin{bmatrix} x \\ y \\ z \end{bmatrix} - 1 = 0$

(d) $[x \quad y \quad z] \begin{bmatrix} 1 & 0 & 0 \\ 0 & 1 & 0 \\ 0 & 0 & -1 \end{bmatrix} \begin{bmatrix} x \\ y \\ z \end{bmatrix} - 7 = 0$

(e) $[x \quad y \quad z] \begin{bmatrix} 0 & 0 & 3/2 \\ 0 & 0 & 0 \\ 3/2 & 0 & 3 \end{bmatrix} \begin{bmatrix} x \\ y \\ z \end{bmatrix} + [0 \quad -14 \quad 0] \begin{bmatrix} x \\ y \\ z \end{bmatrix} + 9 = 0$

(f) $[x \quad y \quad z] \begin{bmatrix} 0 & 0 & 1 \\ 0 & 1 & 0 \\ 1 & 0 & 2 \end{bmatrix} \begin{bmatrix} x \\ y \\ z \end{bmatrix} + [2 \quad -1 \quad 3] \begin{bmatrix} x \\ y \\ z \end{bmatrix} = 0$

4. (a) Ellipsoid. (b) Hyperboloid of one sheet. (c) Hyperboloid of two sheets.

(d) Elliptic cone. (e) Elliptic paraboloid. (f) Hyperbolic paraboloid.

(g) Sphere.

**5.** (a) $9x'^2 + 36y'^2 + 4z'^2 = 36$, ellipsoid

(b) $6x'^2 + 3y'^2 - 2z'^2 = 18$, hyperboloid of one sheet

(c) $3x'^2 - 3y'^2 - z'^2 = 3$, hyperboloid of two sheets

(d) $4x'^2 + 9y'^2 - z'^2 = 0$, elliptic cone

(e) $x'^2 + 16y'^2 - 16z' = 32$, hyperboloid of one sheet

(f) $7x'^2 - 3y'^2 + z' = 0$, hyperbolic paraboloid

(g) $x'^2 + y'^2 + z'^2 = 25$, sphere

**6.** (a) $25x'^2 - 3y'^2 - 50z'^2 - 150 = 0$, hyperboloid of two sheets

(b) $2x'^2 + 2y'^2 + 8z'^2 - 5 = 0$, ellipsoid

(c) $9x'^2 + 4y'^2 - 36z = 0$, elliptic paraboloid

(d) $x'^2 - y'^2 + z' = 0$, hyperbolic paraboloid

**7.** $2x''^2 - y''^2 - z''^2 = 1$, hyperboloid of two sheets

**8.** $x''^2 + y''^2 + 2z''^2 = 4$, ellipsoid

**9.** $x''^2 - y''^2 + z'' = 0$, hyperbolic paraboloid

**10.** $6x''^2 + 3y''^2 - 8\sqrt{2}z'' = 0$, elliptic paraboloid

**EXERCISE SET 7.1 (page 277).**

**1.** (a) $.28 \times 10^1$     (b) $.3452 \times 10^4$     (c) $.3879 \times 10^{-5}$

(d) $-.135 \times 10^0$     (e) $.17921 \times 10^2$     (f) $-.863 \times 10^{-1}$

**2.** (a) $.280 \times 10^1$     (b) $.345 \times 10^4$     (c) $.388 \times 10^{-5}$

(d) $-.135 \times 10^0$     (e) $.179 \times 10^2$     (f) $-.863 \times 10^{-1}$

**3.** (a) $.28 \times 10^1$     (b) $.35 \times 10^4$     (c) $.39 \times 10^{-5}$

(d) $-.14 \times 10^0$     (e) $.18 \times 10^2$     (f) $-.86 \times 10^{-1}$

**4.** $x_1 = -3, x_2 = 7$       **5.** $x_1 = 13/8, x_2 = 14/8, x_3 = 21/8$

**6.** $x_1 = 1, x_2 = 2, x_3 = 3$     **7.** $x_1 = 0, x_2 = 0, x_3 = 1, x_4 = -1$

**8.** $x_1 = .997, x_2 = 1.00$       **9.** $x_1 = -2, x_2 = 0, x_3 = 1$

**10.** $x_1 = 0, x_2 = 1$ (without pivoting); $x_1 = 1, x_2 = 1$ (with pivoting)

**EXERCISE SET 7.2 (page 282).**

**1.** $x_1 \approx 2.81, x_2 \approx .94$; exact solution is $x_1 = 3, x_2 = 1$

**2.** $x_1 \approx .954, x_2 \approx -1.90$; exact solution is $x_1 = 1, x_2 = -2$

**3.** $x_1 \approx -2.99, x_2 \approx -.998$; exact solution is $x_1 = -3, x_2 = -1$

**4.** $x_1 \approx 0.00, x_2 \approx 2.02$; exact solution is $x_1 = 0, x_2 = 2$

**5.** $x_1 \approx 3.03, x_2 \approx 1.02$; exact solution $x_1 = 3, x_2 = 1$

**6.** $x_1 \approx 1.01, x_2 \approx -2.00$; exact solution is $x_1 = 1, x_2 = -2$

**7.** $x_1 \approx -3.00, x_2 \approx -1.00$; exact solution is $x_1 = -3, x_2 = -1$

**8.** $x_1 \approx .05$; $x_2 \approx 2.00$; exact solution is $x_1 = 0, x_2 = 2$

**9.** $x_1 \approx .492, x_2 \approx .006, x_3 \approx -.996$; exact solution is $x_1 = 1/2, x_2 = 0, x_3 = -1$

**10.** $x_1 \approx 1.00, x_2 \approx .998, x_3 = 1.00$; exact solution is $x_1 = 1, x_2 = 1, x_3 = 1$

**11.** $x_1 \approx .499, x_2 \approx .0004, x_3 \approx -1.00$; exact solution is $x_1 = 1/2, x_2 = 0, x_3 = -1$

**12.** $x_1 \approx 1.00, x_2 \approx 1.00, x_3 \approx 1.00$; exact solution is $x_1 = 1, x_2 = 1, x_3 = 1$

**13.** $a, d, e$

## EXERCISE SET 7.3 (page 291).

**1.** (a) $\lambda = 3$    (b) No dominant eigenvalue    (c) $\lambda = 6$    (d) $\lambda = 3$

**2.** (a) $\begin{bmatrix} 1.00 \\ .503 \end{bmatrix}$    (b) 5.02    (c) The dominant eigenvector is $\begin{bmatrix} 1 \\ 1/2 \end{bmatrix}$ the dominant eigenvalue is 5

(d) The percentage error is .4%

**3.** (a) $\begin{bmatrix} 1.00 \\ .750 \end{bmatrix}$    (b) 8.01

(c) The dominant eigenvector is $\begin{bmatrix} 1 \\ 3/4 \end{bmatrix}$; the dominant eigenvalue is 8

(d) The percentage error is .125%

**4.** (a) $\begin{bmatrix} 1.00 \\ -.560 \end{bmatrix}$    (b) $-4.00$

(c) The dominant eigenvector is $\begin{bmatrix} 1 \\ -1/2 \end{bmatrix}$; the dominant eigenvalue is $-4$

(d) The percentage error is 0%

**5.** (a) At the end of two iterations the dominant eigenvalue and eigenvector are approximately $\lambda_1 \approx 20.1$ and $x \approx \begin{bmatrix} 1 \\ .119 \end{bmatrix}$

(b) The exact values of the dominant eigenvalue and eigenvector are $\lambda_1 = 20$ and $x = \begin{bmatrix} 1 \\ 2/17 \end{bmatrix}$

**6. (a)** At the end of three iterations the dominant eigenvalue and eigenvector are approximately $\lambda_1 \approx -9.95$ and $x = \begin{bmatrix} -.978 \\ 1 \end{bmatrix}$

**(b)** The exact values of the dominant eigenvalue and eigenvector are $\lambda_1 = -10$ and $x = \begin{bmatrix} -1 \\ 1 \end{bmatrix}$

**7. (a)** $\begin{bmatrix} .027 \\ .027 \\ 1 \end{bmatrix}$ **(b)** 10.0

**(c)** The dominant eigenvector is $\begin{bmatrix} 0 \\ 0 \\ 1 \end{bmatrix}$; The dominant eigenvalue is 10

## EXERCISE SET 7.4 (page 296).

**1. (a)** $\begin{bmatrix} 1 \\ .509 \end{bmatrix}$ **(b)** 7.00 **(c)** $\lambda_2 \approx 2.00, v_2 \approx \begin{bmatrix} -.51 \\ 1 \end{bmatrix}$

**(d)** Exact eigenvalues 7, 2; exact eigenvectors

$$v_1 = \begin{bmatrix} 1 \\ 1/2 \end{bmatrix}, \quad v_2 = \begin{bmatrix} -1/2 \\ 1 \end{bmatrix}$$

**2. (a)** $\begin{bmatrix} 1 \\ .503 \end{bmatrix}$ **(b)** 12.0 **(c)** $\lambda_2 \approx 2.02, v_2 \approx \begin{bmatrix} -.532 \\ 1 \end{bmatrix}$

**(d)** Exact eigenvalues 12, 2; exact eigenvectors

$$v_1 = \begin{bmatrix} 1 \\ 1/2 \end{bmatrix}, \quad v_2 = \begin{bmatrix} -1/2 \\ 1 \end{bmatrix}$$

# Index